The geology of the southern North Sea

BRITISH GEOLOGICAL SURVEY

United Kingdom Offshore Regional Report

The geology of the southern North Sea

T D J Cameron, A Crosby, P S Balson, D H Jeffery, G K Lott,
J Bulat and D J Harrison

LONDON   HMSO   1992

Production of this report was funded by the Department of Energy and the Natural Environment Research Council

The coastline used on many maps and diagrams in this book is based on Ordnance Survey mapping

Bibliographic reference

CAMERON, T D J, CROSBY, A, BALSON, P S , JEFFERY, D H , LOTT, G K, BULAT, J, AND HARRISON, D J. 1992. *United Kingdom offshore regional report: the geology of the southern North Sea.* (London: HMSO for the British Geological Survey.)

© *NERC copyright 1992*

First published 1992

ISBN 0 11 884492 X

Printed in the United Kingdom for HMSO
Dd 0292035 1/93 C20 531/3 12521

Contents

1 **Introduction** 1
　Offshore research 1
　Geological summary 3
2 **Crustal structure** 5
　Crustal thickness 5
　Crustal structure: the seismic evidence 5
　Crustal structure: the gravity and magnetic evidence 6
3 **Pre-Carboniferous basement** 10
　Lower Palaeozoic 10
　Devonian 11
4 **Post-Devonian basin development** 13
　Carboniferous 13
　Permian to Jurassic 14
　Cretaceous to Quaternary 15
　Structural styles 17
　　Carboniferous section 17
　　Dowsing Fault Zone and Swarte Bank Hinge Zone 17
　　Other Mesozoic fault zones 18
　Halokinesis in the Anglo-Dutch Basin 20
5 **Carboniferous** 23
　Dinantian 24
　　Dinantian sediments in the north 24
　　Dinantian sediments along the flanks of the London–Brabant Massif 26
　Namurian 27
　　Namurian sediments in the northern wells 27
　　Namurian sediments adjacent to the London–Brabant Massif 30
　Westphalian 32
　　Coal-measure facies in the Anglo-Dutch Basin 32
　　Coal-measure facies around the London–Brabant Massif 34
　　Redbed facies of the Anglo-Dutch Basin 34
　　Upper Westphalian facies around the London–Brabant Massif 36
6 **Lower Permian** 38
　Leman Sandstone Formation 39
　　Fluvial facies 39
　　Aeolian facies 40
　　Reworked-sand facies 40
　Silverpit Formation 42
　　Lacustrine facies 42
　　Sabkha and other lake-marginal facies 42
7 **Upper Permian** 43
　Z1 Group 45
　　Kupferschiefer Formation 45
　　Zechsteinkalk Formation 45
　　Werraanhydrit Formation 45
　Z2 Group 47
　　Hauptdolomit Formation 48
　　Basalanhydrit Formation 48
　　Stassfurt Halite Formation 49
　Z3 Group 49
　　Grauer Salzton Formation 49
　　Plattendolomit Formation 50
　　Hauptanhydrit Formation 52

　　Leine Halite Formation 53
　　Roter Salzton Formation 53
　Z4 Group 53
　　Pegmatitanhydrit Formation 53
　　Aller Halite Formation 54
　Z5 Group 54
　　Grenzanhydrit Formation 54
8 **Triassic** 55
　Bacton Group 56
　　Bunter Shale Formation 56
　　Bunter Sandstone Formation 58
　Haisborough Group 59
　　Dowsing Dolomitic Formation 59
　　Dudgeon Saliferous Formation 61
　　Triton Anhydritic Formation 65
　Penarth Group 65
9 **Jurassic** 67
　Lias Group 68
　West Sole Group 72
　Humber Group 76
10 **Cretaceous** 80
　Cromer Knoll Group 83
　　Spilsby Sandstone Formation 83
　　Speeton Clay Formation 84
　　Carstone and the Red Chalk Formation 86
　The Lower Cretaceous of the southern province 86
　Chalk Group 88
11 **Tertiary** 91
　Palaeogene 94
　　Paleocene 94
　　　Danian 94
　　　Thanetian 94
　　Eocene 96
　　　Ypresian 96
　　　Lutetian 97
　　　Bartonian 97
　　　Priabonian 97
　　Oligocene 97
　Neogene 98
　　Miocene 98
　　Pliocene 99
12 **Pleistocene** 101
　Deltaic division 101
　　Element A 103
　　Element B 107
　　Shoreline changes in the early Pleistocene 108
　Nondeltaic division 109
　　Element C 109
　　Element D 111
　　Element E 111
　　Element F 112
　　Element G 113
　　Element H 115
13 **Holocene** 116
　The Holocene sea-level rise 116
　Holocene sediments 119
　Bedforms 120
　　Sand ribbons 120

Sand waves 120
Sandbanks (tidal sand ridges) 122
Sand Hills 122
Norfolk Banks 122
Thames Estuary banks 124
Distribution and sources of sea-bed sediment 125
Gravel 125
Sand 127
Mud 128
Carbonate 129

14 Economic geology 131
Hydrocarbons 131
Carboniferous reservoirs 132
Lower Permian reservoirs 133
Upper Permian reservoirs 133
Triassic reservoirs 134
Nonhydrocarbon resources 134
Sand and gravel aggregates 134
Other minerals 136

References 137
Index 149

FIGURES

1 Location of the report area xii
2 Sub-Pleistocene geology of the report area 2
3 Generalised cross-section from the central part of the report area 23
4 Regional structural setting of the report area 4
5 Interpreted deep-seismic reflection section off the Yorkshire coast 5
6 Deep-seismic reflection section, with interpretation, of a line running obliquely from the Anglo-Dutch Basin across the eastern margin of the Sole Pit Trough 6
7 Migrated deep-seismic reflection section, with interpretation, of a line running across the northern flank of the London–Brabant Massif 7
8 Migrated deep-seismic reflection section, with interpretation, illustrating mid-crustal seismic reflections adjacent to the north coast of East Anglia 8
9 Bouguer gravity anomaly map 8
10 Pseudorelief aeromagnetic anomaly map of the southern North Sea 9
11 Depth to the top of the Lower Palaeozoic 10
12 Postulated maximum extents of continental Devonian and marine Middle Devonian deposits 11
13 Depth to the top of Carboniferous rocks, and the location of principal Mesozoic fault zones 13
14 Seismic section and interpretation showing an example of the complex stratigraphical and structural relationships along the Dowsing Fault Zone 16
15 Seismic section and interpretation showing the complex stratigraphical and structural relationships on the northern flank of the London–Brabant Massif 18
16 Seismic section and interpretation across a Jurassic extensional graben on the southern flank of the Mid North Sea High 19
17 Seismic sections showing examples of halokinetic deformation as a result of the movement of Upper Permian salts 21 and 22
18 Pre-Permian subcrop map of the southern North Sea and eastern England 23
19 Carboniferous structural features, and the locations of wells that have penetrated Dinantian strata, together with a schematic profile through the Dinantian sediments of the southern North Sea 25
20 Correlation of the Dinantian sediments penetrated by well 44/2-1 with the Dinantian sequence of the Northumberland Trough 26
21 The transition from fluviodeltaic to Yoredale facies in the uppermost Dinantian sediments of well 42/10a-1 27
22 Distribution of Namurian sediments, with the locations of offshore wells that penetrate them, and a schematic profile through the Namurian sediments of the southern North Sea 28
23 Namurian sequences in four offshore wells 29
24 Fence diagram illustrating schematically the stratigraphical relationships of Carboniferous rocks on the northern flank of the London–Brabant Massif 30
25 Distribution of Westphalian sediments, and the locations of offshore wells that penetrate them, together with a schematic profile through the Westphalian sediments of the southern North Sea 31
26 Westphalian sequences from five wells in the Anglo-Dutch Basin 33
27 Westphalian sequences on the northern and southern flanks of the London–Brabant Massif 35
28 Seismic section showing the character of the Westphalian coal-measure and redbed facies at the centre of the Anglo-Dutch Basin 36
29 Schematic profile through the Lower Permian sediments of the southern North Sea 38
30 Distribution of the principal Lower Permian facies 39
31 Distribution and thickness of the Leman Sandstone Formation 40
32 Lower Permian sequences from wells in three parts of the southern North Sea 41
33 Distribution and thickness of the Silverpit Formation, showing the distribution of the Silverpit Halite Member 42
34 North–south sections showing successive cycles in the Upper Permian sediments of the southern North Sea 44
35 Correlation of carbonate and anhydrite formations in the first and second Zechstein cycles 46
36 Distribution, thickness and facies of the Zechsteinkalk Formation in the Z1 Group, and equivalent formations in eastern England 47
37 Distribution, thickness and facies of the Werraanhydrit Formation in the Z1 Group, and equivalent formations in eastern England 47
38 Distribution, thickness and facies of the Hauptdolomit Formation of the Z2 Group, and equivalent formations in eastern England 48
39 Distribution, thickness and facies of the Basalanhydrit Formation in the Z2 Group, and equivalent formations in eastern England 49
40 Well correlation of Z2 evaporite cycles on the southern margin of the Zechstein basin, with inset map depicting the distribution and generalised thickness of the Stassfurt Halite Formation, and equivalent formations in eastern England 50
41 Well correlation of formations in the Z3, Z4 and Z5 groups 51
42 Distribution, thickness and facies of the Plattendolomit Formation in the Z3 Group, and equivalent formations in eastern England 52
43 Distribution, thickness and facies of the Hauptanhydrit Formation in the Z3 Group, and the equivalent formation in eastern England 52

44 Distribution, thickness and lithology of the Leine Halite Formation in the Z3 Group, and the equivalent formation in eastern England   53
45 Distribution, thickness and lithology of the Aller Halite Formation in the Z4 Group, and the equivalent formation in eastern England   54
46 Triassic lithostratigraphy of eastern England and the southern North Sea   55
47 Distribution and thickness of the Bunter Shale Formation and the distribution of the Amethyst Member   56
48 Correlation of the Bunter Shale Formation between four offshore wells   57
49 Distribution and thickness of the Bröckelschiefer Member of the Bunter Shale Formation   58
50 Distribution and thickness of the Bunter Sandstone Formation   59
51 The Bunter Sandstone Formation in three offshore wells   59
52 Distribution and thickness of the Dowsing Dolomitic Formation   60
53 Schematic profile through the Dowsing Dolomitic Formation   61
54 Correlation of the Dowsing Dolomitic Formation in four offshore wells   62
55 Distribution, thickness and facies of the Main Röt Halite Member   63
56 Distribution of the Upper Röt Halite Member and contemporary facies   63
57 Distribution and thickness of the Muschelkalk Halite and Muschelkalk Evaporite members   63
58 Distribution and thickness of the Dudgeon Saliferous Formation   63
59 Correlation of the Dudgeon Saliferous and Triton Anhydritic formations between three offshore wells   64
60 Distribution and thickness of the Keuper Halite Member   65
61 Distribution and thickness of the Triton Anhydritic Formation   65
62 Correlation of the uppermost Triassic sediments between four offshore wells   66
63 Distribution and thickness of the Penarth Group   66
64 Distribution and thickness of Jurassic rocks   67
65 Jurassic lithostratigraphy of eastern England and the southern North Sea   69
66 Distribution and thickness of the Lias Group   70
67 Lithostratigraphical correlation of the Lias Group between selected wells on the East Midlands Shelf   71
68 Lithostratigraphical correlation of the Lias Group between two wells off Flamborough Head   72
69 Distribution and thickness of the West Sole Group (Middle Jurassic)   73
70 Lithostratigraphical correlation of the West Sole Group from selected wells   74
71 Model of the Bajocian depositional environment in the southern North Sea   75
72 Distribution and thickness of the Humber Group (Upper Jurassic)   75
73 Model of middle Oxfordian depositional environments in the southern North Sea   76
74 Distribution of middle Oxfordian strata, with the thickness of the oolitic limestones   77
75 Sparker profile showing the unconformity between the Kimmeridge Clay and Speeton Clay formations to the north-west of Flamborough Head   78
76 Stratigraphical correlation of Cretaceous rocks   81
77 Generalised Cretaceous facies and palaeogeographies   82
78 Sub-Cretaceous geology of the southern North Sea   83
79 Distribution of the Lower Cretaceous (Cromer Knoll Group)   84
80 A comparison of the Lower Cretaceous in an onshore borehole with the Cromer Knoll Group in the offshore type well   85
81 Correlation of Cretaceous rocks between selected wells across the southern North Sea and the adjacent land, using gamma-ray and sonic logs   87
82 Distribution and thickness of the Upper Cretaceous (Chalk Group)   88
83 Sparker profile showing discontinuities in the Chalk   90
84 Generalised palaeogeography of north-west Europe during Paleocene to Eocene times   91
85 The distribution of Tertiary sediments, and the depth in metres below sea level to the base of post-Danian Tertiary sediments in the UK sector of the southern North Sea   92
86 Tertiary stratigraphy, sea level and temperature   93
87 Palaeogene stratigraphy in the southern North Sea region   95
88 Summary logs of two BGS boreholes in the southern North Sea which recovered Palaeogene sediments   96
89 BGS high-resolution boomer record showing small-scale faulting within the London Clay Formation   97
90 Neogene stratigraphy in the southern North Sea region   98
91 (a) Pleistocene formations in the southern North Sea referred to in the text
(b) Pollen-zone divisions of the Pleistocene, with palaeomagnetic and absolute timescales, and suggested positions of UK sector offshore formations within this scheme   102
92 Generalised thickness of Pleistocene sediment in the southern North Sea   103
93 Cross-sections showing the dispositions of the Pleistocene formations of the southern North Sea   104
94 Schematic illustration of the relationships of the stratigraphical elements and early Pleistocene formations in the southern North Sea   105
95 Acoustic facies and subfacies within seven of the eight formations of element A   106
96 Distribution and acoustic facies of the Westkapelle Ground Formation and the crags of East Anglia   107
97 Distribution and acoustic facies of the IJmuiden Ground Formation and its correlatives   108
98 Distribution and acoustic facies of the Winterton Shoal Formation   109
99 Distribution and acoustic facies of the Markham's Hole Formation   109
100 Distribution and acoustic facies of the Outer Silver Pit Formation   110
101 Distribution and acoustic facies of the Aurora Formation   110
102 Distribution and generalised thickness of the Yarmouth Roads Formation   110
103 Inferred Cromerian Complex shorelines   111
104 Inferred early Pleistocene shorelines   111
105 Distribution of the Swarte Bank Formation   112
106 Distribution of the Egmond Ground and Sand Hole formations west of 3°E   112
107 Distribution of the Cleaver Bank Formation west of 3°E   113
108 Distribution of the Eem and Brown Bank

formations 113
109 Distribution of deposits of the last glaciation (late Weichselian) 114
110 Generalised palaeogeography of the southern North Sea at the Weichselian ice maximum 114
111 Sea-bed sediment distribution 117
112 Generalised bathymetry 118
113 Maximum surface-current velocities for mean spring tides 119
114 Generalised curve of relative sea level during the Holocene in the North Sea region 120
115 Hypothetical early Holocene coastlines 121
116 Distribution of the Elbow Formation and equivalent sediments 122
117 Distribution of zones with abundant sand waves 123
118 High-resolution seismic profile across the southern tip of Sandettie Bank, showing large sand waves with opposing asymmetry 123
119 Net sand-transport directions in the southern North Sea region 124
120 Distribution of major sandbanks in the southern North Sea 125
121 Morphology of the Norfolk Banks 126
122 Shallow-seismic profile across the Norfolk Banks 127
123 Morphology of nearshore sandbanks near the coast of East Anglia 128
124 Seismic profile (sparker) at the northern end of South Cross Sand, showing dipping internal reflectors and surface sand waves 129
125 Net sand transport directions in the southernmost part of the North Sea 129
126 Locations of gasfields and other significant discovery wells in the southern North Sea 131
127 Gas reserves in the southern North Sea at fields in production or under development 132
128 Areas of the southern North Sea currently licensed for marine aggregate production 135

Foreword

In 1966, the British Geological Survey (then the Institute of Geological Sciences) began to investigate the geology of the United Kingdom (UK) Continental Shelf. At first the work was funded by the Department of Education and Science through the Natural Environment Research Council. Since the mid 1970s the Department of Energy has provided the greater part of the funds. Two principal programmes have been undertaken, namely hydrocarbon studies and regional mapping. This report is a product of the latter programme.

In the regional mapping programme systematic surveys of the UK Continental Shelf were carried out by BGS, leading to the preparation and publication of a major series of geological and geophysical maps at 1:250 000 scale and a series of UK Offshore Regional Reports. This report describes the geology of that part of the north-west European continental shelf which comprises the UK sector of the Southern North Sea Basin. It is an area where, since the early 1960s, there has been considerable exploration for hydrocarbons. The report has been compiled from information derived from published scientific literature, released hydrocarbon wells, and data collected during the BGS regional mapping programme.

The report will be of interest to all geologists and students of the subject who are seeking a general account of the geology of this economically important area.

Peter J Cook, DSc
Director
British Geological Survey

6 July 1992

ACKNOWLEDGEMENTS

Amoco (UK) Exploration Company, Arco British Limited, Conoco (UK) Limited, GECO, Seismograph Services (England) Limited, Western Geophysical and Elsevier Applied Science Publishers have kindly granted permission to reproduce the seismic sections illustrated as Figures 8, 14 to 17 and 28. Dr C J Southworth is thanked for commenting on the Triassic chapter, part of which is based around his researches. Dr C King of Palaeoservices is thanked for identification of samples and constructive comments on the Tertiary chapter.

Responsibilities for production of the report have been as follows:

T D J Cameron — Basin development, Pre-Carboniferous, Carboniferous, Lower Permian, Upper Permian and Triassic, as well as hydrocarbon aspects of the economic geology and contribution to the crustal structure section

A Crosby — Compiler of report and author of the introduction and Cretaceous

P S Balson — Tertiary and Holocene

D H Jeffery — Pleistocene

G K Lott — Jurassic

J Bulat — Crustal structure

D J Harrison — Nonhydrocarbon economic geology

In addition to the named authors, the report draws heavily on the pool of expertise in BGS, not only in the Marine groups but also in the Land Survey, and particularly in the specialist fields of biostratigraphy, sedimentology, cartography and book production. In particular, the following are thanked for constructive comment: R A Chadwick, N G T Fannin, R W Gallois, R Harland, H M Pantin, T C Pharaoh, E R Shepherd-Thorn and colleagues, N J P Smith, S J Stoker and G Warrington.

The Offshore Regional Report Series is co-ordinated by the Marine Geology Group, and is edited by D Evans with the assistance of A G Stevenson. The project has been financed by the Department of Energy/NERC.

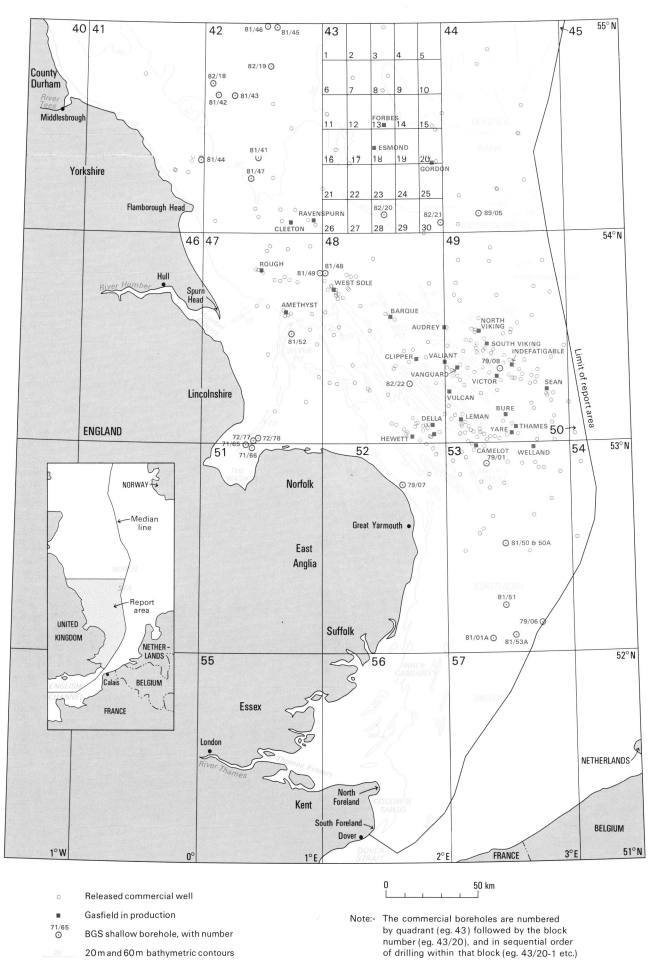

Figure 1 Location of the report area, showing licence quadrants, commercial wells released prior to April 1990, gasfields, BGS shallow boreholes, and simplified bathymetry.

1 Introduction

The southern North Sea report area covers some 62 000 km², and extends from the east coast of England to the median line which divides United Kingdom (UK) waters from those of The Netherlands, Belgium and France. The northern limit lies at latitude 55°N, and the southern limit is marked by a line drawn from near South Foreland on the coast of England to meet the median line in the direction of Calais on the French coast (Figure 1). The report area therefore extends from the Dogger Bank in the north to the Dover Strait in the south, an area of considerable importance to the UK economy. Up to 13 million tonnes per annum of sand and gravel are being dredged from the sea floor (Ardus and Harrison, 1990), and the many gasfields are currently producing about 29 billion cubic metres of gas per annum (Department of Energy, 1990). Proven gas reserves ensure that the southern North Sea will be a major gas-producing area well into the 21st century.

Much of the report area lies within the shallow-water, southern embayment of the North Sea (Figure 1). To the south of 53°N, water depths are generally less than 50 m, and the sea bed shows a range of features including sand waves up to 16 m high. Between about 53° and 54°N, a wide area of shallow water extends north-eastwards from the Norfolk coast to the median line. Numerous tidal sand ridges up to 40 m in amplitude and between 20 and 60 km in length occur in this shallow area, where the crests of the largest ridges rise to less than 10 m below sea level. Elsewhere in this shallow-water area, a gently undulating sea floor is locally incised by narrow deeps such as Sole Pit, Markham's Hole and the Silver Pit. The latter is the deepest in the report area, with water up to 98 m deep, that is, almost 80 m below the surrounding sea bed.

The Dogger Bank consists of a variable thickness of Holocene sands reworked from glacial deposits, and its highest point is less than 15 m below sea level. To the south of the Dogger Bank is the east–west-trending Outer Silver Pit, that was created shortly after the last glaciation. Between the Dogger Bank and the English coast, the sea floor undulates gently, and is mostly 40 to 80 m below sea level. Two systems of moribund tidal sand ridges up to 25 m in height occur to the south-west of the Dogger Bank.

Along the east coast of England, strata of Permian to Cretaceous age occur in the north, whereas in the south, Cretaceous to Tertiary sediments crop out or are buried beneath less than 100 m of Quaternary deposits (Figure 2). To the north of Flamborough Head, coastal erosion of resistant Permian limestones, Jurassic sandstones and limestones, and Upper Cretaceous chalk has created a mainly rocky coastline with many high sea cliffs (see front cover). South of Flamborough Head, the south Yorkshire coastline is formed by lower cliffs cut into late Pleistocene glacial deposits. By contrast, the coastline between the Humber Estuary and The Wash is generally low lying and sandy, with sea walls to protect many parts from inundation and erosion during severe storms. Much of the East Anglian coast is formed of stretches of cliffs cut into Pleistocene marine, fluvial and glacial deposits, separated by low-lying areas with marshes. Tertiary deposits crop out in places on the topographically subdued coasts of the Thames Estuary, and in the extreme south of the area, the sea cliffs of North Foreland and South Foreland are composed of chalk.

OFFSHORE RESEARCH

Before the search for hydrocarbons began in the 1960s, there was comparatively little information on the pre-Holocene geology of the southern North Sea. Early reconnaissance surveys (Donovan, 1963; Donovan and Dingle, 1965; Donovan, 1968; Dingle, 1971) established that many of the Mesozoic formations exposed in the sea cliffs of north-east England continue at the sea bed, or are buried beneath only a thin cover of Quaternary sediments, for up to 100 km offshore. These early surveys enabled the near-surface expression of some of the more deep-seated geological structures to be mapped. Elsewhere, the Mesozoic formations are mantled by an eastward-thickening wedge of Tertiary and Quaternary sediments (Figure 3), so that early attempts at determining the regional structure relied on gravity and aeromagnetic surveys (Collette, 1960; Donovan, 1963; Fleischer, 1963; Cook, 1965). These geophysical methods established the configuration of the principal sedimentary basins, and revealed the presence of halokinetic structures.

Since 1964, more than a thousand wells have been drilled to a maximum depth of 5000 m, and many hundred thousand kilometres of digitally recorded seismic-reflection data have been acquired by companies exploring for hydrocarbons beneath the southern North Sea. Following the pioneering work of Kent (1967), the information collected during these surveys has resulted in the publication of many hundreds of papers describing various aspects of the geology. Consequently, the structural setting and stratigraphy of the Late Carboniferous to Tertiary succession has been well established for the area north of 52°30'N. Only three wells have penetrated pre-Carboniferous strata, but the interpretation of deep-seismic reflection surveys shot and processed for the British Institutions Reflection Profiling Syndicate (BIRPS) has provided some information on the deeper geological structure of the area (Blundell et al., 1991).

Much of the exploration database remains unpublished, but some is available for scientific research because Acts of Parliament require those companies licensed to explore for hydrocarbons on the UK Continental Shelf to provide the Secretary of State for Energy with copies of all geophysical and well data acquired during their exploration programmes (Brooks, 1983; Walmsley, 1983). Under the terms of the Acts, all well data are released by the Department of Energy five years after receipt by the Secretary of State (Fannin, 1989).

The search for hydrocarbons has yielded only limited information on the near-surface Tertiary and Quaternary geology of the offshore area. BGS (British Geological Survey, previously the Institute of Geological Sciences, IGS) carried out shallow geological surveys of the Humber Estuary (McQuillin et al., 1969) and The Wash (Wingfield et al., 1978) in the late 1960s and early 1970s. In 1979, IGS was commissioned by the Department of Energy to complete a reconnaissance geological survey of the whole of the UK Continental Shelf

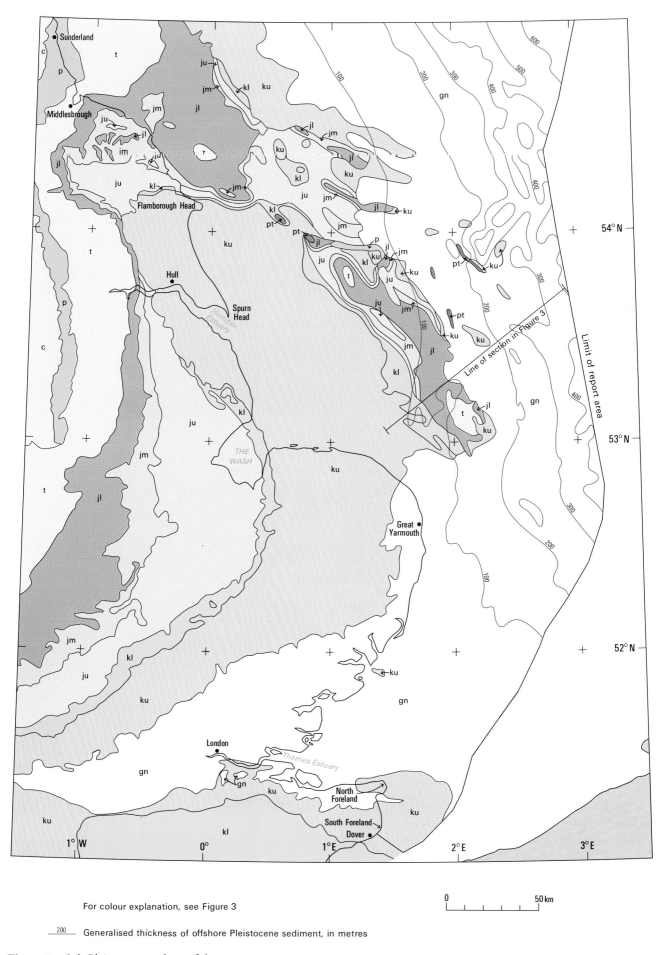

For colour explanation, see Figure 3

⎯200⎯ Generalised thickness of offshore Pleistocene sediment, in metres

Figure 2 Sub-Pleistocene geology of the report area.

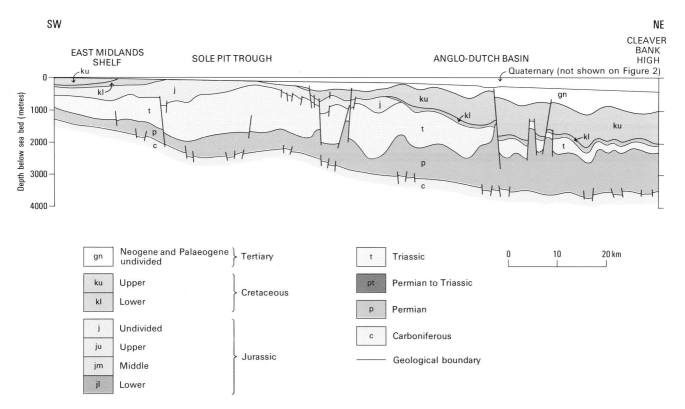

Figure 3 Generalised cross-section from the central part of the report area. For location see Figure 2.

using geophysical techniques, sea-bed sampling, and drilling (Fannin, 1989). The survey of the southern North Sea was carried out between 1979 and 1984, and has resulted in the production of maps of the solid and Quaternary geology, sea-bed sediments, Bouguer gravity anomaly and aeromagnetic anomaly at a scale of 1:250 000 (see inside back cover). In the report area, those maps which straddle the median line have been published jointly with the Rijks Geologische Dienst (Geological Survey of The Netherlands) and the Belgische Geologische Dienst (Geological Survey of Belgium). The BGS maps are based on the interpretation of more than 23 000 km of shallow-seismic data, calibrated using released commercial well data and the results of 29 rotary-cored shallow boreholes drilled by BGS to a maximum depth of 206 m below sea bed. The surface geology has been sampled by 6 m vibrocorer at 1232 sites, by gravity corer at a further 2284 locations, and by grab at 5814 sites.

GEOLOGICAL SUMMARY

More than 200 years of research into the stratigraphy and sedimentology of the Upper Palaeozoic, Mesozoic and Cenozoic formations of eastern England have provided a valuable basis for interpretation of the offshore deposits. Hydrocarbon exploration and BGS surveys have revealed that the post-Carboniferous sequences in the centre of the Southern North Sea Basin are generally thicker and more complete than those to the west (Figure 3). This is because eastern England and its adjacent offshore area have, since the Carboniferous, been situated largely on the periphery of the basin (Rhys, 1974). Despite this, the geology of the onshore and offshore areas have many features in common; the inverted Cleveland Basin and the East Midlands Shelf continue offshore from Yorkshire and Lincolnshire respectively (Figure 4). The London–Brabant Massif, extending from East Anglia across the Southern Bight of the North Sea into Belgium, was a basin-marginal upland area from Early Carboniferous to Early Cretaceous times.

Relatively little is known about the Devonian and older rocks in the area, but the Carboniferous fluviodeltaic, coal-measure and redbed sequences offshore are very similar to those of central and eastern England. They are buried beneath more than 4000 m of Permian, Mesozoic and Cenozoic sediments in the deepest parts of the Southern North Sea Basin. As in England, these Carboniferous rocks were gently folded and faulted during the Variscan orogeny.

Throughout Permian and Triassic times, most of the southern North Sea lay within a gently subsiding Variscan foreland basin which extended from eastern England through northern Germany into Poland (Ziegler, 1982). Aeolian, fluvial and desert-lake sediments accumulated between the London–Brabant Massif and the Mid North Sea High during the Early Permian. There were five, short-lived, marine transgressions across the basin during the Late Permian, producing a complex sequence of marine and evaporite deposits which are locally more than 1000 m thick. The evaporites have been deforming by halokinesis intermittently since Middle Triassic times, leading to the widespread growth of salt pillows and diapirs across the central offshore area. Triassic sediments are up to 1650 m thick, and dominated by reddish brown mudstones with subsidiary sandstones and evaporites; these were deposited in a range of playa-lake, floodplain, fluvial and quasimarine environments (Fisher, 1986).

The fully marine conditions which extended across the southern North Sea late in the Triassic Period have continued intermittently until the present day. The Sole Pit Trough and the Cleveland Basin were the principal depocentres during the Jurassic, accumulating up to 1000 m of marine mudstones with subsidiary sandstones and limestones before undergoing erosion at the end of the Jurassic Period, followed by post-Jurassic inversion (Glennie and Boegner, 1981; Van Hoorn, 1987). The Lower Cretaceous sediments are dominantly argillaceous, and are more than 800 m thick adjacent to contemporary growth faults in the Dowsing Fault Zone.

Upper Cretaceous, Tertiary and Quaternary sediments are all thickest near the UK–Netherlands median line. Upper Cretaceous pelagic carbonates of the Chalk Sea are locally more than 1000 m thick, and following their deposition there was widespread uplift and regression prior to the deposition of up to 800 m of mainly argillaceous marine sediments during the Palaeogene. Neogene sediments are generally absent due to nondeposition and/or erosion. The early Quaternary was dominated by the north-westward expansion across the southern North Sea of peripheral delta systems from continental Europe, and is now represented by some 600 m of sediment where best developed. Later Pleistocene history has been dominated by glacial erosion and deposition prior to the establishment of the present-day, strongly tidal, marine environment between 7000 and 10 000 years ago.

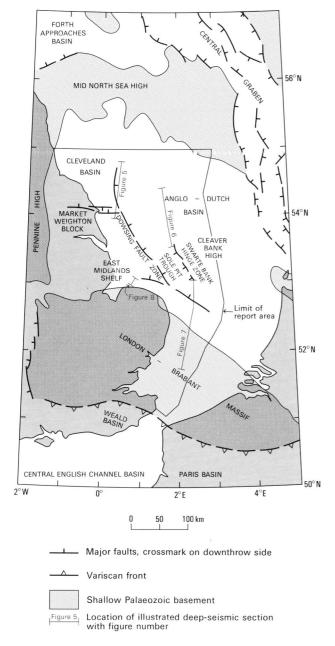

Figure 4 Regional structural setting of the report area. Locations of deep-seismic reflection sections are also shown.

2 Crustal structure

CRUSTAL THICKNESS

The boundary between the Earth's crust and mantle is marked by a major change in seismic velocity from about 7 to 8 km/s; this boundary has become known as the Mohorovičić Discontinuity, or Moho. It is commonly observed in seismic refraction experiments, and in certain circumstances on deep-seismic reflection sections. Both refraction and reflection data indicate that the base of the crust is at an average depth of 30 ± 5 km beneath the British Isles (Meissner et al., 1986).

There is only limited seismic information concerning the thickness of the Earth's crust beneath the southern North Sea. On a deep-seismic reflection line shot for BIRPS off Yorkshire (Figure 5) there are numerous, discontinuous, high-amplitude seismic reflectors below 7 s two-way travel time (TWTT). Close to the coast these have a well-defined flat base at between 9.5 and 10 s TWTT (Freeman et al., 1988), deepening to 11 s farther offshore (Figures 5 and 6). Elsewhere in the North Sea, the base of similar lower-crustal reflections has been found to correlate with the Moho as deduced from gravity and refraction data (Donato and Tully, 1981; Barton et al., 1984; Holliger and Klemperer, 1989).

Blundell et al. (1991) have demonstrated that beneath the Mid North Sea High, the reflection Moho coincides with that identified by seismic-refraction methods. Using an average crustal velocity of 6 km/s (Freeman et al., 1988), the reflection-seismic data suggest that the Moho is between 28.5 and 33 km deep east of the Yorkshire coast.

The lower-crustal reflectors are rather indistinct in the centre of the southern North Sea (Figure 6), but a conspicuous high-amplitude event between 11 and 12 s TWTT may correspond to the reflection Moho. This event appears to be at depths of more than 30 km, perhaps as much as 35 km, and is deepest where the strongly layered and relatively low-velocity sediments are thickest. Gravity modelling (Holliger and Klemperer, 1990) predicts that the base of the crust occurs at 32 ± 2 km in the central offshore area.

On the southern margin of the North Sea Basin there is a broad zone of lower-crustal reflectors parallel to the northern flank of the London–Brabant Massif (Blundell et al., 1991). These reflectors have a well-defined base at almost 11 s TWTT (Figure 7), again recording the reflection Moho at a depth of about 33 km. The reflection Moho is lost southwards beneath the London–Brabant Massif, but to the south of the massif in northern France, it reappears at a depth of about 38 km (Bois et al., 1987).

CRUSTAL STRUCTURE: THE SEISMIC EVIDENCE

Deep-seismic reflection sections around the British Isles have produced images of the continental crust which have a number of features in common. The idealised British profile has been referred to by Matthews and Cheadle (1986) as the 'typical BIRP'; in the southern North Sea the reflection sequences off the Yorkshire coast conform most closely to it (Figure 5). Strong reflections from post-Carboniferous sediments occur on this profile down to 1.8 s TWTT, 3 km below sea level. The middle crust, represented by a relatively featureless seismic zone extending down to a depth of about 7 s TWTT, is essentially nonreflective because it includes few significant velocity contrasts, and may contain a structure that is too complex to be imaged. On BIRPS data there are discontinuous, weak, seismic reflectors towards the top of the middle crust above 3 s TWTT. These may image layering within Carboniferous sediments, and are more clearly resolved on commercial deep-seismic profiles across the area. The lower crust is represented by discontinuous, high-amplitude reflectors between 7 and 11 s TWTT. Current hypotheses on the origin of this type of lower-crustal seismic layering suggest that it may represent mafic sills and layered igneous intrusions; extensional and compressional strain-banding and mylonite zones; laminae of hydrated rocks; or fluid-filled cracks (Klemperer, 1989).

The post-Carboniferous sediments are up to 4 km thick in the centre of the southern North Sea. Figure 6 illustrates Mesozoic and Tertiary sediments gently folded over northwest-trending Upper Permian salt pillows. Although the reflection Moho is locally clearly defined on the section, there are relatively few seismic events within the lower crust, and there is no clear boundary between the middle and lower crust. It seems that seismic resolution of the lower-crustal

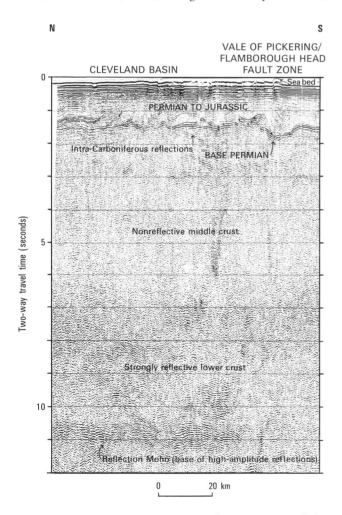

Figure 5 Interpreted deep-seismic reflection section off the Yorkshire coast. For location see Figure 4.

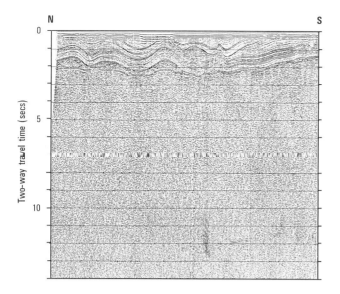

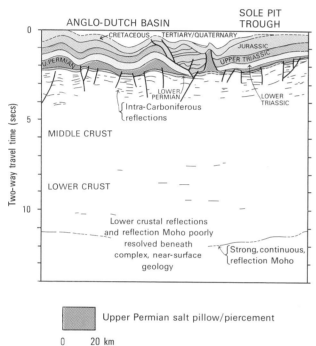

Figure 6 Migrated deep-seismic reflection section, with interpretation, of a line running obliquely from the Anglo-Dutch Basin across the eastern margin of the Sole Pit Trough. For location see Figure 4.

structure in this central area is adversely affected by distortion of the seismic wavefront as it passes through the strongly anisotropic, folded sediments of the upper crust.

Sedimentary cover upon the London–Brabant Massif is relatively thin. On the deep-seismic section in Figure 7, the top of the Lower Carboniferous limestones can be traced southwards from 2 s TWTT at its deepest to its subcrop limit beneath Permian and thin Westphalian sediments on the flank of the massif. Stratigraphical relationships demonstrate that this area was already established as a high by Carboniferous times. The lower-crustal structure beneath the massif is mostly masked by high-amplitude, mid-crustal, dipping reflections which can be traced from about 3.5 s to more than 7 s TWTT. Reston and Blundell (1987) have used a network of commercial seismic-reflection profiles to determine that these reflecting surfaces are stacked in an imbricate manner, and dip at 20° towards the south-west. They can be mapped for about 30 km along the northern flank of the London–Brabant Massif. Reston and Blundell (1987) have interpreted these reflections as mid-crustal shears which developed to accommodate differential Mesozoic extension along a transitional zone between the stable crust of the massif and the relatively mobile crust beneath the North Sea Basin. Alternatively, they may be relicts of earlier imbricate thrusting initiated during Caledonian compression (Blundell et al., 1991). The reflections themselves may represent lithological contrasts or syndeformational magmatism within the shear zones (Blundell et al., 1991).

Shallower, high-amplitude, mid-crustal reflections have been observed dipping at about 30° north-eastwards from 1.9 to 3.8 s TWTT on commercial seismic-reflection profiles adjacent to the north coast of East Anglia. These appear to terminate at depth against the south-westerly dipping imbricate events (Figure 8). Nearby boreholes onshore have proved that the shallow, dipping reflections originate from within folded Lower Palaeozoic turbidites and shales (Pharaoh et al., 1987). Thus they most likely image a thrust zone within a Caledonian fold belt that has been inferred to occur along the northern flank of the London–Brabant Massif (Pharaoh et al., 1987).

Seismic and borehole evidence suggests that beneath the sedimentary cover of the London–Brabant Massif lie volcanic and intrusive rocks; these are interbedded with, and intruded into, Lower Palaeozoic basinal sediments, and were deformed within Caledonian shear zones. The deep structure of the massif has affinities with many of the ancient Precambrian cratonic areas of the world (Chadwick et al., 1989). A reflection Moho is absent beneath the North American and Baltic cratons (Brown et al., 1987; Lund and Heikkinen, 1987), and Hirn et al. (1987) have deduced that there is a Precambrian cratonic basement beneath the London–Brabant Massif in northern France, where there are also no lower-crustal reflectors. Basement xenocrysts within Ordovician volcanic rocks in Belgium have been dated as pre-Cadomian in age, at least 870 Ma old (André and Deutsch, 1984), and were probably derived from the lower crust.

CRUSTAL STRUCTURE: THE GRAVITY AND MAGNETIC EVIDENCE

Many of the features on Bouguer gravity-anomaly and the smoothed pseudorelief aeromagnetic anomaly maps of the southern North Sea (Figures 9 and 10) have been produced by post-Carboniferous structures. Despite this, all gravity anomalies are elongated in a north-westerly direction, which is the Precambrian Charnoid structural trend in southern Britain. The sedimentary cover has its greatest thickness in the centre of the Anglo-Dutch Basin, coinciding with a regional negative gravity anomaly (Figure 9), and with low-amplitude magnetic anomalies which straddle the median line between UK and Dutch waters. The relatively thin sedimentary cover above the London–Brabant Massif is represented by a large positive gravity anomaly extending south-eastwards from East Anglia; the negative gravity anomaly to the south of the massif may indicate a concealed Devonian–Carboniferous sedimentary basin. Two large positive gravity anomalies adjacent to north-east England and north-east of East Anglia (Figure 9) correspond to the mid-Triassic to Jurassic sedimentary basins of the Sole Pit Trough and the Cleveland Basin, which have been inverted. Residual high Bouguer gravity-anomaly values indicate that these sediments are now overcompacted, having been uplifted by between 2 and 2.5 km since Jurassic times (Bulat and Stoker, 1987).

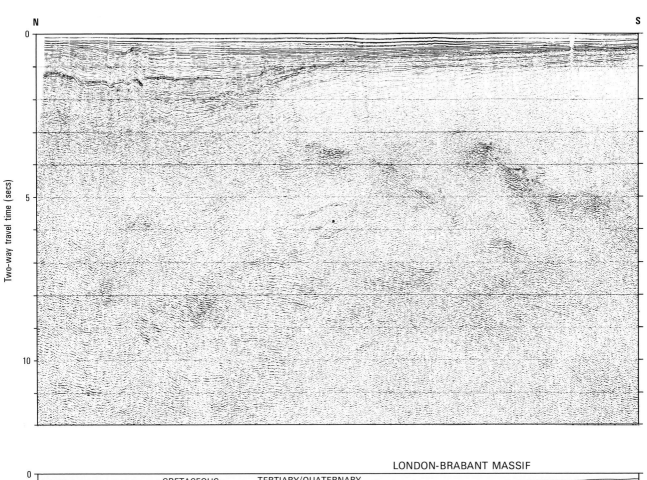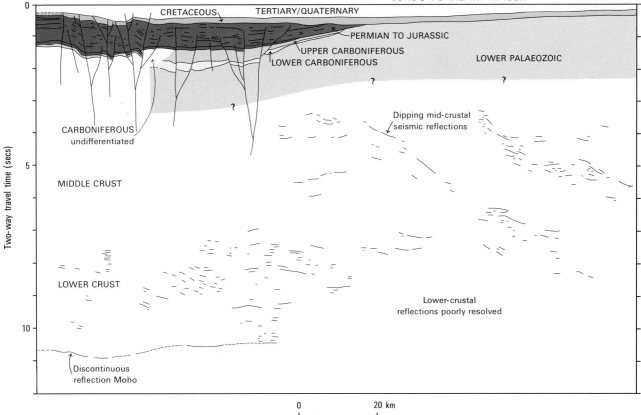

Figure 7 Migrated deep-seismic reflection section, with interpretation, of a line running across the northern flank of the London–Brabant Massif. For location see Figure 4.

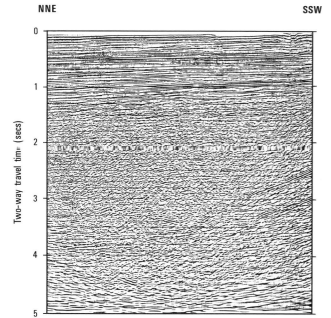

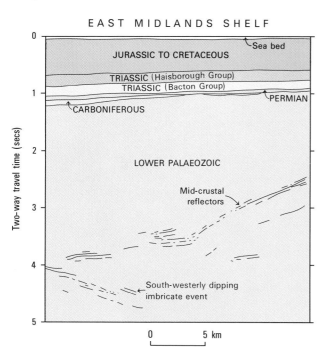

Figure 8 Migrated deep-seismic reflection section, with interpretation, illustrating mid-crustal seismic reflections adjacent to the north coast of East Anglia. For location see Figure 4. Reproduced by courtesy of Western Geophysical.

The most conspicuous features on the pseudorelief aeromagnetic anomaly map (Figure 10) are the west-north-west-trending, discontinuous but markedly linear anomalies that contrast with the magnetically subdued zones associated with the thick sedimentary basins. Two of these anomalies extend up to 150 km off north-east England; the southern anomaly has been correlated by Kirton and Donato (1985) with the Blyth swarm of Tertiary basaltic dykes. Though individual dykes are up to 21 m wide onshore (Land, 1974), they are presumed to be larger or more numerous offshore, as they do not generate a large magnetic anomaly in north-east England (Kirton and Donato, 1985). Along strike to the east-southeast from the Blyth swarm, a group of anomalies passes along the trace of a linear basement fault close to the Murdoch

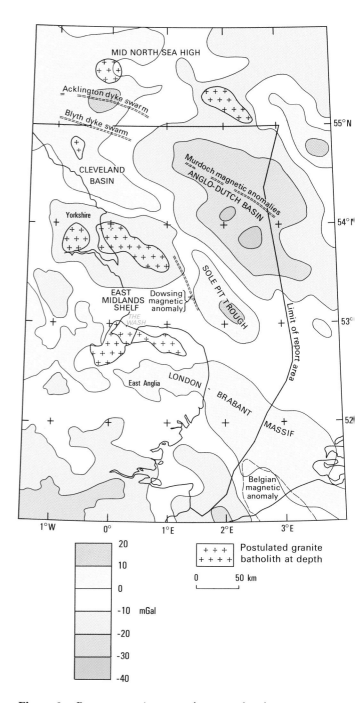

Figure 9 Bouguer gravity anomaly map. Also shown are linear features interpreted from aeromagnetic anomalies, and postulated granites.

Most of the remaining anomalies have probably been generated by features within the middle crust. One of the most conspicuous is the negative Bouguer gravity anomaly extending offshore from Yorkshire. Donato and Megson (1990) have deduced from gravity modelling that this represents a late Caledonian granite batholith about 140 km long and 40 km wide buried beneath 3 to 4 km of mainly post-Devonian sediments. There may be a smaller, deeply buried granite 130 km to the north; this is one of three separate batholiths postulated by Donato et al. (1983) beneath and adjacent to the Mid North Sea High (Figure 9). The crest of another likely Caledonian batholith which extends offshore from East Anglia beneath The Wash may be less than 1.5 km below sea level (Allsop, 1987).

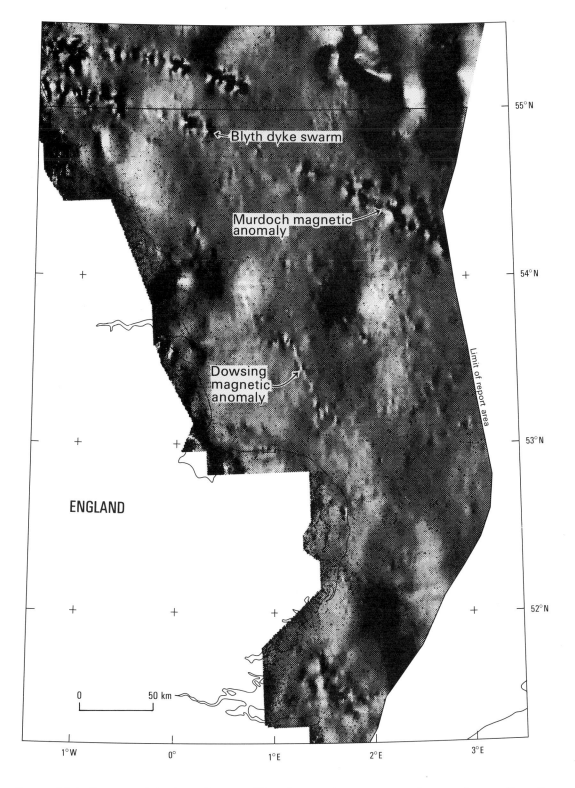

Figure 10 Pseudorelief aeromagnetic anomaly map of the southern North Sea, based on data acquired by Canadian Aeroservice. The map was produced by treating the data as a topographic relief surface illuminated from the east at an elevation of 30°.

Gasfield. The anomalies could indicate an injection of Tertiary dykes along this line of weakness, but they are relatively low-amplitude features, and intrusions have not been observed on the seismic sections. Another group of magnetic anomalies trends north-north-westwards along the trace of the Dowsing Fault Zone; these are higher-amplitude features, but it is unclear as to whether they represent another swarm of dykes, large-scale mineralisation, or were generated by crustal magnetic anisotropy along this line of basement weakness.

There is a very large aeromagnetic anomaly extending offshore from Belgium into the south-eastern North Sea (Figure 9). Similar anomalies occur onshore in Belgium where Cambrian or Precambrian, mildly metamorphosed sediments are in the shallow subsurface (Service Géologique de Belgique, 1964; 1968). The amplitude of the magnetic anomaly diminishes north-westwards, suggesting that its source is either absent or more deeply buried beneath the UK sector of the North Sea.

3 Pre-Carboniferous basement

Lower Palaeozoic and Devonian rocks are buried beneath many kilometres of Carboniferous, Mesozoic and Cenozoic sediments in most of the southern North Sea and eastern England (Figure 11). Geophysical evidence suggests that there may be Precambrian cratonic rocks beneath the London–Brabant Massif (Chapter 2), underlying the Early Palaeozoic basinal-marine sediments that have been penetrated by boreholes on the southern margin of the North Sea Basin. A patchy cover of Devonian alluvial and marine beds exists beneath south-east England, and Upper Devonian fluvial sediments approach to less than 2.5 km below sea level on the southern flank of the Mid North Sea High in the extreme north of the report area, where they have been sampled in one well.

LOWER PALAEOZOIC

In Early Palaeozoic times, the southern North Sea area, Wales, England and Belgium formed part of a late Proterozoic terrane, termed Eastern Avalonia by Soper et al. (1987). By the Early Ordovician, Eastern Avalonia lay 40° to 60° south of the equator (Cocks and Fortey, 1982). The Scottish Highlands formed part of the Laurentian Precambrian shield in equatorial latitudes, that was separated from Eastern Avalonia by many thousands of kilometres of Iapetus Ocean (Harland and Gayer, 1972). Scandinavia lay within the Baltic shield, separated from Eastern Avalonia by the Tornquist Sea, an arm of the Iapetus Ocean until it closed late in Ordovician times (Cocks and Fortey, 1982; Ziegler, 1982).

The southern North Sea now lies to the south-east of the Iapetus suture, the line of eventual closure of the Iapetus Ocean. Its Lower Palaeozoic basement may comprise mainly basinal-marine and volcaniclastic sediments similar to those in the Lake District and North Wales, and was likewise de-

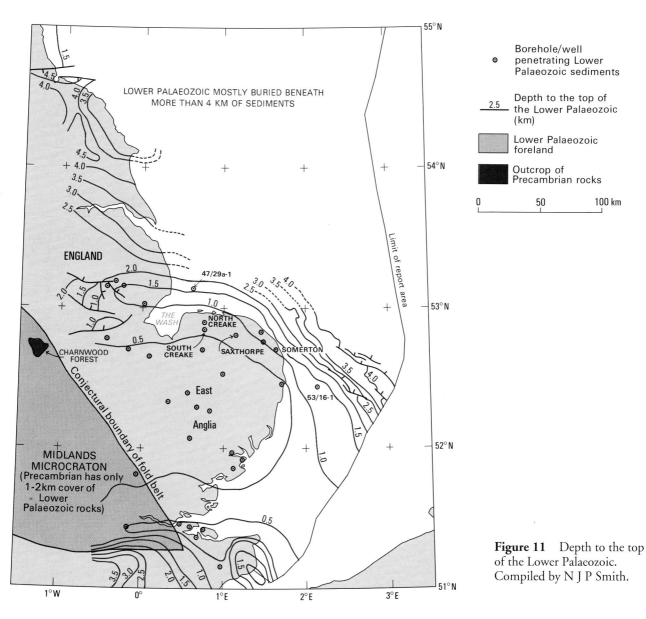

Figure 11 Depth to the top of the Lower Palaeozoic. Compiled by N J P Smith.

formed during the Caledonian orogeny. To the south-west, the Lower Palaeozoic rocks overlying the wedge-shaped Midlands Microcraton (Figure 11) are comparatively unaffected by folding and metamorphism (Pharaoh et al., 1987), and are relatively thin (Smith, 1987). The Midlands Microcraton acted as a foreland to the southern North Sea deformation (Turner, 1949; Smith, 1987).

Lower Palaeozoic basinal-marine sediments have been penetrated by two offshore wells on the northern flank of the London–Brabant Massif (Figure 11). Well 47/29a-1 terminated in a sequence of 115 m of grey Ordovician turbidites composed of alternating siltstones, silty shales and silica-cemented, very fine-grained, micaceous sandstones. Well 53/16-1 penetrated 229 m of lithologically similar, unfossiliferous, reddened rocks beneath the base of the Carboniferous; these were ascribed to the Lower Devonian by the well-site geologist, but further work has shown that they have closer affinity with the Early Palaeozoic turbidite facies of the London–Brabant Massif.

Early Palaeozoic, possibly Silurian, turbidite lithologies have been encountered by boreholes in East Anglia at Saxthorpe, Somerton and South Creake (Figure 11). The shales at South Creake have a penetrative slaty cleavage defined by parallel microfractures and calcite veins. The nearby North Creake borehole recorded 60 m of weakly metamorphosed tuffs and agglomerates directly beneath the Triassic. Phemister, in Kent (1947), interpreted these as Precambrian in age, based on their supposed petrographical affinity with the Charnian felsic volcanic suites exposed at Charnwood Forest in the English Midlands (Figure 11). However, Pharaoh et al. (1987) have reported Ordovician Rb-Sr whole-rock ages for comparable volcanic rocks lying between Charnwood Forest and East Anglia, and geochemical evidence supports their inference that the North Creake rocks are Ordovician (Pharaoh et al., 1991). Precambrian rocks in the shallow subsurface are therefore likely to be confined to the Midlands Microcraton. Le Bas (1972) has speculated that calcalkaline suites of Ordovician volcanic rocks are abundant in the subsurface of the London–Brabant Massif from Wales through central England and East Anglia to Belgium.

Lower Palaeozoic sediments may be many kilometres thick beneath the southern North Sea, and their base is not defined on deep-seismic sections. Shallow-marine and basinal Cambrian to Lower Ordovician sediments are more than 3000 m thick on the southern flank of the London–Brabant Massif in Belgium (Walter, 1980), and may be similarly thick at the base of the Lower Palaeozoic sequence beneath the Dover Strait. Pharaoh et al. (1987) have shown that the metamorphic grade of the Lower Palaeozoic basement increases from diagenetic conditions in the English Midlands, to lower greenschist facies in the vicinity of The Wash. Higher metamorphic grades are likely to be represented in the centre of the southern North Sea. Lower Palaeozoic rocks have a north-westerly trending structural grain in East Anglia (Turner, 1949; Smith, 1987; Pharaoh et al., 1987). If the trends of the Variscan faults in the offshore area have been influenced by the structural grain of the basement, Caledonian deformation will have mainly west-north-west and north-west trends beneath the southern North Sea.

DEVONIAN

By the Devonian, the closure of the Iapetus Ocean had welded together northern and southern lithospheric plates to form a Laurasian 'Old Red Sandstone' continent (Anderton et al., 1979). Within this continent, interior drainage basins such as the Midland Valley of Scotland accumulated thick, post-orogenic, molasse sequences, mainly by denudation of adjacent Caledonian mountain chains. Coastal plains around the southern margin of the continent in southern Ireland, South Wales and southern England accumulated relatively thin sequences of alluvial sediments (Anderton et al., 1979).

Lower Devonian sediments have not been recorded from the Northumberland Trough in northern England (House et al., 1977), and are unlikely to be extensive in the northern part of the report area. Seismic-reflection evidence suggests that younger Devonian sediments may be more than 1000 km thick adjacent to the Mid North Sea High. Farther south, those wells which have penetrated through the base of the Carboniferous on the northern flank of the London–.Brabant Massif have not recorded Devonian sediments overlying the Early Palaeozoic basement (Figure 12). It seems likely that the Devonian sediments have a thin and patchy distribution over most of the central and southern offshore area.

On seismic sections running through the site of well 44/2-1, a strong and continuous reflector occurs about 900 m beneath the top of the Devonian. A similarly continuous, high-amplitude reflector at the Argyll Oilfield of the central North Sea images marine, Middle Devonian dolomites with shales and a thick, fossiliferous limestone (Pennington, 1975). Ziegler (1982) suggested that these sediments were deposited in a temporary embayment of the Proto-Tethys Ocean of northern Europe, along a line adopted later by the Central Graben of the North Sea. If the southern North Sea seismic reflector is imaging an equivalent sequence of marine carbonate deposits, then the Middle Devonian seas may have ex-

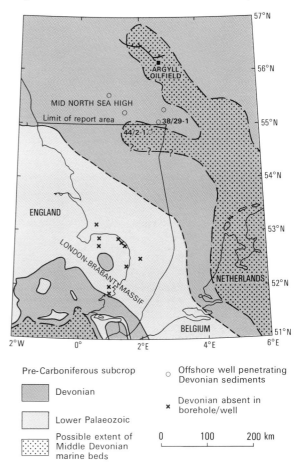

Figure 12 Postulated maximum extents of continental Devonian and marine Middle Devonian deposits. Adapted from Ziegler (1982) and Smith (1985).

tended around the south-eastern flank of the Mid North Sea High (Figure 12).

Fluvial sediments were deposited extensively over the crest and southern flank of the Mid North Sea High, at least during Late Devonian times; these may be continuous with Upper Devonian sediments in the Northumberland Trough, which were deposited in a palaeodrainage system directed eastwards towards the North Sea (Simon and Bluck, 1982). Upper Devonian rocks have been penetrated by only one offshore well in the report area, well 44/2-1 (Figure 12) which terminated in a 58 m-thick sequence of Famennian alluvial sandstones and shales. Nearby well 38/29-1 proved a minimum of 352 m of Late Devonian, fine-grained, red, fluvial sandstones with bands of medium- and coarse-grained, pebbly sandstone and red or brown, micaceous shales. Three other wells which penetrated through the base of the Carboniferous on the crest of the Mid North Sea High (Figure 12) have recorded similar facies of probable Late Devonian age.

Allsop (1985) has used borehole and geophysical data to map the extent of two Middle to Late Devonian basins overlying the crest and southern flank of the London–Brabant Massif in East Anglia and south-east England (Figure 12). There may be up to 1400 m of Devonian sediments in the larger southern basin, which Bouguer gravity-anomaly values suggest may extend beneath the Dover Strait. The Devonian sediments in this southern province are dominantly alluvial, but spores obtained from some boreholes indicate that there may have been marine incursions into south-east England during the Middle Devonian (House et al., 1977).

4 Post-Devonian basin development

The geology of the southern North Sea is the result of a long and complex history of basinal subsidence, punctuated by discrete episodes of uplift and widespread erosion during end-Silurian, Late Carboniferous, Late Jurassic, Late Cretaceous and mid-Tertiary times. These crustal movements can be identified across much of north-west Europe, and structural contours on the top of the Carboniferous (Figure 13) illustrate their net effect. The main area of subsidence has been the Anglo-Dutch Basin.

Lower Palaeozoic sediments are likely to be many kilometres thick beneath most or all of the southern North Sea. They were mildly deformed and intruded by granite batholiths during the Caledonian orogeny of Late Silurian to Early Devonian times, but the effects of this deformation offshore are deeply buried beneath thick sequences of Late Palaeozoic and younger sediments. Nevertheless, the dominant west-north-westerly and north-westerly trends of the faults displacing these sediments may be inherited from the structural grain of the Early Palaeozoic basement.

CARBONIFEROUS

Following Caledonian uplift, most of the southern North Sea remained an upland area of net erosion in Devonian times. It lay on the southern margin of the supercontinent formed by the collision of the Laurentian and Fennoscandian landmasses. Crustal extension initiated the break-up of this part of the supercontinent early in the Carboniferous by forming zones of rapidly subsiding grabens and half-grabens; these are separated by horsts on which condensed sequences of sediments accumulated (Leeder, 1982; Grayson and Oldham, 1987). The London–Brabant Massif became established as a stable upland area early in the Dinantian, and formed the

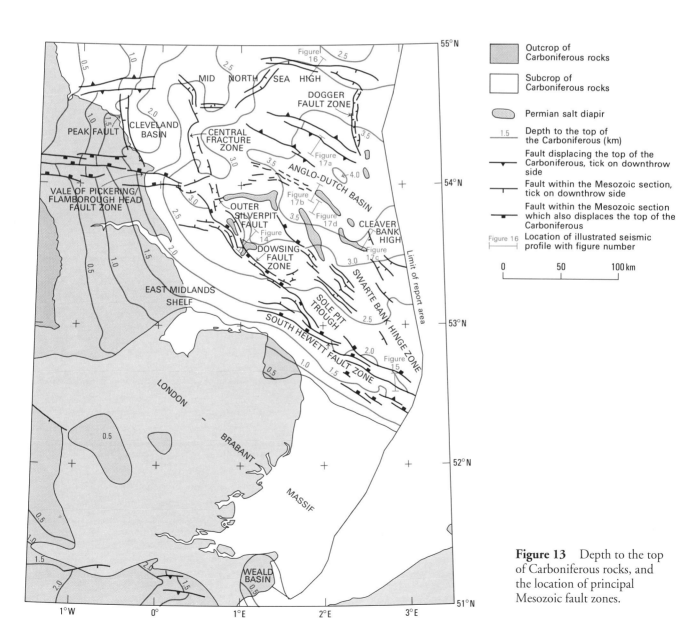

Figure 13 Depth to the top of Carboniferous rocks, and the location of principal Mesozoic fault zones.

southern boundary of the Southern North Sea Basin until mid-Cretaceous times. Up to 4000 m of deep-water or deltaic sediments were deposited in the Dinantian grabens, whereas some of the horsts remained as positive features until late in the Dinantian. The positions of the horsts and subsiding basins in northern England have been relatively well established, and it appears that some of the most persistent highs there were kept buoyant by their Caledonian granite cores (Bott, 1967). The Dinantian palaeogeography of the southern North Sea is less well known, but some or all of the granite batholiths inferred from their Bouguer gravity anomalies (Figure 9) may lie beneath Early Carboniferous horsts. Bott (1987) has calculated that the grabens and horsts of northern England were formed by between 5 and 10 per cent extension of the uppermost crust during the Early Carboniferous, although there may have been as much as 19 per cent extension across the Northumberland Trough (Kimbell et al., 1989). The ductile lower crust may have stretched by around 20 per cent to accommodate the regional Dinantian subsidence (Bott, 1987).

Leeder (1982, 1987) has invoked a transition from lithospheric extension to thermal subsidence to account for the establishment of relatively uniform, deltaic sedimentary facies across northern and central England by mid-Namurian times. These deltaic facies also extend across the southern North Sea, where Upper Carboniferous sediments are locally more than 3000 m thick. Bott (1987) pointed out that thermal subsidence alone cannot account for the much greater thickness of Upper Carboniferous than underlying Dinantian sediments in parts of northern England. Instead, he suggested that crustal extension continued spasmodically until at least Duckmantian (Westphalian B) times. Stratigraphical evidence contradicts this, and suggests that if there was continuing crustal extension, it was largely confined to the lower crust.

From mid-Westphalian times, sedimentation in the southern North Sea was increasingly affected by the approach of the Variscan deformation front from the south (Besly, 1988). Regional uplift caused the poorly drained, deltaic facies to be superseded by better-drained, alluvial-plain sediments. Contemporary uplift of the London–Brabant Massif is indicated by mid-Westphalian unconformities on both its southern and northern flanks (Ramsbottom et al., 1978; Tubb et al., 1986).

During the Variscan orogeny, the southern North Sea formed a foreland area close to the Variscan mountains that stretched from south-west England through France into eastern Europe. Differential Late Carboniferous and Early Permian uplift accompanied by peneplanation of the foreland led to the erosion of more than 1500 m of Upper Carboniferous sediments from those areas of the southern North Sea where Namurian or Dinantian rocks underlie the base-Permian unconformity. On a regional scale, the pre-Permian subcrop map (see Figure 18) demonstrates that transpression associated with north–south compression imposed mainly west-north-westerly trending Variscan structures in the south and west of the offshore area. In the east-central part of the report area, Bolsovian (Westphalian C) and Westphalian D sediments are preserved in north-north-west-trending Variscan synclines, whereas Variscan structures are mainly aligned east–west along the southern flank of the Mid North Sea High and in northern England.

PERMIAN TO JURASSIC

Following cessation of Variscan compression, the southern North Sea began to subside once more during the Early Permian. Most of the area lies within a Variscan foreland, postorogenic-collapse basin which, until Late Triassic times, extended from eastern England through northern Germany into Poland (Ziegler, 1982). The Pennine High (Figure 4) became a positive feature for the first time during the Early Permian, separating this north European basin from a contemporary Irish Sea basin. The London–Brabant Massif continued to form the southern boundary of the Southern North Sea Basin. The Mid North Sea High formed its northern boundary and accumulated condensed sequences of Permian and Triassic sediments.

More than 2700 m and perhaps as much as 3000 m of sediments were deposited in the centre of the southern North Sea during the Permian and Triassic. Isopachs of these sediments (see Chapters 6 to 8) demonstrate that the Sole Pit Trough temporarily became an important depocentre within the Variscan foreland basin during the Early Permian, and then more continuously from mid-Triassic times. Increased Early Permian subsidence in the trough was initiated by late Variscan wrench reactivation of its basement faults in an east–west-oriented, transtensional stress regime (Glennie and Boegner, 1981). The Cleaver Bank High and the northern part of the East Midlands Shelf (Figure 13) formed contemporary intrabasinal highs, but the isopachs of the Lower Permian sediments may to some extent be imaging a relict post-Variscan topography.

There are no indications of differential intrabasinal subsidence in the isopachs of the Upper Permian or basal Triassic formations (Chapters 7 and 8). In mid-Triassic times, the Sole Pit Trough was reactivated as a depocentre and extended temporarily around the northern flank of the East Midlands Shelf. Contemporary uplift of peripheral source areas caused the coarse-grained, clastic deposits of the Bunter Sandstone Formation to spread rapidly across the offshore area (Fisher, 1986). These source areas were quickly peneplaned, and the Upper Triassic sequence is dominated by argillaceous sediments.

Outside the report area, Late Triassic and Early Jurassic crustal extension led to the development of the Viking and Central grabens along the central axis of the North Sea (Ziegler, 1982) in response to regional east–west transtensional stress. There is a rise in the Moho beneath both of these graben systems, but the reflection Moho beneath the southern North Sea has a generally flat disposition, indicating that there was only limited contemporary crustal extension here. Glennie and Boegner (1981) have instead related the enhanced subsidence of the Sole Pit Trough to renewed wrench reactivation of its Variscan basement faults in this east–west transtensional stress regime. These tectonic movements triggered the earliest, mid-Triassic halokinesis of the Upper Permian salts in the centre of the offshore area (Brunstrom and Walmsley, 1969).

Fully marine conditions were established in the southern North Sea and in most of northern Europe at the end of the Triassic Period, as crustal extension along the central axis of the North Sea allowed the Tethys Ocean to link to the Boreal Ocean for the first time (Ziegler, 1982). In the southern North Sea, differential subsidence of the Sole Pit Trough was accentuated by the development of growth faults along its western margin. These remained active until Middle Jurassic times, and account for the major thickness and facies changes of the Lower and Middle Jurassic sediments between the Sole Pit Trough and the East Midlands Shelf. Simultaneously, the Cleveland Basin became established as a new depocentre in east Yorkshire (Kent, 1980a), linking offshore to the Sole Pit Trough around the north of the East Midlands Shelf. The London–Brabant Massif had retained positive relief throughout the Permo-Triassic, but by Early Jurassic times had reduced in size sufficiently for sedimentation to resume on its southern flank.

Along the central axis of the North Sea, subsidence related to crustal extension beneath the major graben systems was superseded in Middle Jurassic times by domal uplift and widespread erosion. In the southern North Sea, this resulted in an erosional surface commonly referred to as the late-Cimmerian Unconformity. More than 1000 m, and perhaps as much as 2000 m of Jurassic and Triassic strata were eroded from the Cleaver Bank High at this time (Glennie, 1986a). Several hundred metres of sediments were eroded from the southern part of the Sole Pit Trough towards the end of the Jurassic (Glennie and Boegner, 1981), whereas sedimentation was continuous on the nearby East Midlands Shelf (Kent, 1980a). The increased tectonic activity and destabilising of the overburden triggered a widespread pulse of diapirism in the southern and western parts of the Permian halite basin (Glennie, 1986b).

Isopachs of Upper Jurassic sediments (see Figure 72) demonstrate that the East Midlands Shelf replaced the Sole Pit Trough as the principal depocentre before the end of the Jurassic; up to 350 m of marine sediments were deposited there during the Late Jurassic.

CRETACEOUS TO QUATERNARY

By Late Jurassic times, regional doming along the central axis of the North Sea had been superseded by renewed rifting and rapid subsidence within its major graben systems. More widespread, but gentle subsidence allowed marine sedimentation to resume across most of the southern North Sea early in Cretaceous times. The Lower Cretaceous sediments are typically less than 200 m thick, but growth faulting (Kirby and Swallow, 1987) enabled accumulation of up to 1000 m of mainly argillaceous sediments in parts of the Vale of Pickering–Flamborough Head Fault Zone (Figure 13). Similarly, about 1000 m of Lower Cretaceous sediments are present locally within the Dowsing Fault Zone. Conversely, contemporary uplift caused erosion of up to 200 m of basal Lower Cretaceous and Jurassic sediments in the adjacent northern part of the East Midlands Shelf (Kent, 1980a). These observations demonstrate that sinistral transcurrent crustal movements occurred within and beneath the fault zones during Early Cretaceous times (Kirby and Swallow, 1987). South of the London–Brabant Massif, the Weald Basin was subsiding rapidly, and there may be more than 800 m of Early Cretaceous alluvial, lagoonal and lacustrine sediments beneath Kent and the Dover Strait.

Lithospheric extension and localised rifting along the central axis of the North Sea finally ceased around the middle of the Cretaceous Period, to be replaced by thermal subsidence and gentle, regional, crustal downwarping (Barton and Wood, 1984). A contemporary global rise in sea level allowed pelagic carbonate sedimentation to extend across most of the southern North Sea in the Late Cretaceous. The London–Brabant Massif was submerged for the first time since the early Palaeozoic, though the Upper Cretaceous sediments of the Chalk Group are markedly condensed over its crest. Local transpression caused some areas at the southern end of the Dowsing Fault Zone to retain positive relief until near the end of the Cretaceous (Glennie and Boegner, 1981), although local transtension farther north along the same fault zone allowed more than 800 m of chalky limestones to accumulate close to active growth faults. Hancock and Scholle (1975) have demonstrated that the Sole Pit Trough was a zone of relatively slow subsidence. The Cleaver Bank High, which had been subsiding since the Early Cretaceous, was established as the main depocentre in the offshore area, and accumulated more than 1000 m of Upper Cretaceous sediments. The Chalk Group is similarly thick in rim synclines between contemporaneously active salt swells, and is strongly condensed over the crests of these swells.

A major phase of basin inversion during, or at the end of, the Late Cretaceous affected many basins in north-west Europe (Ziegler, 1987), including the Sole Pit Trough and Cleveland Basin (Glennie and Boegner, 1981; Kent, 1980a; Van Hoorn, 1987). In the southern North Sea, erosion was less widespread than during the Late Jurassic regional uplift, although there is uncertainty as to the amount and precise timing of uplift of the inverted basins, for the original thickness of Cretaceous sediments is unknown. Kent (1980a) used stratigraphical evidence to demonstrate that the Cleveland Basin was inverted by between 0.5 and 1.3 km at this time. The Sole Pit Trough has been inverted by between 1.2 and 1.8 km since the Early Cretaceous (Marie, 1975), but some of this may have taken place during mid-Tertiary regional uplift of the south-western North Sea (Bulat and Stoker, 1987). Whatever its amount, the Late Cretaceous inversion in the southern North Sea can be interpreted as resulting from a phase of regional compression that reactivated basement faults by sinistral strike-slip. The displacements along individual faults were very small, and have been likened by Glennie and Boegner (1981, p. 119) to 'deformation of a rectangle into a parallelogram by movement along closely spaced parallel lines'.

An unconformity separates the Chalk Group from overlying Palaeogene sediments across almost all of the southern North Sea. This unconformity was caused by widespread uplift and marine regression from the British Isles and much of the surrounding continental shelf around the end of the Cretaceous Period. Jenyon (1984) has demonstrated that karstic topography became established at the top of the Chalk Group over large areas off East Anglia.

Tertiary and Quaternary subsidence in the North Sea has been dominated by broad, synclinal downwarping towards a depositional axis that extends from the Viking Graben through the Central Graben towards The Netherlands (Parker, 1975; Caston, 1977). Because of this, the depocentres for the major Cenozoic sedimentary units in the southern North Sea lie in Dutch waters. The Palaeogene sequence comprises up to 800 m of dominantly argillaceous, marine sediments.

A second phase of tectonic activity affected the southern North Sea in Oligocene to Miocene times. Many of the basement faults were reactivated by dextral strike-slip (Glennie and Boegner, 1981), which triggered another major phase of halokinetic activity (Van Hoorn, 1987). Most of the salt swells and pillows north of 54°N were initiated in mid-Tertiary times. Seismic interpretation suggests that contemporary regional uplift of the western half of the offshore area caused erosion of between 200 and 400 m of Palaeogene sediments from the inverted Sole Pit Trough. Glennie (1986a) has related this deformation to regional north–south compression caused by continental convergence and collision during the Alpine orogeny. South of the London–Brabant Massif, the Weald Basin was also inverted by Alpine compression during the Oligocene and Miocene (Lake and Karner, 1987).

There are very few Miocene deposits, and only local outliers of Pliocene sediments, in the UK sector of the southern North Sea. Subsidence and sedimentation became more widespread early in the Quaternary; up to 600 m of lower

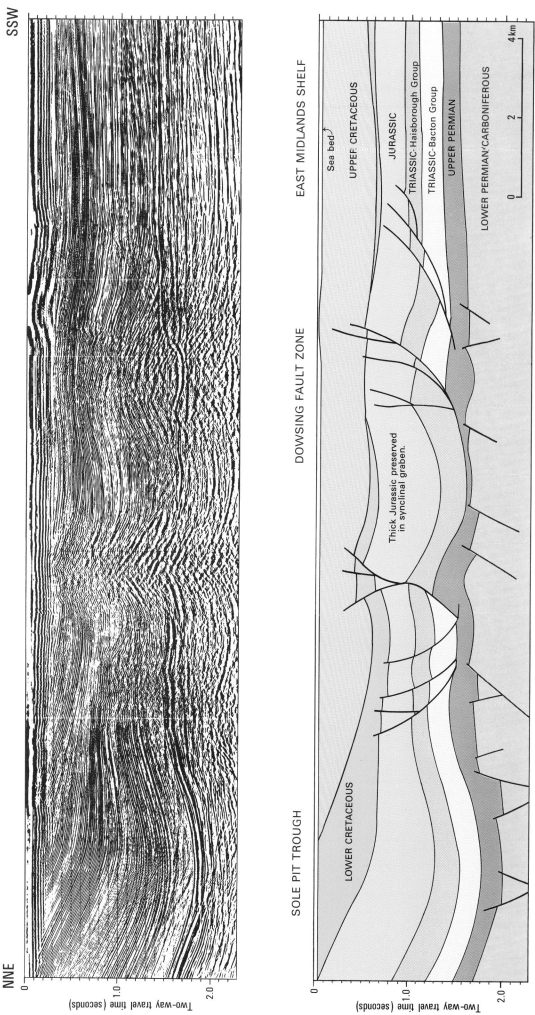

Figure 14 Seismic section and interpretation showing an example of the complex stratigraphical and structural relationships along the Dowsing Fault Zone. Seismic section courtesy of Conoco (UK) Ltd and Arco British Ltd. For location see Figure 13.

and middle Pleistocene sediments record the progressive north-westward expansion of peripheral delta systems across the southern North Sea (Cameron et al., 1987). Sedimentation in the last 400 000 years has been dominated by glacial erosional and depositional processes.

STRUCTURAL STYLES

Post-Devonian basinal subsidence in the southern North Sea has been punctuated by regional episodes of deformation during Late Carboniferous, Late Jurassic, Late Cretaceous and mid-Tertiary times. The earliest of these established a complex system of basement-cutting faults in the Carboniferous and older Palaeozoic rocks. Subsequent episodes of deformation reactivated many of these faults, markedly influencing the Permian and overlying sediments. In the Anglo-Dutch Basin, the style of post-Carboniferous deformation is dominated by halokinetic activity within the Upper Permian salts. Where the halites of the second Upper Permian evaporite cycle are more than 130 m thick (Taylor, 1986), their deformation has effectively decoupled reactivation of the basement faults from deformation of the overlying post-Permian sediments. The Anglo-Dutch Basin is separated from the relatively undeformed basin-marginal areas of the East Midlands Shelf and London–Brabant Massif by the Vale of Pickering–Flamborough Head, Dowsing, and South Hewett fault zones (Figure 13). Stratigraphical relationships adjacent to these and other Mesozoic fault zones indicate that they have had a long and very complex movement history in response to alternating transtensional and transpressional regional stresses.

Carboniferous section

The Carboniferous sediments of the southern North Sea are displaced by numerous west-north-westerly or north-westerly trending faults, almost all of which continue through the top of the Carboniferous into the overlying Permian strata. Detailed mapping from seismic-reflection surveys has revealed that many of the faults developed with wrench or normal components during and soon after the end-Carboniferous Variscan orogeny (Glennie and Boegner, 1981); these may have been reactivated to displace the Permian and younger sediments during subsequent episodes of deformation. Glennie (1986a) and Walker and Cooper (1987) have observed that swarms of faults are separated by strike-parallel zones in which relatively few faults have developed; the fault traces describe an *en-échelon* pattern in some parts, but elsewhere form complex, anastomosing patterns which can be traced for up to 100 km along the direction of strike (Walker and Cooper, 1987). Both of these observations suggest that the fault swarms link to deep-seated crustal fractures, and may have formed in response to wrench reactivation of the Early Carboniferous growth faults during Variscan compression. Many of these faults have a reverse component in the north of the area, which suffered the greatest Variscan uplift and erosion. Some faults change their sense of throw from normal to reverse along strike, confirming that they include a wrench component in their history. The East Midlands Shelf and London–Brabant Massif are cut by relatively few faults, and were stable areas during most of the Carboniferous.

Dowsing Fault Zone and Swarte Bank Hinge Zone

The Dowsing Fault Zone and Swarte Bank Hinge Zone form complementary areas of structural complexity on the south-western and north-eastern margins of the Sole Pit Trough (Figure 13); they have many features in common.

In the north, upward-branching fault patterns (Harding and Lowell, 1979) dominate at structural levels above the Permian; these appear to describe a mainly synclinal, extensional geometry (Glennie and Boegner, 1981; Walker and Cooper, 1987). The Mesozoic sediments are dissected by systems of *en-échelon* asymmetric grabens that trend north-west to south-east; these started to develop during the Late Triassic in the Dowsing Fault Zone, and during the Late Jurassic in the Swarte Bank Hinge Zone. Each graben is bounded by listric normal faults which sole out in the Upper Permian halites (Figure 14); halokinetic deformation of these, effectively decoupled the faulting in the Mesozoic cover from contemporary deformation of the pre-Permian basement. In the Swarte Bank Hinge Zone, the master faults for each graben dip away from the Sole Pit Inversion axis (Walker and Cooper, 1987). Antithetic faults and subsidiary listric faults occur outside the main grabens, and link to the master faults via *décollements* on the Triassic halite members of the Haisborough Group. The Upper Permian halites are strongly attenuated beneath some of the listric, fault-bounded grabens. Jenyon and Cresswell (1987) suggested that withdrawal of the salt was directly responsible for the formation of extensional grabens, but identical Mesozoic structures have formed farther south where the salt is thin or absent. It seems more likely that the salt took a passive role in the deformation (Gibbs, 1986).

Southward stratigraphical thinning of Upper Permian halites causes a change in structural style within both fault zones towards the London–Brabant Massif. Beyond the limit of halite deposition, upward-branching anticlinal 'flower' geometry is common within the Mesozoic section, but synclinal grabens are also present and there are many more subsidiary faults between the Dowsing and Swarte Bank fault zones. All faults within the Mesozoic section here can be traced downwards through the Permian strata into their root zones in the Carboniferous. Many faults display components of apparent dip-slip and reverse-slip movement, which Badley et al. (1989) have related to extension followed by inversion (Figure 15). Tubb et al. (1986) have demonstrated that there are significant changes in the thickness of Lower Carboniferous limestones across some of the deep-seated faults on the northern flank of the London–Brabant Massif; the faulting within the Mesozoic section was caused mainly by Cretaceous reactivation of these Early Carboniferous growth faults.

Glennie and Boegner (1981) deduced from fault geometry that there have been strike-slip movements along the Dowsing Fault Zone over a very long time, in response to both transtensional and transpressional local crustal stresses. By contrast, Walker and Cooper (1987) have invoked no significant post-Variscan strike-slip or compression along the northern part of the Swarte Bank Hinge Zone. Instead, they interpreted the deformation as resulting from extensional, gravitational creep of the post-Permian sediments off the rising, north-eastern flank of the Sole Pit Trough, mainly during Early Cretaceous times. The *en-échelon* arrangement of the Mesozoic grabens may indicate a component of oblique slip in their underlying basement faults.

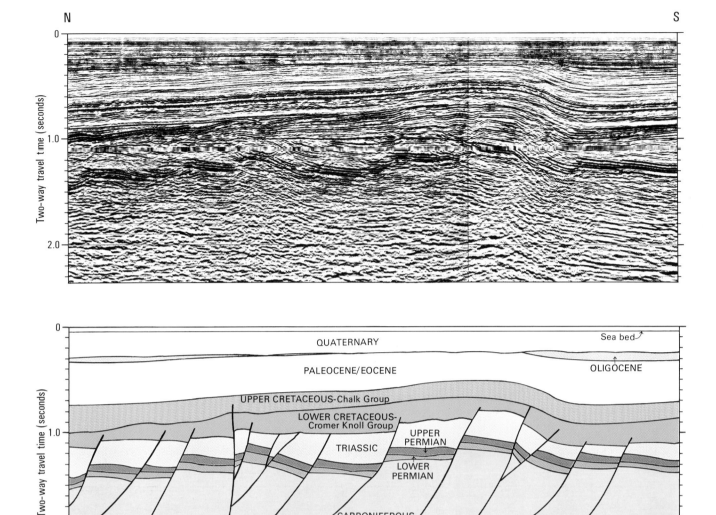

Figure 15 Seismic section and interpretation showing the complex stratigraphical and structural relationships on the northern flank of the London–Brabant Massif. For discussion of the interpretation see Badley et al. (1989). Seismic section courtesy of GECO. For location see Figure 13.

Other Mesozoic fault zones

Away from the Sole Pit Trough, faulting in the Mesozoic sediments of the Anglo-Dutch Basin is localised to a few discrete zones trending approximately east–west or north–south, oblique to the regional basement structure (Figure 13). Three of these fault systems meet the Dowsing Fault Zone along the north-eastern margin of the East Midlands Shelf; these are the Vale of Pickering–Flamborough Head Fault Zone, the Outer Silver Pit Fault and the Central Fracture Zone. Where these fault zones intersect, steeply dipping faults extend from the basement through the Upper Permian halites to displace the overlying Mesozoic strata. Otherwise, the halites have effectively decoupled the post-Variscan deformation of the basement from faulting of the Mesozoic sediments across the whole of this region.

The east–west-trending graben which dominates the Vale of Pickering–Flamborough Head Fault Zone was initiated during the Late Jurassic (Kirby and Swallow, 1987) by failure along the hinge zone between the East Midlands Shelf and the Cleveland Basin. Continuing movement on its listric, southward-dipping master fault and associated antithetic faults caused an upper crustal extension of about 25 per cent across the graben by mid-Cretaceous times; the listric fault soles out in Upper Permian evaporites. Kirby and Swallow (1987) have speculated that the north–south extension was caused by sinistral transcurrent movements along deep-seated faults in the basement.

The Outer Silver Pit Fault was invoked by Glennie and Boegner (1981) to account for a diapiric salt wall extending 30 km east from the Dowsing Fault Zone (Figure 13). There is undoubtedly a large, steeply dipping fault running into the diapir from its western end; this fault extends from the basement through Upper Permian halites to effect an apparent southerly downthrow of more than 300 m in the Triassic sediments. Its displacement of the Mesozoic section dies out quickly to the east. The continuation of the fault beneath the diapir is poorly resolved on seismic-reflection sections, and there is no east–west fault displacing the top of the Carboniferous to the east of the diapir. If the Outer Silver Pit Fault continues to meet the Swarte Bank Hinge Zone, as suggested by Glennie and Boegner (1981), then the dislocation is confined to deeper levels in the basement.

A southerly dipping listric normal fault trends east–west for 75 km along the southern flank of the Mid North Sea

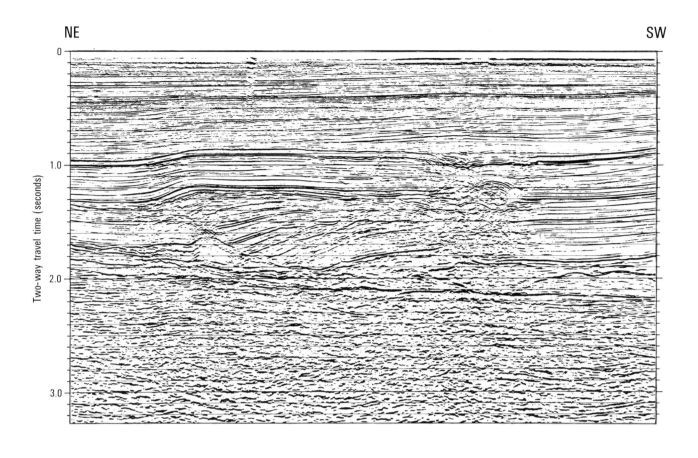

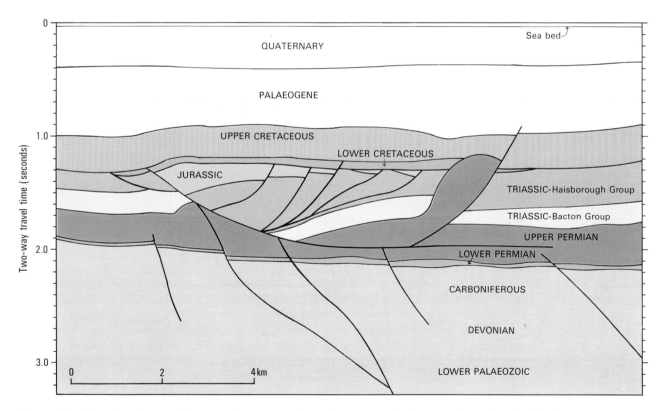

Figure 16 Seismic section and interpretation across a Jurassic extensional graben on the southern flank of the Mid North Sea High. Adapted from Jenyon (1985). Seismic section courtesy of Seismograph Services (England) Limited. For location see Figure 13.

High. The fault has developed along a line of anisotropy within Upper Permian sediments, marking the northern limit of halokinetic deformation of the thick, basinal salts. More than 600 m of Upper Triassic and perhaps Jurassic sediments that are preserved in its hanging-wall prism, which is about 6 km wide (Figure 16), had been completely eroded from its footwall by Early Cretaceous times (Jenyon, 1985). Attenuation of Lower Triassic sediments above the fault zone indicates Jurassic upper-crustal extension of about 1.5 km; this was accompanied by migration of Upper Permian salt to form diapirs in the hanging wall. Later anticlinal folding in many parts of the fault zone (Jenyon, 1985) was caused by post-Cretaceous reverse reactivation of the listric normal fault in a compressive, probably mid-Tertiary, stress regime. By analogy with the Vale of Pickering–Flamborough Head Fault Zone, the structure may have developed above sinistral transcurrent movement on a deep basement lineament, and then was reactivated by dextral transpression during the Tertiary.

Stratigraphical and structural relationships are more complex within the north–south-trending Mesozoic fault systems (Figure 13). The Central Fracture Zone (Kirby and Swallow, 1987) is a northward continuation of the Dowsing Fault Zone, but is strongly oblique to its underlying west-north-westerly trending basement faults. The Dogger Fault Zone (S J Stoker, written communication, 1988) comprises a train of asymmetric grabens above a linear Upper Permian salt swell. It is oblique to the dominant north-north-west trend of basement faults in this region, and the grabens face alternately east and west along the length of the fault zone. There are major thickness changes in the Middle and Upper Triassic, Jurassic, Upper Cretaceous and Tertiary sediments across individual faults, indicating a strike-slip component in the movement history. There is a 1 km sinistral offset of a crosscutting minor graben in the Chalk near the southern end of the fault zone. There are indications that the fault zone developed by Late Triassic and Jurassic growth faulting, and then reactivated by strike-slip in response to Late Cretaceous or mid-Tertiary transpression.

HALOKINESIS IN THE ANGLO-DUTCH BASIN

Mesozoic and early Tertiary sediments have been gently folded over Upper Permian salt swells and pillows across most of the offshore area north of 53°20'N. Many of the halokinetic structures are several hundred metres in amplitude, but there are fewer large salt walls and diapirs in the southern North Sea than in the equivalent evaporite sequence of northern Germany (Brunstrom and Walmsley, 1969).

Sannemann (1968) and Jenyon and Cresswell (1987) have suggested that after deposition, the Upper Permian salt remained in a metastable, potentially mobile state because of its low density compared with the overlying sediments. Halokinetic motion was then triggered by changes in the regional stress regime. The earliest salt movements in the southern North Sea occurred during the Middle Triassic along the Mesozoic fault zones bordering the Sole Pit Trough (Brunstrom and Walmsley, 1969; Walker and Cooper, 1987). Salt structures then developed more widely in this trough throughout the Jurassic in response to the differential stresses generated as it subsided. This led to the formation of primary peripheral sinks containing anomalously thick Jurassic sediments between the active salt swells.

Almost all the salt swells are aligned parallel to the structural grain of the pre-Permian basement. This led Jenyon (1986a) to speculate that the salt structures of the southern North Sea were most active during regional orogenic episodes, when halokinesis was initiated by movements along the many faults which continue upwards from the basement to terminate within the Upper Permian section. Glennie (1986b) inferred that the timing of the salt movements could be used to date the principal movements on the faults.

Many seismic-reflection sections give the impression that brittle deformation of the basement was translated into folding of the overlying sediment by ductile deformation of the relatively mobile salt layer. Brunstrom and Walmsley (1969) observed that most of the largest salt walls and diapirs in the North Sea occur to the south of 54°N, where the majority were active intermittently between Jurassic and mid-Tertiary times. However, the depocentre of the Permian salt basin lay to the north of 54°N, outside the major Mesozoic fault zones, where deformation was delayed until mid-Tertiary times. This implies that in the north either the salt was less mobile, or the shear stresses generated in the salt layer by deformation of the basement were less pervasive until the mid-Tertiary.

Once set into motion, the salt structures of the southern North Sea have adopted a very wide variety of shapes and sizes (Jenyon, 1986b); examples of some are illustrated in Figure 17. The geometry of most of the salt structures demonstrates that differential loading became an increasingly important factor as they continued to develop, causing lateral migration of halite into the growing salt swells. Areas of salt depletion have been preserved as salt withdrawal synclines. Migration may have been particularly rapid at the end of the Jurassic and around the middle of the Tertiary, when the uppermost crust was destabilised by widespread uplift and planation (Jenyon, 1986a).

Suites of uplift-induced normal faults and fractures developed over the crests of many of the largest salt pillows (Jenyon, 1988). As some of these pillows continued to grow, salt was injected along their crestal fractures causing piercement of the overburden, and forming salt walls and diapirs. Where the injected salt encountered a circulatory system of undersaturated water, halite in the piercement was dissolved. Eventually the upper parts of these salt swells were dissolved, leading to collapse grabens in the overburden (Jenyon, 1986b), which remained active for as long as the salt continued to dissolve. Some collapse grabens have displaced the base of the Quaternary by up to 100 m, and late Pleistocene sediments have subsided by up to 10 m (Balson and Cameron, 1985). Many of these grabens may still be active.

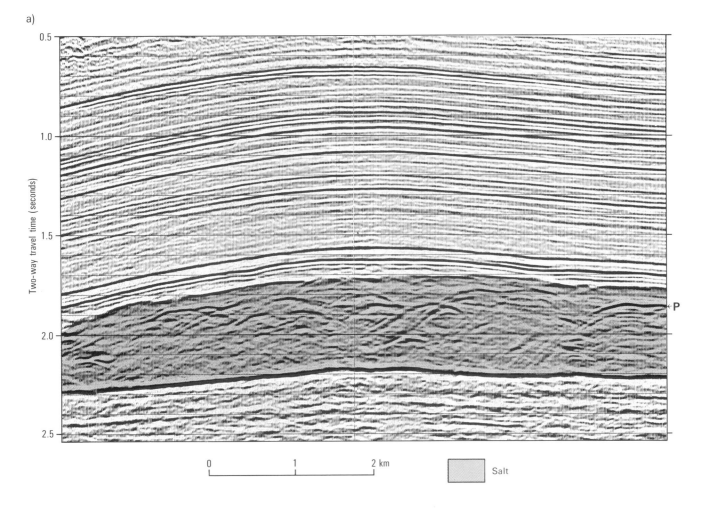

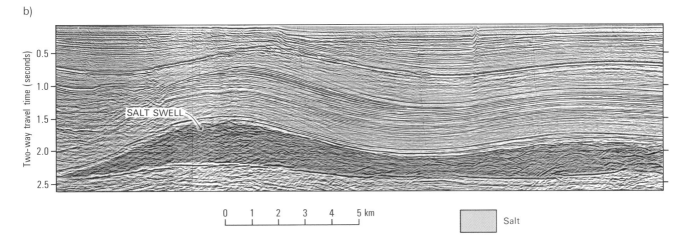

Figure 17 Seismic sections showing examples of halokinetic deformation as a result of the movement of Upper Permian salts.

For locations see Figure 13. (a) illustrates the disruption of a competent limestone bed (P) due to lateral migration of salt. Reproduced after Jenyon and Cresswell (1987) courtesy of Seismograph Services (England) Ltd (SS(E)L). (b) to (d) are from Jenyon (1986b), reproduced courtesy of Elsevier Applied Science Publishers and SS(E)L. The salt swells depicted in (b) are typical of many of those formed in mid-Tertiary times in the centre of the Anglo-Dutch Basin. (c) illustrates a collapse graben formed by minor dissolution of salt at the crest of a large salt pillow. The asymmetric salt diapir depicted in (d) was formed by piercement of the Triassic to Tertiary overburden in Late Jurassic and mid-Tertiary times above an active basement fault.

c)

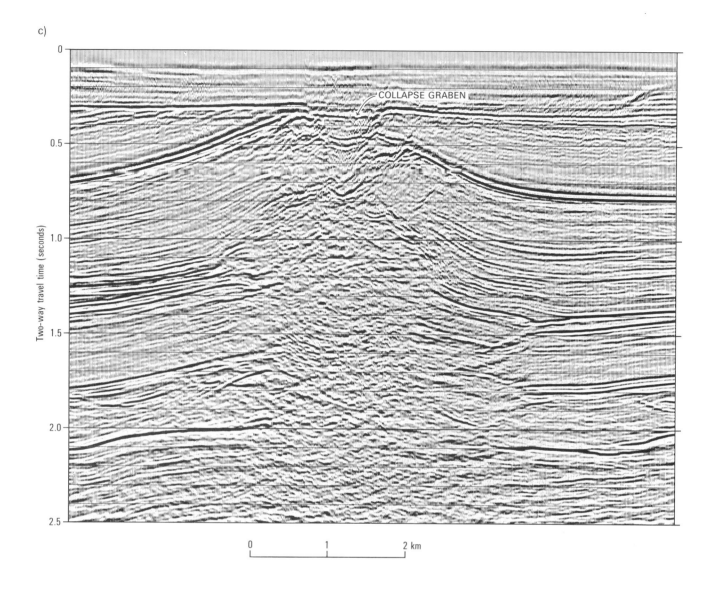

d)

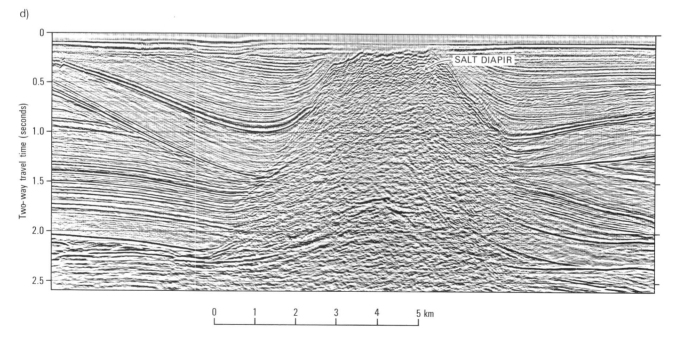

Figure 17 *continued* See page 21 for captions. Note salt features cannot be accurately defined on these diagrams.

5 Carboniferous

By the end of the Devonian, the relief of the Old Red Sandstone continent had been severely reduced by erosion. Early in the Carboniferous, crustal extension caused regional subsidence and allowed the sea to transgress northwards across the substantially peneplaned surface of England; it reached as far as the Northumberland Trough by early Dinantian times. Marine shales, sandstones and limestones then continued to be deposited across most of eastern England and the offshore area between the London–Brabant Massif and the Mid North Sea High during the remainder of the Dinantian. By the middle of the Dinantian, periodic southward advances of a delta system to the north caused a regression to dominantly fluviodeltaic conditions in the Northumberland Trough and its eastward offshore continuation. More widespread delta advances during the Namurian then extended fluvial sandstone deposition and fluviodeltaic conditions southward across the remainder of eastern England and the southern North Sea. Most of the overlying early Westphalian sequence is composed of shale-dominated coal measures deposited in an upper delta-plain environment. Contemporaneously, relatively sandy, redbed-facies sediments were being deposited locally on the flanks of the London–Brabant Massif, and with improved drainage and increasing aridity, these extended across the remainder of the area during the late Westphalian. Stephanian sediments have not been proven in the UK sector.

Both sedimentary facies and palaeomagnetic data demonstrate that the slow drift of the Laurasian continent during the Carboniferous Period carried the southern North Sea from southern, subtropical, arid climes through humid, equatorial conditions into relatively arid, equatorial latitudes (Glennie, 1986b; Besly, 1990). Variscan deformation and uplift in the Late Carboniferous, accompanied by erosion, has caused a variety of Dinantian, Namurian and Westphalian rocks to subcrop beneath the regional unconformity at the base of the Permian in the southern North Sea (Figure 18).

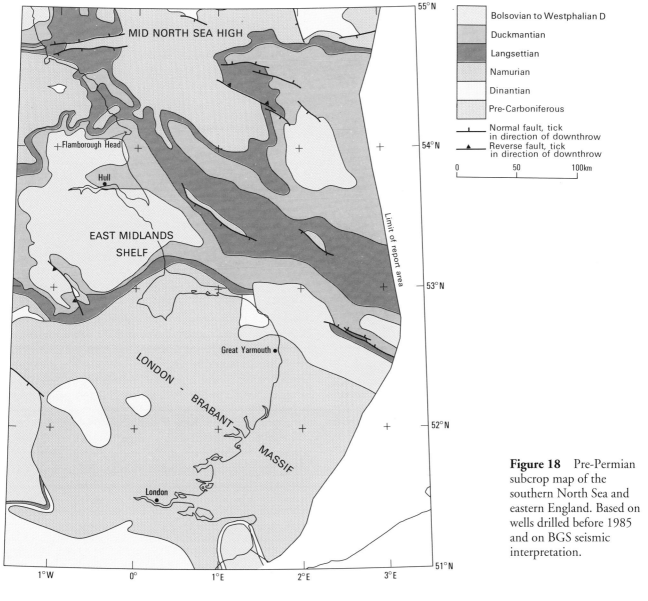

Figure 18 Pre-Permian subcrop map of the southern North Sea and eastern England. Based on wells drilled before 1985 and on BGS seismic interpretation.

Carboniferous strata have been penetrated by more than 95 per cent of the released exploration wells offshore; their present-day depth of burial beneath Permian, Mesozoic and Cenozoic sediments increases from 800 m on the northern flank of the London–Brabant Massif to more than 4000 m beneath the centre of the Anglo-Dutch Basin (Figure 13). The Carboniferous sediments achieve their maximum thickness, perhaps more than 5000 m in central parts of the offshore area, and they are absent only over the crest of the London–Brabant Massif.

Relatively few early wells penetrated more than 100 m of the Carboniferous strata. In many wells drilled before 1980, the beds are described as 'Carboniferous Undifferentiated' on company records, but since the discovery of significant volumes of gas in Westphalian sandstones beneath the Anglo-Dutch Basin, many wells have been drilled more than 500 m into the Carboniferous section. This has provided valuable new data on Westphalian and Namurian stratigraphy (Figure 18); late Westphalian strata are now known to be less extensive beneath the Anglo-Dutch Basin than in the earlier assessments of Eames (1975), Glennie and Boegner (1981), and Smith (1985).

DINANTIAN

Both the thickness and facies of Dinantian (Tournaisian and Viséan) sediments in eastern England vary dramatically across syndepositional faults (Grayson and Oldham, 1987; Smith and Smith, 1989). These faults bound highs which were either persistent land throughout most of the Early Carboniferous or are overlain by condensed or incomplete marine sequences. Basins and half-grabens accumulated thick successions of marine or fluviodeltaic sequences between the highs (Grayson and Oldham, 1987). The London–Brabant Massif was one such high; during late Viséan times, extensive carbonate platforms became established on the East Midlands Shelf and beneath the Kent Coalfield, on its northern and southern flank respectively. Two other highs, the Alston and Askrigg blocks (Figure 19), were not finally submerged by the Carboniferous sea until late in the Dinantian, and are capped by shallow-water carbonates. The basins included the Gainsborough Trough and the Edale and Widmerpool gulfs of central England, where thick sequences of turbidites and deep-water shales accumulated. The Northumberland Trough was a much larger basin, comprising an asymmetrical, southward-deepening half-graben; it contains more than 4000 m of Dinantian sediments (Kimbell et al., 1989).

Delta-related Yoredale sedimentation cycles encroached into the Northumberland Trough at various intervals throughout the Dinantian, and extended farther south across the highs and lows of northern England before the end of the Dinantian (Taylor et al., 1971). Each Yoredale cycle records a major phase of delta progradation followed by re-establishment of marine conditions. A typical cycle comprises a basal marine limestone overlain by a variety of delta-related sedimentary facies, including thick sandstones deposited in distributary channels, and thin but widespread coals representing abandonment of delta lobes (Elliott, 1975). Many cycles are completed by post-abandonment shoreline deposits.

Most models now invoked to explain the wide range of Dinantian sedimentary environments in northern and central England involve the formation of the small, fault-bounded basins and half-grabens by north–south crustal extension (Leeder, 1982), perhaps accompanied by transtensional shear (Grayson and Oldham, 1987). Locally, thickness and facies were also influenced by Early Carboniferous eustatic changes in sea level, as well as by differential rates of subsidence (Grayson and Oldham, 1987). Bott (1967; 1987) suggested that two of the most persistent highs, the Alston and Askrigg blocks, were kept particularly buoyant by their Caledonian granite cores. Similarly, the offshore continuation of the East Midlands Shelf was a zone of relatively slow Dinantian subsidence (Leeder and Hardman, 1990) above a large subsurface granite pluton (Donato and Megson, 1990).

Offshore, Dinantian sediments have been sampled in only a small number of wells around the periphery of the Anglo-Dutch Basin (Figure 19). These have confirmed that late Viséan platform carbonates extend from East Anglia along the northern flank of the London–Brabant Massif, and that the marine and fluviodeltaic deposits of the Northumberland Trough continue offshore along the southern margin of the Mid North Sea High. Between these areas, it seems likely that the complex pattern of highs and lows with syndepositional faults continues from eastern England across the centre of the southern North Sea, but there are insufficient data to locate these in most of the offshore area. It does seem possible, however, that the northern limit of carbonate shelf deposition on the flank of the London–Brabant Massif was determined by one or more such faults, perhaps located at the Dowsing Fault Zone and its south-eastwards continuation across the Flemish Bight. Figure 19 offers a schematic profile through the Dinantian sediments of the southern North Sea.

Dinantian sediments in the north

There is a striking lithological similarity between the lower Dinantian sequence penetrated in well 44/2-1 (Figure 20) and that of the Northumberland Trough, 260 km to the west. The Dinantian sediments are 660 m thick in the well, and are overlain unconformably by Permian strata; perhaps as much as 1000 m of uppermost Dinantian sediments have been removed by pre-Permian erosion. The Devonian–Carboniferous boundary was penetrated near the base of well 44/2-1, and the lowermost Tournaisian fluvial sandstones and shales rest conformably on Devonian fluvial sediments (Besly, 1990). A dolomite bed about 60 m above the base of the Carboniferous heralds a transition to a more complex sequence of dolomitic limestones, sandstones and shales. There are comparable dolomites containing a restricted fauna of ostracods, bivalves and gastropods within the equivalent Cementstone Group of Northumberland, which Leeder (1974) has interpreted as indicating deposition in a periodically hypersaline, probably lagoonal, environment. Onshore, the dolomites become more fully marine westwards along the axis of the Northumberland Trough.

An influx of increasingly thick sandstone beds occurs near the top of the Tournaisian in well 44/2-1 (Figure 20), slightly earlier than in Northumberland. These and the overlying lower Viséan sediments are similar in lithology to the Fell Sandstone Group of northern Northumberland, which Hodgson (1978) interpreted as the sheet-sand deposits of braided-river courses that flowed south or south-eastwards across a fluviodeltaic alluvial plain. Interbedded sandstones and shales with thin coal seams occur at the top of the Dinantian section in well 44/2-1, and in nearby well 44/7-1. These were deposited contemporaneously with the Scremerston Coal Group of Northumberland during a major regression of the braided-river facies that allowed fluviodeltaic sedimentation to become established during mid-Dinantian times.

In northern England, the remainder of the Dinantian sequence is dominated by Yoredale cycles. Delta-related or marine sandstones and shales predominate, but there are also

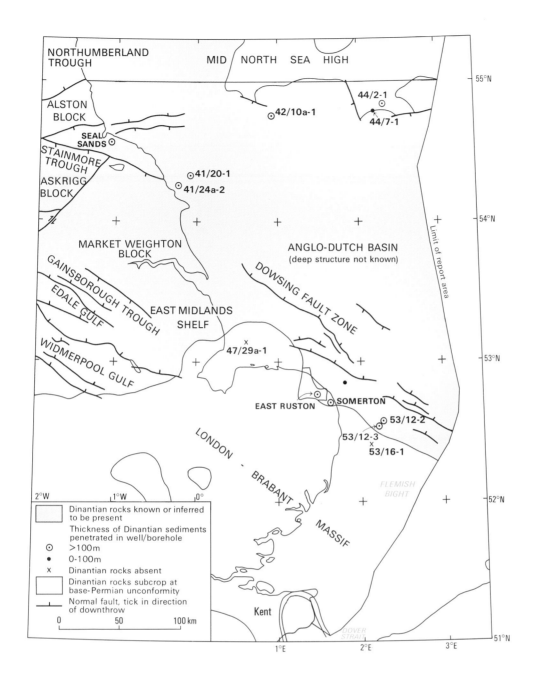
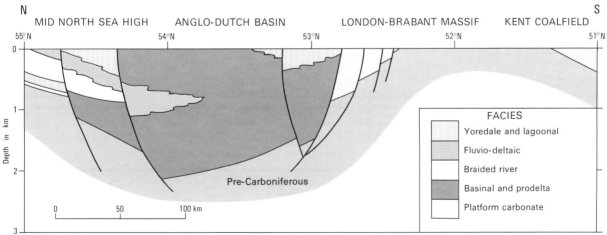

Figure 19 Carboniferous structural features, and the locations of wells that have penetrated Dinantian strata, together with a schematic profile through the Dinantian sediments of the southern North Sea in which the datum is the base of the Namurian. Structural interpretation after Plant et al. (1988).

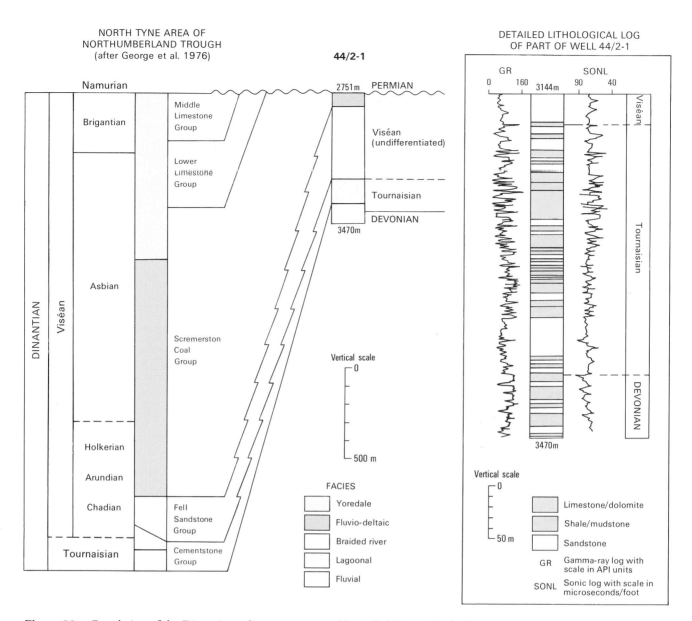

Figure 20 Correlation of the Dinantian sediments penetrated by well 44/2-1 with the Dinantian sequence of the Northumberland Trough, with detail of the transition from fluvial to lagoonal facies within the Tournaisian sequence at the base of well 44/2-1. For location of well see Figure 19.

thin seams of coal and beds of commonly shelly, marine, limestone. The Seal Sands borehole (Figure 19) has proved 1200 m of uppermost Dinantian sediments composed of the Yoredale cycles on the east coast of England (Dunham and Wilson, 1985). Carbonates and coal seams are intercalated with sandstones and shales in the Dinantian sequences penetrated by wells 41/20-1, 41/24a-2 and 42/10a-1 (Figures 19 and 21), suggesting that shallow-marine and deltaic facies had extended over most of the northern offshore area by the end of the Dinantian.

There is a broad zone beneath the Anglo-Dutch Basin in which the Dinantian sediments are deeply buried and have not been penetrated by exploration wells. Onshore, the Yoredale facies in the Seal Sands borehole (which terminated above the base of the Dinantian in sediments of early Viséan age) is underlain by 2249 m of mudstones, siltstones and sandstones with a few thin limestones (Dunham and Wilson, 1985). Similar sequences of marine sediments may underlie beds of the Yoredale facies beneath the northern part of the Anglo-Dutch Basin, but only adjacent to Dinantian syndepositional faults are they likely to achieve such remarkable thicknesses. It seems more likely that the total Dinantian sequence is generally between 1000 and 2500 m thick in the central offshore area (Figure 19), perhaps thinning to less than 1000 m over structural highs.

Dinantian sediments along the flanks of the London–Brabant Massif

Adjacent to the London–Brabant Massif, well 53/12-2 penetrated the uppermost 450 m of Dinantian limestones in an area where seismic-reflection sections indicate that these may be locally more than 1 km thick. Dinantian syndepositional faults can be observed dipping towards the north-east on the seismic sections, and Tubb et al. (1986) have speculated that reef limestones may have accumulated on the footwalls of such faults.

In well 53/12-2, the Dinantian comprises a lower dolomitic limestone, in places argillaceous, overlain by about 300 m of limestone containing oolite beds. Fossiliferous beds containing abundant corals, brachiopods, bivalves, bryozoans and crinoids occur within equivalent sediments at Somerton

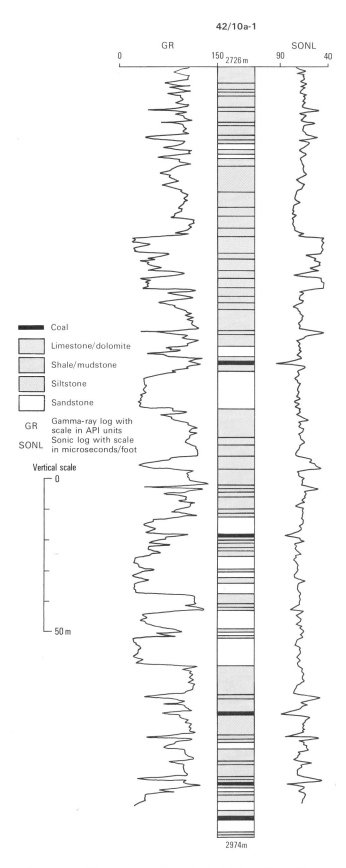

Figure 21 The transition from fluviodeltaic to Yoredale facies in the uppermost Dinantian sediments of well 42/10a-1. For location see Figure 19.

and East Ruston in East Anglia (Figure 19); these are likely to be intercalated locally within the Lower Carboniferous limestones offshore.

To the south-east, Dinantian carbonates are up to 750 m thick in The Netherlands, where they were deposited in an open-marine to restricted, sublagoonal environment with practically no siliciclastic input (van Staalduinen et al., 1979). There are widespread gaps in the sedimentary sequence, with locally condensed sequences, and reworked Tournaisian fossils within the upper Viséan limestones. This suggests that there was contemporary, intermittent uplift and erosion along the northern flank of the London–Brabant Massif (van Staalduinen et al., 1979). The limestones are notably absent in wells 47/29a-1 and 53/16-1 (Figure 19), where Upper Carboniferous rocks rest directly on Lower Palaeozoic basement; it seems that the limestones had already been eroded to their present southern limit by mid-Carboniferous times, and that they may never have extended over the crest of the London–Brabant Massif.

South of the London–Brabant Massif, up to 300 m of Dinantian, bituminous, pyritic, oolitic or finely crystalline limestones and calcite mudstones have been proved beneath Langsettian (Westphalian A) coal measures in the Kent Coalfield (George et al., 1976). Mid-Viséan rocks underlie the Westphalian throughout the coalfield, and Tournaisian faunas have been described from near the base of the limestones in two boreholes (Mitchell, 1981). If later Viséan carbonates were deposited in this area, they have been completely removed by pre-Westphalian erosion. Seismic sections offshore confirm that the Dinantian limestones of Kent continue beneath the Dover Strait (Figure 19), and south-east of Dover they may be as much as 900 m thick. Their erosional limit along the southern flank of the London–Brabant Massif is uncertain.

NAMURIAN

The differential bathymetry inherited from the Dinantian continued to play a major role in controlling depositional patterns in central England during Namurian times (Collinson, 1988). Basinal mudstones, turbidites and turbidite-fronted delta sequences infilled most lows during the early Namurian, before a succession of stacked, sheet-delta sequences extended diachronously across the whole of northern and central England during the remainder of the Namurian. These sediments display an overall cyclic pattern; they are mainly nonmarine, but Ramsbottom (1977) interpreted six major, and several minor, eustatic rises in sea level on the basis of widespread goniatite-bearing marine shales.

Namurian rocks have been penetrated by more than 20 wells in the southern North Sea (Figure 22); most were terminated before reaching Dinantian sediments, and the Namurian sequence has been truncated beneath the base-Permian unconformity in many of the remainder. Despite this, the sediments encountered are likely to be representative of much of the offshore Namurian, and a schematic profile has been constructed to illustrate the lateral relationships of the principal Namurian sedimentary facies (Figure 22). The Namurian sequence appears to be attenuated in well 41/20-1 (Figure 23) relative to wells in the centre of the Anglo-Dutch Basin. As in central England (Collinson, 1988), basinal tectonics and local subsidence patterns are likely to have affected bathymetry and the style of sedimentation offshore until at least mid-Namurian times.

Namurian sediments in the northern wells

Well 43/25-1 (Figure 22) has proved that Namurian sediments are locally more than 1250 m thick in the north-central part of the Anglo-Dutch Basin. Its basal sediments were assigned a possible very early Namurian age by the well oper-

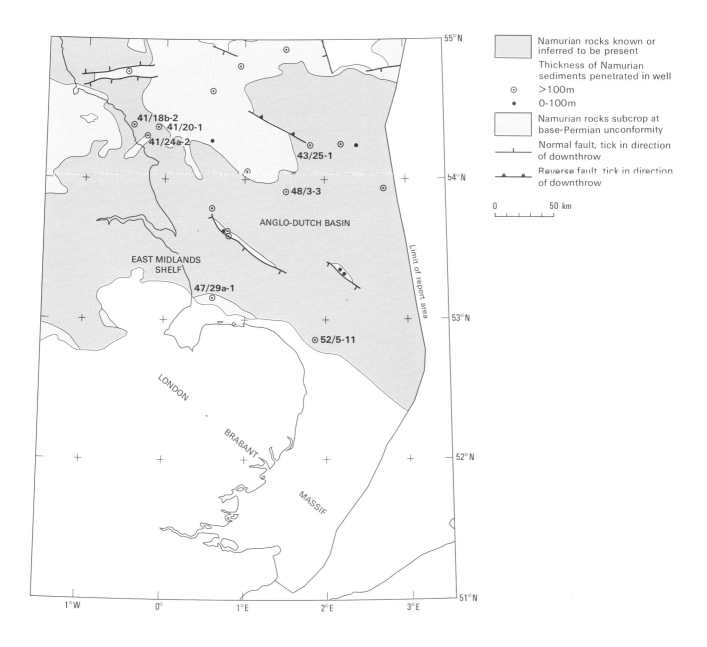

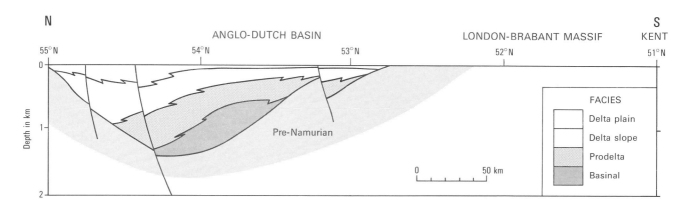

Figure 22 Distribution of Namurian sediments, with the locations of offshore wells that penetrate them, and a schematic profile through the Namurian sediments of the southern North Sea in which the datum is the base of the Westphalian.

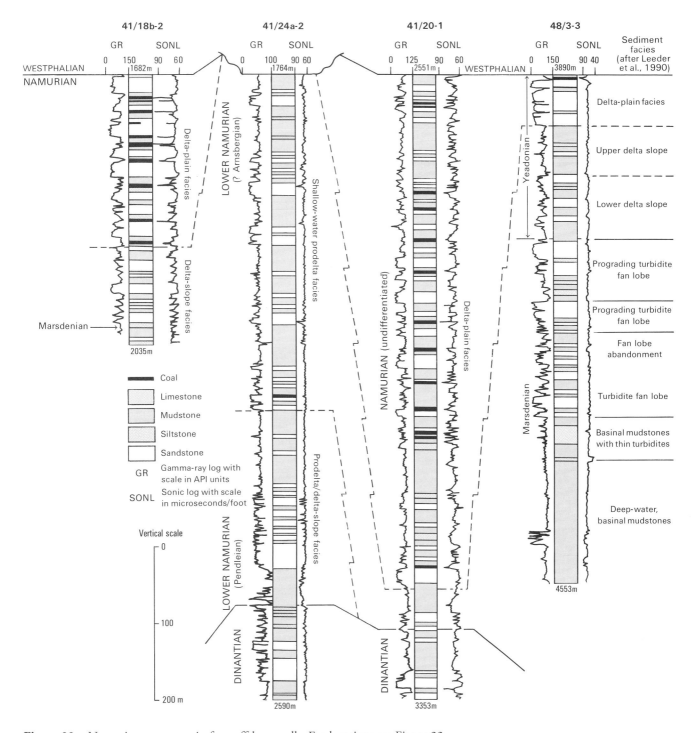

Figure 23 Namurian sequences in four offshore wells. For locations see Figure 22.

ators, suggesting that the well terminated close to the Dinantian–Namurian boundary. Leeder and Hardman (1990) have observed that there may be locally more than 2000 m of Namurian sediments above the Dinantian depocentres. For comparison, Smith and Smith (1989) have noted that the Namurian attains a maximum thickness of 1400 m in north-eastern England.

Leeder and Hardman (1990) have used Collinson's (1988) observations in northern England to deduce that the Namurian sediments of the southern North Sea bear the imprint of repeated progradation of major fluviodeltaic systems across both areas. Successive advances of these fluviodeltaic systems caused the area covered by delta-plain sedimentation to expand gradually southwards across the North Sea. As a result, the Namurian sediments penetrated by wells in the centre of the Anglo-Dutch Basin display a large-scale upward-shallowing sequence of sedimentary facies.

The middle and upper Namurian (Marsdenian to Yeadonian) in well 48/3-3 (Figure 23) provides a good illustration of the progressive infill of a deep basin as the fluviodeltaic systems encroached (Leeder et al., 1990). The 140 m-thick unit at the base of the well is dominated by black mudstones that were deposited when the basin was still remote from coarse-grained, clastic, sediment supply. Beds of turbidite siltstone are increasingly intercalated within the mudstones in the overlying 75 m, and the thin sandstones which occur within the 90 m section above are prodelta turbidites deposited during the progradation of a turbidite fan lobe into the basin. Two further turbidite fan lobes are represented within the overlying 140 m sequence of sandstones, siltstones and mudstones, and each of these culminates in a

20 m-thick, multistorey, coarse-grained, fan-channel, sandstone bed. The remainder of the Namurian section is probably Yeadonian in age (Leeder et al., 1990). The rapidly alternating sandstones and mudstones at the base of the Yeadonian comprise a lower delta-slope facies, and are succeeded by an upper delta-slope sequence of mudstones and thin sandstones. The fluviodeltaic deposits at the top of the Namurian include thin coal seams and a 28 m-thick, medium- to very coarse-grained sandstone bed that was deposited in a distributary channel. Leeder et al. (1990) have suggested that the thickest sandstone intervals in well 48/3-3 may correlate with the sandy sheet-delta sediments of the youngest of the 'Millstone Grit' cycles of northern England.

Nearer to the English coast, 704 m of basal Namurian sediments were penetrated by well 41/24a-2 (Figures 22 and 23). This section and the basal 300 m of the Namurian in nearby well 41/20-1 are mudstone dominated, but siltstones and thin beds of fine-, medium- and coarse-grained sandstone occur throughout, and are locally abundant. There are carbonates intercalated with deltaic deposits in the underlying Dinantian section in well 41/24a-2, suggesting that the Namurian sediments were deposited in shallow-water, prodelta and delta-plain, fluviodeltaic environments. Deep-water, basinal facies may have become restricted mainly to the centre of the Anglo-Dutch Basin by early Namurian times.

Well 41/18b-2 records an upward transition from mudstone-dominated to relatively sandy deltaic facies within its 353 m-thick Namurian section (Figure 23). The sediments near the base of the well have been dated as late Namurian, perhaps Marsdenian, on the basis of goniatites and foraminifera (N J Riley, written communication, 1989). The fluviodeltaic sandstone beds are up to 23 m thick; their gamma-ray profiles are interpreted as displaying deposits from a wide range of major and minor distributary channels, prograding crevasse splays, and minor mouth-bars (cf Steele, 1988). Coal seams are unusually abundant within the upper Namurian in well 41/18b-2, and also occur within the uppermost 475 m of the Namurian in well 41/20-1 (Figure 23). Thick, interbedded, distributary-channel sandstones are notably absent in the latter well, suggesting that a shallow-water delta or lower-delta, floodplain environment prevailed here through the late Namurian. Only the fluviodeltaic facies occurs in all wells which penetrated upper Namurian strata in the south-central part of the Anglo-Dutch Basin.

Namurian sediments adjacent to the London–Brabant Massif

Three wells appear to provide conflicting evidence on the basal relationships of the southernmost Namurian deposits in

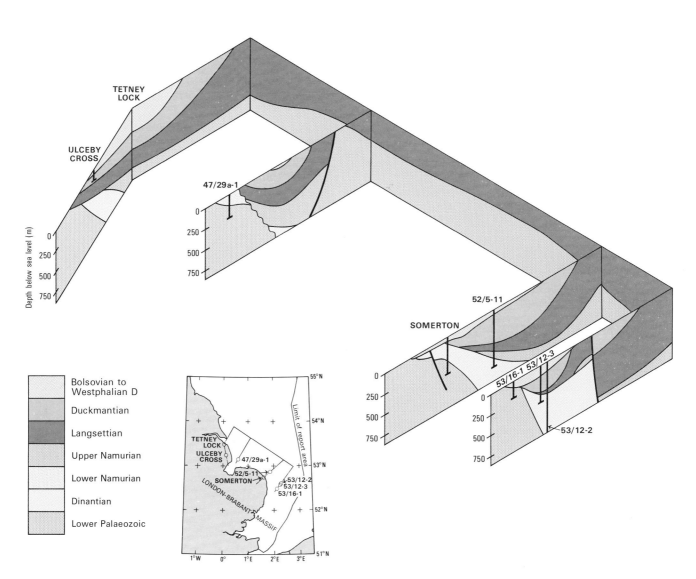

Figure 24 Fence diagram illustrating schematically the stratigraphical relationships of Carboniferous rocks on the northern flank of the London–Brabant Massif.

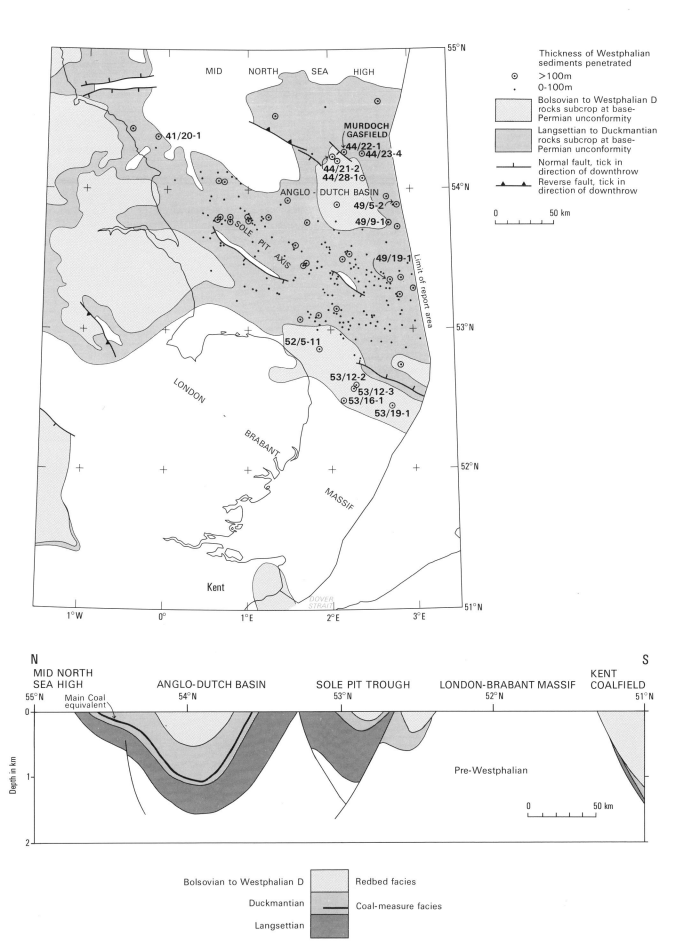

Figure 25 Distribution of Westphalian sediments, and the locations of offshore wells that penetrate them, together with a schematic profile through the Westphalian sediments of the southern North Sea in which the datum is the base of the Permian.

the North Sea. In well 52/5-11 (Figure 24), approximately 130 m of Yeadonian (late Namurian) sandstones and mudstones rest unconformably on Dinantian limestones. This led Ramsbottom et al. (1978) to speculate that progressively younger Namurian beds onlap southwards on to Dinantian or older rocks along the northern flank of the London–Brabant Massif, as around other upland areas of Britain and Ireland. Some 90 km along strike to the west, well 47/29a-1 proved 169 m of lower Namurian sandstones and shales resting on Ordovician basement; later Namurian and Westphalian deposits are notably absent, and may have been eroded beneath the base-Permian unconformity. Farther along strike, Namurian rocks are completely absent in south Lincolnshire, where Langsettian (Westphalian A) coal measures rest on Dinantian limestones, as they do to the south of well 52/5-11.

If the dating by the well operators is correct, the evidence from well 47/29a-1 suggests that a transgression soon after the end of the Dinantian extended sedimentation locally beyond the Dinantian limestones on to the London–Brabant Massif (Figure 24). Regional mid-Namurian uplift of the massif then caused regression and erosion of lower Namurian deposits over most of this southern area. A second transgression subsequently extended late Namurian and Westphalian sedimentation progressively southwards across the flank of the London–Brabant Massif. The stratigraphic onlap envisaged by Ramsbottom et al. (1978) was therefore delayed until the late Namurian, and where both lower and upper Namurian deposits are preserved to the south of the Anglo-Dutch Basin, they are likely to be separated by an unconformity. Namurian rocks have not been recorded from the Kent Coalfield (Ramsbottom et al., 1978), and are unlikely to occur beneath the Dover Strait.

WESTPHALIAN

Three comparable and widespread marine bands divide the Westphalian deposits of central England into four chronostratigraphical units. Deltaic sedimentation from Langsettian (Westphalian A) through Duckmantian (Westphalian B) to early Bolsovian (Westphalian C) times deposited the coal-bearing facies of the Coal Measures. Primary redbed-facies sediments were deposited in northern and central England during later Bolsovian and Westphalian D times.

Guion and Fielding (1988) deduced that most of the deltaic sedimentation took place on a low-lying, paralic plain that was occupied by swamps with brackish and freshwater lagoons in which muds, silts and layers of plant debris accumulated. Major river distributaries brought channel sands on to the plains, and breaching of these channels allowed crevasse-channel sands and crevasse-splay sands to be deposited in the interdistributary lagoonal areas. Irregular subsidence resulted in cyclic sequences of deposition; these show much variation in thickness, with lateral splitting and coalescing of beds. The coal-bearing facies was superseded by a primary redbed facies composed mainly of mud-dominated alluvial and lacustrine sediments, but there are many intercalations of minor meandering channel deposits and widespread pedogenic textures (Besly, 1990).

In England, both the coal-bearing and primary redbed facies of the Westphalian have been traditionally referred to the Coal Measures (Ramsbottom et al., 1978), although there are no coals within most of the redbed sequence. There is no formal stratigraphical nomenclature for the offshore deposits, although the redbeds have been variously described as 'Barren Red Beds' (Rhys, 1974), 'Barren Measures' (Tubb et al., 1986), or the 'Barren Red Group' (Besly, 1990). In this report, the informal terms coal measures and redbeds will be used, and these are described separately.

More than 250 exploration wells have penetrated Westphalian strata in the southern North Sea, and almost all have encountered the Langsettian and Duckmantian coal measures (Figure 25). The redbeds are preserved only in the centre of the Anglo-Dutch Basin and along the flanks of the London–Brabant Massif (Figure 25). Where the Langsettian and Duckmantian coal measures subcrop at the base Permian unconformity, they have been reddened by secondary oxidation, locally for more than 50 m beneath their top. This oxidation has altered thin coal seams directly beneath the unconformity to dolomite in at least one well off the north-east coast of East Anglia.

In northern England, there are 1525 m of coal-bearing Langsettian and Duckmantian sediments in the centre of the Pennine Basin (Calver, 1969). The overlying redbed deposits are also 1525 m thick in the Lancashire and North Staffordshire coalfields, where they may range up to Stephanian in age. The maximum combined thickness of Westphalian and Stephanian deposits onshore is about 3050 m (Calver, 1969). The offshore wells indicate that up to 1600 m of Westphalian sediments are preserved in the centre of the Anglo-Dutch Basin, of which about 1030 m are coal bearing. Elsewhere, relatively few of the wells have penetrated to depths sufficient to deduce Westphalian thickness. Glennie (1986b) has estimated that the coal-bearing sequence preserved beneath the base-Permian unconformity is about 1200 m thick along a Sole Pit axis of late Westphalian inversion. There are indications that the Langsettian sequence may be appreciably thicker here than beneath the centre of the Anglo-Dutch Basin.

Coal-measure facies in the Anglo-Dutch Basin

As in northern England, the coal measures of the Anglo-Dutch Basin are dominated by grey mudstones and silty mudstones, with varying proportions of sandstone, coal and seat earth. Calver (1969) counted about 80 coal seams in the Pennine Basin, and Glennie (1986a) has estimated that coals form between 3 and 4 per cent of the total thickness of Westphalian deposits offshore. Fine- to coarse-grained sandstone beds, generally less than 10 m thick, make up less than 20 per cent of the coal-bearing sequence in the centre of the Anglo-Dutch Basin. To the south, the coal measures appear to become more argillaceous, and contain about 10 per cent sandstone in wells adjacent to the London–Brabant Massif. Most of the thickest and coarsest-grained sandstones were deposited in fluvial distributary channels, and although these account for less than 5 per cent of the total sequence, one forms the main reservoir for the Murdoch Gasfield and is 40 m thick in nearby well 44/23-4 (Figure 25). Similar coarse-grained sandstones are up to 30 m thick onshore in the Durham Coalfield, where they occur in belts between 15 and 20 km wide that can be traced for tens of kilometres in the direction of channel flow (Fielding, 1986).

The stratigraphy of the coal measures is best established offshore in the north-central part of the Anglo-Dutch Basin, where several wells have each penetrated more than 500 m of Westphalian section. The Subcrenatum Marine Band, which defines the Namurian–Westphalian boundary in England and Wales (Ramsbottom et al., 1978), occurs within thick shale sequences in wells 44/22-1 and 44/23-4 (Figure 26). The well operators placed the Namurian–Westphalian boundary at the base of a distributary channel sandstone in both wells, but the marine band is more likely to coincide with the highest natural gamma-ray peaks within the underly-

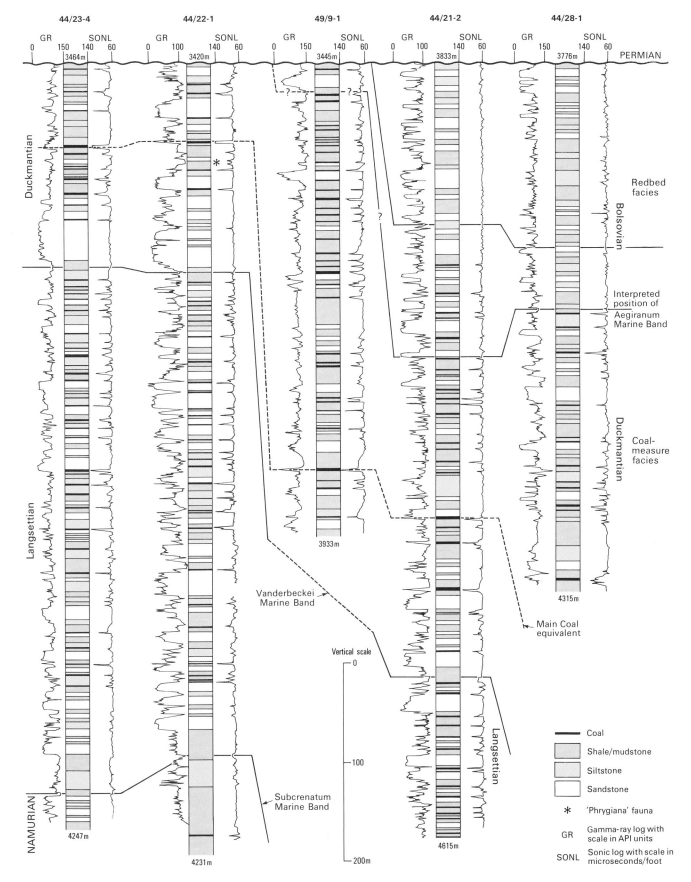

Figure 26 Westphalian sequences from five wells in the Anglo-Dutch Basin. For locations see Figure 25.

ing shales, as in Figure 26. High gamma-ray peaks followed by a gradual upward decrease in natural radioactivity have been correlated with marine bands in the Yorkshire and Derbyshire coalfields (Knowles, 1964).

The Vanderbeckei Marine Band defines the base of the Duckmantian in England (Ramsbottom et al., 1978), and has been positively identified within shale horizons between 460 and 515 m above the Subcrenatum Marine Band in three offshore wells (Figure 26). The Aegiranum Marine Band, defining the Duckmantian–Bolsovian boundary in England, occurs close above the highest coal seam in well 44/28-1. The sequence penetrated by well 49/9-1 (Figure 26) suggests that the Duckmantian and early Bolsovian coal measures have a combined thickness of about 550 m in the central offshore area.

Guion and Fielding (1988) envisaged that a lower delta-plain or shallow-water environment persisted from the late Namurian into early Langsettian times in the Pennine Basin. The offshore data confirm that this environment was continuous across most or all of the southern North Sea. As in the Pennine Basin, there are relatively few thin coals for about 200 m above the base of the Westphalian in wells 44/22-1 and 44/23-4 (Figure 26). In northern England, there are many thin marine bands within this part of the Langsettian sequence, and sandstone beds are particularly extensive (Calver, 1969). The wells drilled so far, suggest that the equivalent sandstones are thinner and finer-grained offshore, implying a less extensive network of fluvial distributary channels than in northern and central England.

Coal seams within the remainder of the Langsettian and most of the Duckmantian coal measures of northern England are many and thick; sandstones are less abundant, there are few marine bands, and freshwater mussel beds are common. Guion and Fielding (1988) have interpreted these deposits as an upper delta-plain facies, perhaps reflecting major delta progradation southwards across the Pennine Basin during early Langsettian times. The coal seams are also thickest and most numerous within the upper Langsettian and Duckmantian deposits of the southern North Sea, yet in some offshore wells there are thicker distributary channel sandstone beds intercalated within these than within the underlying lower delta-plain facies. Sandstones are most abundant in the middle part of the Langsettian in well 41/20-1 off Yorkshire, and become more numerous and coarser-grained upwards through the Langsettian in wells 49/5-2 and 49/19-1 near the median line (Figure 25). They are thickest and coarsest close above the base of the Duckmantian adjacent to the Murdoch Gasfield (Figure 25). None of the other wells which have penetrated thick Westphalian sections display significant variations in the proportion of sandstone in the coal measures.

Perhaps the most easily identified marker bed offshore is an unusually thick coal seam between 90 and 150 m above the base of the Duckmantian (Figure 26), which forms a conspicuous seismic marker in the north-central part of the Anglo-Dutch Basin. A freshwater mussel band 18 m below this coal seam in well 44/22-1 has yielded a fauna typical of the '*Phrygiana*' faunal belt onshore (N J Riley, written communication, 1989). In Durham, the '*Phrygiana*' faunal belt is overlain by the Main Coal (Ramsbottom et al., 1978), which is about 130 m above the base of the Duckmantian. A thick coal seam occurs at a similar stratigraphical level in most of the coalfields of England and Wales (Ramsbottom et al., 1978), and the thick offshore seam is believed to be a close time equivalent of the Main Coal.

Marine bands are relatively numerous within the uppermost Duckmantian and basal Bolsovian of northern England (Calver, 1969); Guion and Fielding (1988) have speculated that the deltas may have receded to allow a return to lower delta-plain conditions late in Duckmantian times. There are insufficient wells penetrating this part of the sequence to determine whether a comparable facies change occurred offshore.

Coal-measure facies around the London–Brabant Massif

Wells off the coast of East Anglia provide good evidence that Langsettian and Duckmantian coal measures here overstep Namurian and Dinantian on to Lower Palaeozoic basement. In well 52/5-11 (Figure 27), these coal measures have a combined thickness of 464 m, and rest conformably on Namurian sediments. Some 40 km to the south-east, the Namurian is absent in well 53/12-3, and there are only 26 m of uppermost Langsettian and 200 m of Duckmantian beds lying unconformably on Dinantian limestones (Figure 24). In both wells, the Duckmantian sediments may have been partially eroded beneath a mid-Westphalian unconformity (Tubb et al., 1986). The Main Coal marker is less clearly defined in the attenuated Duckmantian sequence off East Anglia than in the Anglo-Dutch Basin.

The Langsettian and Duckmantian coal measures have a combined thickness of less than 220 m in Kent on the southern flank of the London–Brabant Massif (Bisson et al., 1967), and, as in well 53/12-3, there are no basal Langsettian sediments. No exploration wells have been drilled offshore, but seismic profiles suggest that early and late Westphalian sediments may expand from their maximum combined thickness of almost 900 m in Kent to more than 1600 m beneath the Dover Strait. If so, the coal measures may be appreciably thicker offshore than in Kent.

On both flanks of the London–Brabant Massif, the coal measures comprise mudstone-dominated sequences. Thick distributary-channel sandstones are only developed locally, and the sequences were deposited in a deltaic environment (Read, 1979; Tubb et al., 1986). Coal seams are present throughout the Kent Coalfield, and there are relatively few marine bands, suggesting deposition in an upper delta-plain environment (cf. Guion and Fielding, 1988). Nevertheless, both the Vanderbeckei and Aegiranum marine bands are widespread in Kent, where they contain some of the richest marine faunas in Britain at these levels (Ramsbottom et al., 1978). The Kent No. 11 seam may correlate with the Main Coal (Ramsbottom et al., 1978).

Redbed facies of the Anglo-Dutch Basin

On seismic sections across parts of the Anglo-Dutch Basin, a transparent seismic zone overlies the high-amplitude, continuous, parallel reflectors typical of the coal-bearing sequence (Figure 28). The transparent zone is characteristic of a late Westphalian redbed facies, whose lower boundary appears to be conformable. The lowest redbeds occur 157 m above the base of the Bolsovian in well 44/21-2, and are only 87 m lower in well 44/28-1, 30 km to the south-east (Figure 26). Besly and Turner (1983) have noted that there is a complex, laterally diachronous relationship between the coal measures and the redbeds in central England; the transition may also be locally diachronous offshore (Leeder and Hardman, 1990). Well 44/28-1 penetrated the basal 220 m of the primary redbeds, which seismic interpretation suggests are up to 550 m thick along the axes of Variscan synclines (Figure 25). Westphalian D and Stephanian beds have been proved locally beneath the base-Permian unconformity in the Dutch sector of the North Sea (NAM and RGD, 1980), and also in parts

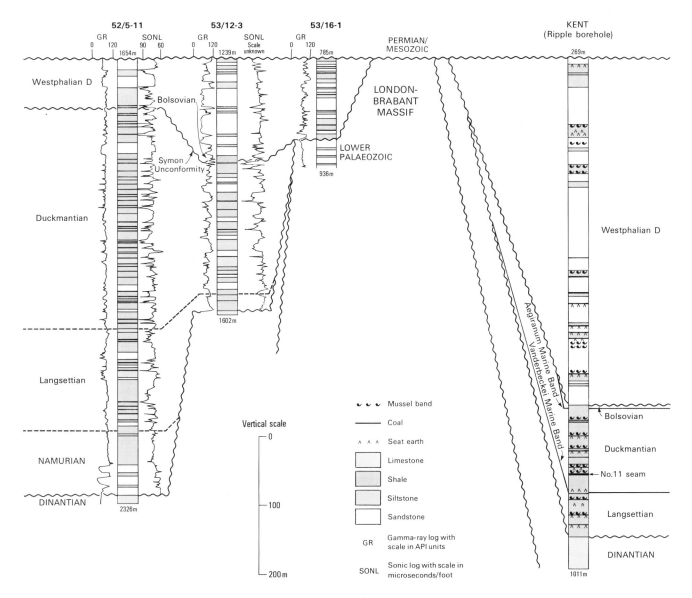

Figure 27 Westphalian sequences on the northern and southern flanks of the London–Brabant Massif. For locations see Figure 25. The Kent sequence is after Bisson et al. (1967) and Ramsbottom et al. (1978).

of central England, where the redbeds are up to 1400 m thick (Besly and Turner, 1983). As the redbeds are relatively thin in the UK sector of the Anglo-Dutch Basin, they may be mainly of Bolsovian age.

The transition from the delta-plain sediments of the coal measures to the alluvial-floodplain facies of the redbeds records a regional improvement in the drainage of the Anglo-Dutch Basin early in Bolsovian times (Besly, 1990). The alluvial floodplains in northern England and the Anglo-Dutch Basin became extensive earlier than those in central England, which remained a predominantly freshwater, lacustrine area until late Westphalian D times (Besly, 1988).

The upper Westphalian sequence is composed of alternating grey, red, brown and purple mudstones, silty mudstones, and grey or reddish purple, fine- to coarse-grained, pebbly sandstones. There are no coal seams. The sandstone beds comprise up to 25 per cent of the total sequence, and are more numerous and thinner than in the coal measures; most are less than 10 m thick, and comprise upward-fining units. The sandstones have been interpreted as fluvial-channel and overbank sheet deposits that were laid down on an alluvial floodplain (Besly, 1990). The floodplain was frequently waterlogged, which allowed the fine-grained sediments to accumulate in poorly drained marshes and lakes.

The massive red mudstones that predominate offshore and in equivalent alluvial Westphalian facies of central England were oxidised by repeated post-depositional lowering of the local water table (Besly and Turner, 1983). Nodules of siderite have been altered to haematite in many of the oxidised beds. Some cyclothems in the Anglo-Dutch Basin have been only partially oxidised, and in cores of these, grey seat earths are overlain by reddened mudstones containing remnant traces of unoxidised rootlets (S J Stoker, written communication, 1989).

There are mature palaeosols intercalated within the massive red mudstones in well 44/28-1; these are regarded as diagnostic of primary redbed deposition, in contrast to secondary oxidation beneath the base-Permian unconformity in central England (Besly, 1990). In the cores, each palaeosol is represented by a grey, leached, upper zone overlying red and grey, vertically veined and mottled beds (S J Stoker, written communication, 1989). Polyphase palaeosol profiles have been observed; each palaeosol formed in relatively well-drained areas of the floodplain. Calcrete horizons are present towards the top of the redbed sequence in well 44/28-1, and dessication cracks suggest that arid conditions had begun to occur intermittently before the end of Carboniferous time.

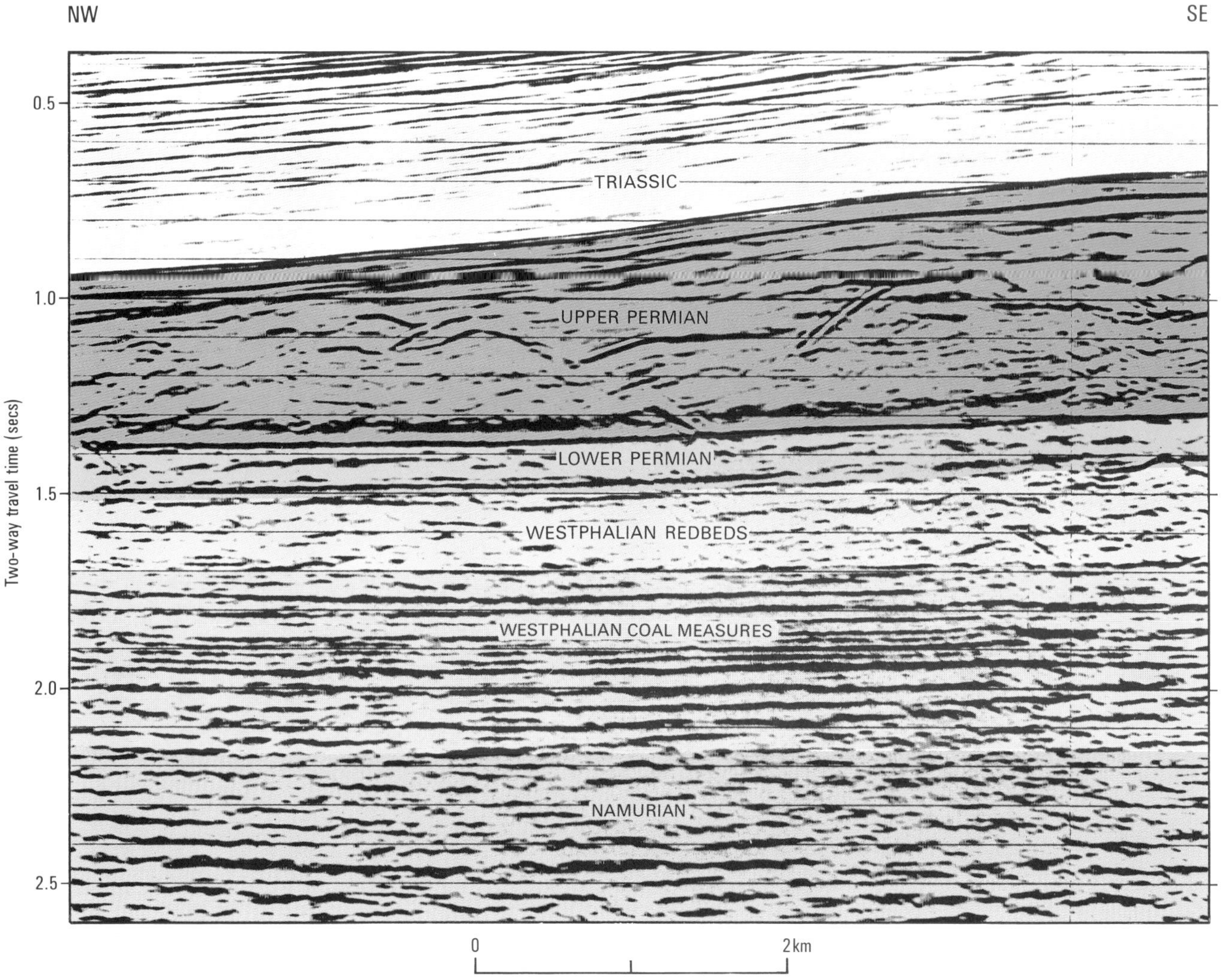

Figure 28 Seismic section showing the character of the Westphalian coal-measure and redbed facies at the centre of the Anglo-Dutch Basin. Section courtesy of Amoco (UK) Exploration Company, interpretation by BGS.

Upper Westphalian facies around the London–Brabant Massif

An angular unconformity separates lower Westphalian coal measures from upper Westphalian redbeds on both the northern and southern flanks of the London–Brabant Massif (Figures 24, 25 and 27). In the north, Tubb et al. (1986) have termed this the Symon Unconformity, and have noted that thin coals intercalated just above the base of the redbeds in some wells are of late Bolsovian age. The sediments directly above the unconformity in well 52/5-11 (Figure 27) are Westphalian D in age (Ramsbottom et al. 1978). Tubb et al. (1986) infer that lower Bolsovian coal measures may be preserved beneath the unconformity in well 53/19-1.

On the northern flank of the London–Brabant Massif, up to 500 m of sediments may be preserved locally between the Symon and base-Permian unconformities. A late Bolsovian resumption of sedimentation introduced a transitional, coal-bearing, alluvial facies into this basin-marginal area (Besly, 1988); all 64 m of upper Westphalian sediments in well 53/16-1 can be referred to this basal facies of the redbeds. The coal-bearing sediments were quickly superseded by an alluvial-floodplain facies comparable to that of the Anglo-Dutch Basin, similarly indicating evolution towards a better-drained environment in Late Carboniferous times (Besly, 1988). The fluvial sandstones are up to 30 m thick, and comprise between 15 and 50 per cent of these primary redbeds (Tubb et al., 1986).

South of the London–Brabant Massif, the Symon Unconformity occurs close above the base of the Bolsovian in Kent. Faunal evidence suggests that most of Bolsovian time is represented by a stratigraphical hiatus (Ramsbottom et al., 1978) between the thin Bolsovian coal measures and an overlying sandy, but coal-bearing, Westphalian D sequence which is up to 700 m thick in the south-east of the Kent Coalfield (Bisson et al., 1967). Offshore, there may be Bolsovian and Westphalian D deposits above this unconformity in the Dover Strait, for the Bolsovian sequence is virtually complete in north-west France (Bless et al., 1977).

In this southern part of the report area, Read (1979) has described both coal-bearing and coal-free late Westphalian alluvial facies from the western Kent Coalfield. In the coal-bearing facies, fluvial sandstones up to 25 m thick comprise up to 40 per cent of the sequence, mudstones are common and relatively thick, and seat earths are thick, argillaceous and typically associated with coal. In the overlying coal-free facies, mudstones are uncommon and thin, seat earths are generally silty and sandy, and coals are absent; more than 70 per cent of the sequence is composed of sandstone. Read (1979) interpreted the coal-bearing facies as the deposits of a river-dominated deltaic environment in which delta-building sedimentary processes were interrupted frequently by the erosive emplacement of distributary-channel sandstones. This was superseded by an alluvial-plain environment in which deposition was dominated by fluvial processes. Unlike the areas north of the massif, none of the overbank deposits have been oxidised, implying a constantly high water table on the floodplain. Boreholes in east Kent (Bisson et al., 1967) suggest that there may be some interdigitation of the upper Westphalian coal-bearing and sandstone-dominated sequences beneath the Dover Strait.

6 Lower Permian

By the end of the Carboniferous Period, Variscan compression had established a chain of mountains extending from south-west England across northern France into eastern Europe. The southern North Sea lay in a rain shadow to the north of these mountains, and its Lower Permian sediments were deposited in a gently subsiding, east–west-trending foreland basin that extended continuously from eastern England across northern Germany into Poland (Ziegler, 1982). This Southern Permian Basin (Glennie, 1986b) was separated from the contemporary, but smaller, Northern Permian Basin by the newly emergent Mid North Sea High. In the southern North Sea, up to 340 m of aeolian and fluvial sands, lacustrine clays and halites were deposited during the Early Permian in an arid or semiarid, tropical-desert environment. Zechstein seas flooded the basin repeatedly later in the Permian, and the base of the lowest marine clay defines the boundary between the Lower and Upper Permian deposits both in eastern England (Smith et al., 1974) and offshore (Rhys, 1974).

The Lower Permian sediments accumulated in an asymmetric basin which sloped gently from south to north (Glennie, 1986c). The base of the Permian in the southern North Sea forms an angular unconformity, for Variscan compression had gently folded and faulted the pre-Permian sediments, and caused widespread uplift and erosion. By the mid-Early Permian onset of deposition, the post-Variscan surface was essentially a peneplain, but the Lower Permian sediments show minor thickness changes across some Variscan faults beneath the Sole Pit Trough (Glennie and Boegner, 1981). It seems that fault scarps inherited from the Variscan orogeny may have influenced sedimentation patterns locally.

The Lower Permian sandstones form the principal reservoir for gas in the Southern North Sea Basin. Many hundreds of wells have confirmed that the generally unfossiliferous Lower Permian rocks are buried beneath Upper Permian, Mesozoic and Cenozoic sediments that locally exceed 3700 m in thickness in the centre of the Anglo-Dutch Basin. Lower Permian rocks have a depth of burial of only 700 m on the northern flank of the London–Brabant Massif, and are absent over its crest; they are also absent in Kent (Smith et al., 1974), and are unlikely to occur beneath the Dover Strait.

Eastern England has its maximum thickness of Lower Permian sediments in County Durham, where up to 60 m of yellow, partially reworked, dune sands are locally underlain by a thin, desert breccia (Smith et al., 1974). Similar dune sands and piedmont breccias have been recorded from boreholes in Yorkshire and Lincolnshire. These Lower Permian deposits are less continuous than they are offshore, and have not been assigned a formal lithostratigraphical name (Smith et al., 1974).

Lower Permian sediments in the UK sector of the southern North Sea have been designated the Rotliegendes Group by Rhys (1974), and those in the Dutch sector the Upper Rotliegend Group (NAM and RGD, 1980). Aeolian and flu-

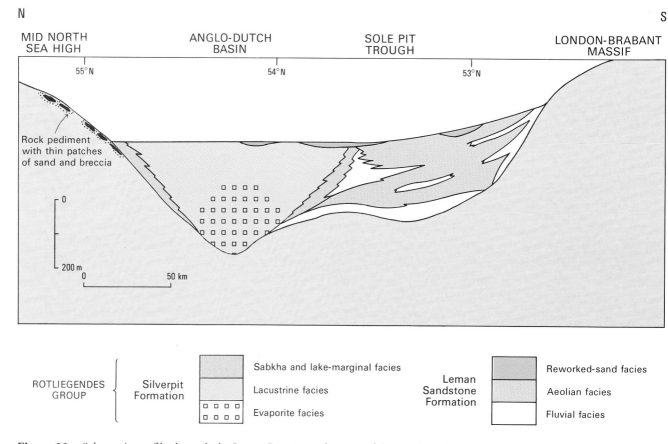

Figure 29 Schematic profile through the Lower Permian sediments of the southern North Sea.

vial, dominantly sandy sediments of the Rotliegendes Group (Figure 29) have been assigned to the Leman Sandstone Formation (Rhys, 1974); they occur mainly to the south of 54°N and along the western margin of the basin. These sandstones pass northwards (Figures 29 and 30) into clays and halites of the Silverpit Formation (Rhys, 1974) that were deposited in a major east–west-trending desert lake which occupied most of the area to the north of 54°N until mid-Permian times. Sabkha facies accumulated around the shore of this lake (Glennie, 1972). Basinal subsidence exceeded the rate of sedimentation such that the desert lake may have been as much as 250 m below global sea level by mid-Permian times (Smith, 1979).

A feather edge of the Leman Sandstone Formation continues beneath the Silverpit Formation for up to 55 km beyond the southern limit of the desert lake (Figure 29). It seems that the desert lake expanded southwards through the Early Permian, and suffered less extreme evaporation and reduced fluctuation in salinity with time, for the beds of halite are concentrated towards the base of the Silverpit Formation. A thin and discontinuous tongue of sandstones also extends northwards over the top of the Silverpit Formation (Figure 29). These reworked aeolian and fluvial sands were transported beyond the shoreline of the lake during the initial stages of the Zechstein marine transgression, but were also assigned by Rhys (1975) to the Leman Sandstone Formation.

The Permian basin of the southern North Sea was bounded to the south by the London–Brabant Massif, and to the west and north by the newly emergent Pennine High and Mid North Sea High respectively (Figure 30). The Sole Pit Trough became an important depocentre in the southern half of this basin during the Early Permian (Glennie and Boegner, 1981), whereas there are only condensed sequences of Lower Permian sediments along the northern margin of the East Midlands Shelf, and to the east of the Sole Pit Trough. The Rotliegendes Group has a second depocentre in the east-central part of the Anglo-Dutch Basin, where the Silverpit Formation desert-lake sediments were deposited. Farther north, there are only thin and patchy developments of Lower Permian piedmont breccias preserved on the southern flank of the Mid North Sea High. This high was not a major source of clastic sediment to the basin during either Permian or Triassic times.

The character of the Lower Permian facies indicates that the southern North Sea had passed from the relatively humid, equatorial conditions of Late Carboniferous times to an arid climatic regime; this was mainly due to northerly drift of Laurasia (Glennie, 1986b). North-west Europe had entered the trade-wind desert belt, and foresets in the aeolian sandstones show that the Early Permian winds blew from the east and north-east across the southern North Sea towards England (Glennie, 1983). The desert lake was fed mainly by ephemeral rivers flowing from the Variscan highlands to the south (Marie, 1975), and the size of the lake may have fluctuated considerably if its water supply was seasonal.

Many sand dunes in the Early Permian desert attained a height of 50 m or more (Glennie, 1983). Today's winds in tropical latitudes are not capable of building and sustaining dunes of more than 10 m in amplitude; seif dunes up to 100 m high in the modern deserts of Arabia and North Africa were probably constructed during Pleistocene glacial intervals. Glennie (1983) inferred that the wind systems which built both the Early Permian and Pleistocene dunes persisted for a much greater part of the year than in the present tropical-desert climatic regime, and had a higher average velocity. Glennie (1983) implied that there was a generally higher subtropical barometric gradient caused by compression of climatic belts around the Equator during Permian and Pleistocene glaciations. The strong, very dry winds which constructed the Permian dunes would have enhanced evaporation, and may have been the cause of the desert lake reducing in size sufficiently for halite deposition to take place at intervals during the early part of the Permian (Glennie, 1972). Conversely, weaker wind systems would have prevailed between glacial episodes when reduced evaporation may have been a factor in the expansion of the desert lake during the latter part of the Early Permian.

LEMAN SANDSTONE FORMATION

The Leman Sandstone Formation is composed almost entirely of sediments of aeolian and fluvial origin up to 290 m thick (Figure 31). Both facies have characteristic suites of sedimentary structures in core sections (Glennie, 1972), and they can sometimes be distinguished on the wireline logs where core is not available. A third distinctive facies composed of grey, relatively structureless, reworked sandstones is present at the top of the formation in the Leman Gasfield (van Veen, 1975) and elsewhere.

Fluvial facies

Fluvial sequences of conglomerates with laminated or homogeneous, red-brown sandstones and dark red clays are interdigitated with, or occur beneath, aeolian sediments in many offshore wells. The most substantial belt of fluvial sediments extends north from the East Anglian coast to the margin of the desert lake (Marie, 1975), but even here they are less than 100 m thick and are intercalated with aeolian sandstones

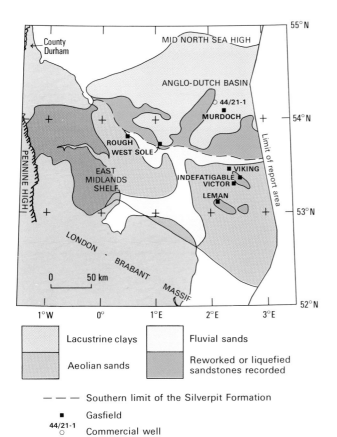

Figure 30 Distribution of the principal Lower Permian facies.

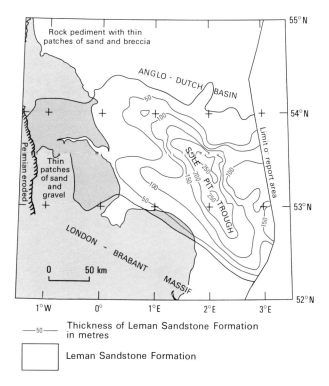

Figure 31 Distribution and thickness of the Leman Sandstone Formation.

(Figures 29 and 30). Elsewhere, the fluvial sediments are generally less than 50 m thick, and Glennie (1986c) has noted that they are better developed in The Netherlands, Germany and eastern Europe than in the UK sector of the North Sea. Alluvial-fan, sheet-flood sands and conglomerates have been described from the Rough Gasfield by Goodchild and Bryant (1986); these are probably typical of the fluvial facies on the western and southern margins of the basin.

Many of the sedimentary structures observed in cores indicate that the fluvial sequences were deposited in temporary streams or wadis (Glennie, 1972). The sandstones are either structureless or have gently inclined or discontinuous laminae, suggesting deposition by flowing water. They are commonly slightly argillaceous and contain rare pebbles and rip-up clasts of red clay. These clasts were eroded from bedded claystones deposited in ephemeral lakes within the wadi system. Locally, especially around the margins of the basin, the pebbles and clasts are sufficiently numerous to define beds of conglomerate; these are up to 1.5 m thick at the Rough Gasfield (Goodchild and Bryant, 1986). Many of the clay flakes are cracked or curled convex-side up, indicating subaerial desiccation (Glennie, 1972). The fluvial sandstones tend to be cemented by dolomite where evaporation of floodwaters was relatively slow, but the basin-margin sands at the Rough Gasfield are less tightly cemented as they were deposited well above the regional water table.

Marie (1975) deduced that the fluvial sediments of the southern North Sea were deposited in alluvial plains and floodplains by rivers flowing northwards from the Variscan highlands towards the desert lake in the centre of the basin. The rainfall was probably seasonal, occurring in violent storms when the floodwaters followed the lines of lowest relief between sand dunes, eroding and reworking aeolian and previously deposited fluvial sediments in their paths. During the dry season, aeolian sediments would have begun to encroach across the wadi surface, but often these would not have become sufficiently well established to divert floodwaters during the following wet season. Aeolian sandstones predominate over the fluvial facies towards the top of the Leman Sandstone Formation (Marie, 1975); the major rivers which fed the gradual expansion of the desert lake at that time evidently did not flow through the UK sector of the North Sea.

Aeolian facies

Stacked dune sequences of red, aeolian sandstones are up to 200 m thick in the Sole Pit Trough (van Veen, 1975). They constitute most or all of the Leman Sandstone Formation in the east and south-east of the offshore area (Marie, 1975), and are intercalated with, or overlie, fluvial sediments to the west of the Sole Pit Trough (Figure 29). The aeolian sandstones are typically clean with common frosted grains, and range between very fine and medium grained. Coarser-grained sandstones occur locally above intraformational unconformities (Glennie, 1972). Unlike the fluvial sediments, the aeolian sandstones do not contain mica flakes.

In cores, the aeolian facies is represented by both planar-bedded and trough-bedded sandstones (Figure 32) in which adjacent laminae commonly show sharp grain-size differences between very fine and coarser sand grades (Glennie, 1986c). The laminae are characteristically inclined; in each dune set their inclinations increase upwards to about 20° or 25° before they are truncated by the nearly horizontal, fine laminae at the base of the succeeding dune. Only the basal few metres of many of the migrating dunes are preserved beneath the truncation defining the base of the next dune in the sequence (Glennie, 1986c). By plotting their bedding inclinations, Glennie (1972) deduced that both transverse and seif dunes were formed; most of the dunes were transverse in the Sole Pit Trough and farther east, but seif dunes may have predominated around the margins of the basin.

Some offshore wells have penetrated thin packages of poorly sorted, slightly silty sand within the dune sequence. These occur in irregular wavy laminations which Glennie (1972) has interpreted as adhesion ripples; they are formed in modern deserts where windblown sand adheres to a slightly dampened surface. The Early Permian adhesion ripples are mostly very localised, and were probably formed where windblown sand was deposited over those interdune areas where the water table was at, or close to, the surface (Glennie, 1972). Unusually, adhesion-rippled sands form a 20 m-thick unit near the top of the dune sequence in the Victor Gasfield, and Conway (1986) has suggested that these record a local rise in the water table due to either a temporary climatic change or a local tectonic event.

Reworked-sand facies

Up to 50 m of structureless or faintly laminated sands are patchily distributed at the top of the Leman Sandstone Formation in many of the gasfields of the southern North Sea, and overlie part of the Silverpit Formation (Figure 29). They are present in Leman and its adjacent gasfields (van Veen, 1975; Arthur et al., 1986), at West Sole (Butler, 1975) and Indefatigable (France, 1975), and have been noted in eastern England (Smith, 1974). They are notably absent in the Viking, Victor and Rough gasfields (Gray, 1975; Conway, 1986; Goodchild and Bryant, 1986). Where the sandstones are present, they are grey and moderately or poorly sorted, contain some clay pebbles, and show irregular or slumped laminations. The sandstones are less porous than the fluvial and aeolian sediments; this partly reflects their tighter grain packing, but is mainly due to cementation by dolomite, which was probably precipitated by porewaters percolating from the overlying Zechstein carbonates (Glennie, 1986c).

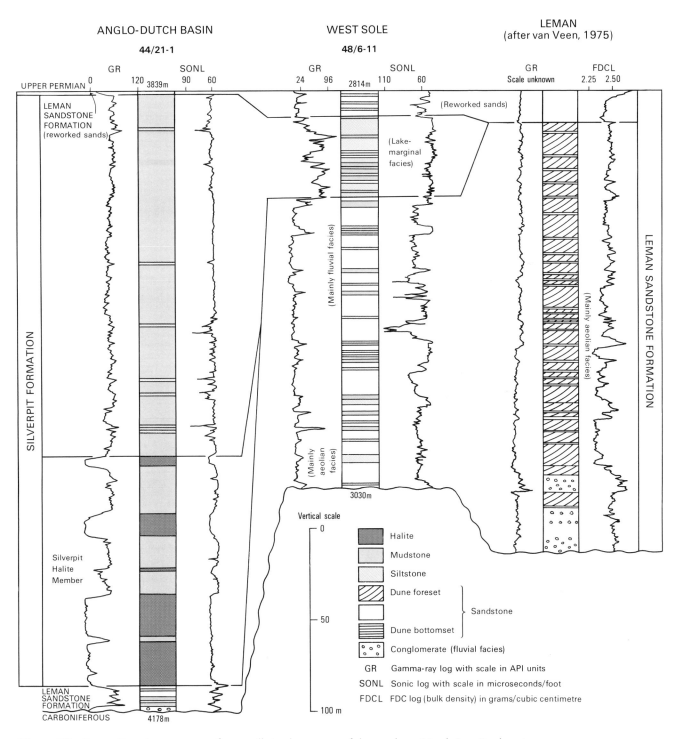

Figure 32 Lower Permian sequences from wells in three parts of the southern North Sea. For locations see Figure 30.

Van Veen (1975) and Glennie and Buller (1983) have observed that the reworked facies of the Leman Sandstone Formation is very similar to the Weissliegend facies in the Lower Permian deposits of Germany and eastern Europe. In many of the North Sea cores, as in Europe, there is a sharp boundary between the laminated aeolian sands and the relatively structureless sandstones which commonly contain zones of faint, contorted lamination and patches of clay enrichment alternating with layers of blurred and indistinct lamination. In some examples, the loss of structure intensifies upwards. Glennie and Buller (1983) have explained these, and deformation structures observed in other parts of the Permo-Triassic basin, as resulting from liquefaction of unconsolidated dune sands, perhaps related to expulsion of air from the sediment shortly after inundation by the Zechstein seas.

In the Leman Gasfield (van Veen, 1975), there appears to be an erosion surface separating the grey sandstones from the underlying aeolian sediments (Figure 32). Here, the upper unit comprises aeolian sands which have been homogenised by subaqueous reworking during rapid transgression of the Zechstein seas across the area. Farther north, this transgression spread the thin, and perhaps discontinuous sheet of sand which overlies the Silverpit Formation for up to 55 km beyond the southern margin of the desert lake (Figure 30).

SILVERPIT FORMATION

In the north-central part of the Southern Permian Basin, the Rotliegendes Group is composed entirely of the lacustrine clays and halites of the Silverpit Formation. These sediments were deposited in a desert lake which during the Early Permian extended about 1200 km eastwards from the North Sea through Germany into Poland, and had its greatest width of over 200 km shortly before the Zechstein transgression (Glennie, 1986c). The Silverpit Formation has a maximum thickness of 340 m in UK waters, but it does not extend as far west as the English coast (Figure 33). Near the northern feather edge of the Leman Sandstone Formation, the basal claystones are interbedded with siltstones and fine-grained sandstones of sabkha and shoreline facies. Despite this, the vertical transition between the Silverpit and Leman Sandstone formations is relatively sharp (Figure 32).

Lacustrine facies

In the centre of the Permian desert lake, the Silverpit Formation is dominated by a monotonous sequence of red-brown, silty, anhydritic mudstones with only a few thin beds of grey-brown, argillaceous siltstone (Figure 32). Interbedded halites occur at the base of the mudstones, and are sufficiently numerous and widespread for the lower part of the sequence to be designated the Silverpit Halite Member (Figures 32 and 33). In well 44/21-1, the type well for the Silverpit Formation (Rhys, 1974), the basal halite interval is 50 m thick (Figure 32), but the halite beds become progressively thinner and less numerous both towards the top of the Silverpit Halite Member, and laterally towards the margin of the lake; most are less than 10 m thick. Halites occur up to 210 m above the base of the Permian in some wells in the Murdoch Gasfield, but even here the halite beds have a combined thickness of less than 70 m.

It seems that the desert lake was relatively shallow and was subject to more extreme fluctuations in salinity during the earliest part of the Permian. At first there were lengthy periods during which the seasonal supply of freshwater was insufficient to prevent widespread deposition of halites by evaporation in the centre of the lake. As the volumes of freshwater reaching the basin gradually increased, or evaporation reduced, the lake expanded southwards over the Leman Sandstone Formation. As it expanded, long-term fluctuations in the salinity of the lake became less extreme, and there are no halite intervals in the topmost 100 m of the Silverpit Formation in any offshore wells.

Sabkha and other lake-marginal facies

The lake-marginal facies of the Silverpit Formation comprises a complex interdigitation of lacustrine with sabkha and aeolian sediments; fluvial sediments are intercalated where major wadis discharged into the desert lake. There are up to 50 m of interbedded claystones, siltstones and sandstones at the base of the Silverpit Formation around the margins of the Permian desert lake (Figure 29). At the West Sole Gasfield, these comprise upward-fining cycles which are about 1 m thick, and

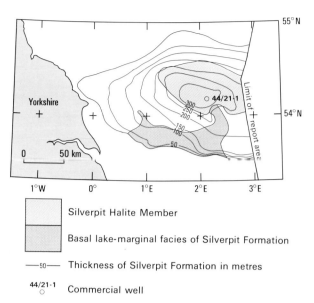

Figure 33 Distribution and thickness of the Silverpit Formation, showing the distribution of the Silverpit Halite Member.

there is good correlation of the cycles between adjacent wells (Butler, 1975) in a zone which was close to where a major north-flowing wadi system entered the lake (Marie, 1975). The sandstones are likely to have been deposited by distal sheet floods in a shallow-water, nearshore environment. The extent of this lake-marginal facies in UK waters is depicted in Figure 33.

Shoreline sabkha facies are extensive around the margins of the desert lake in The Netherlands, where they are characterised by the presence of adhesion-rippled sands and nodular anhydrite within claystones and siltstones (Glennie, 1972). There are only a few interbedded sandstones in which sedimentary structures indicate deposition under aeolian or fluvial conditions during prolonged intervals of lake retreat. As the transition between sabkha and fully lacustrine facies is gradational, the lateral limits of the sabkha facies in the UK sector are not easily determined.

Long-term expansion of the Early Permian desert lake is represented by the migration of the lacustrine facies over shoreline sequences, and of the Silverpit Formation over the Leman Sandstone Formation, in many areas around the southern margin of the lake (Butler, 1975). Short-term fluctuations are also likely to be represented in the lake-marginal sequences; lacustrine clays would have been deposited in the coastal area during intervals of high-water level in the lake, whereas sabkha facies were deposited in the same area during lake retreat. If the latter were sufficiently prolonged, wind-blown sand would have formed adhesion ripples on the surface of the sabkha, and dune sands may have encroached temporarily. The shoreline facies thus comprises a complex interdigitation of lacustrine with sabkha and aeolian facies, intercalated with fluvial sediments where the major wadis discharged into the desert lake.

7 Upper Permian

From Early Permian to Late Triassic times, the southern North Sea lay in a postorogenic Variscan foreland basin extending from eastern England through Germany into eastern Europe. Aeolian, fluvial and lacustrine sediments had partially filled this basin during the Early Permian. An abrupt and basinwide facies transition defines the base of the relatively complex Upper Permian sequence of marine and evaporitic sediments which comprise the Zechstein Supergroup (Rhys, 1974). This major facies change was caused by the first of five widespread Late Permian marine transgressions across the basin. Each transgression records the opening of a temporary seaway connecting the Permian basins of northern Europe with the Boreal Ocean north of Norway along a proto-Atlantic fracture zone (Ziegler, 1982). The primary controls affecting the opening and closing of this seaway were eustatic changes in sea level within the Boreal Ocean (Smith, 1980), and crustal movements within the proto-Atlantic fracture system.

The configuration of the Variscan foreland basin had already been established during the Early Permian, and persisted through Late Permian times. The London–Brabant Massif and Pennine High continued to form the southern and western margins of the basin, but its northern margin became less clearly defined. The Mid North Sea High was inundated by the first of the Zechstein transgressions, and remained partially or wholly submerged as an area of reduced subsidence throughout the remainder of the Permian.

More than 1000 m of Upper Permian sediment accumulated in the centre of the southern North Sea; the sediments are mainly evaporites, and have been extensively deformed by post-Permian halokinesis. Because of this, the thickness of the Upper Permian varies from less than 50 m in zones of extreme salt withdrawal to more than 2500 m in some of the major salt diapirs. Many of these diapirs have pierced through the Triassic section into Jurassic and Cretaceous rocks, and two of the largest have crests within 100 m of the sea bed. Upper Permian rocks crop out at sea bed near the coast to the north of Middlesbrough (Figure 2), but they are buried by between 700 and 2500 m of younger sediments in most of the offshore area (Figures 2 and 3).

The Upper Permian sequences around the margins of the basin are dominated by carbonate and anhydrite. Traced beyond the palaeoshoreline, the carbonate formations pass laterally into fluvial sandstones, and the anhydrites grade into red mudstones and siltstones (Figure 34). Where they occur, these continental deposits are generally less than 50 m thick, but are not easily distinguished from similar Lower Permian and Triassic sediments. The entire Permo-Triassic succession has been eroded over the crest of the London–Brabant Massif, and Upper Permian sediments are unlikely to be present farther south in Kent (Smith et al., 1974) or beneath the Dover Strait.

The effects of the Late Permian marine transgressions divide the Zechstein Supergroup of the southern North Sea into five basinwide evaporite cycles. Each of the first three cycles has a basal sequence of marine sediments deposited when oceanic waters circulated freely in the basin. The second and successive cycles are dominated by evaporites deposited after each constriction of the temporary seaway to the Boreal Ocean. These evaporites generally show an upward transition from anhydrite to halite and then to magnesium or potassium salts, reflecting a gradual increase in the salinity of the basin due to evaporation. In detail, most of the Zechstein cycles contain many departures from this idealised evaporation sequence; these can be mostly attributed to changes in salinity within the basin resulting from the access of oceanic water and the supply of freshwater runoff (Taylor, 1986). Parts of the evaporite sequences have been further complicated by diagenetic alteration of their primary mineral constituents.

The topography inherited from the Early Permian continued to influence the regional distribution of the sedimentary facies during the first and second Zechstein cycles. When the Zechstein seas flooded the area for the first time, most of the basin was already more than 100 m (Taylor and Colter, 1975), and perhaps as much as 250 m, below global sea level (Smith, 1979). Up to 300 m of carbonate and anhydrite were deposited around the margins of the basin during the first Zechstein cycle, but the time-equivalent sediments across the basin floor are less than 50 m thick. This allowed most of the basin to remain well below wave base throughout the second Zechstein transgression. It was the precipitation of thick second-cycle evaporites in the centre of the basin that brought most of the area close to sea level throughout the remainder of the Permian (Figure 34), although subsidence continued to be greatest in the central offshore area.

All five Zechstein cycles display lateral facies and thickness variations from the margins towards the centre of the basin (Figure 34). In the first three cycles, carbonate production exceeded subsidence in the warm, shallow, well-aerated waters around the basin margins. This caused a range of barrier, lagoonal, intertidal and locally emergent environments to develop in a broad coastal shelf extending basinwards between 20 and 75 km from the shoreline of the subtropical seas (Taylor, 1986). The thickest carbonates accumulated on the submarine slope separating these nearshore environments from anoxic conditions which prevailed below wave base in the centre of the basin.

The anhydrites in the first four cycles display lenticular geometry, and are thickest directly basinwards of the carbonate slope facies in the lowest cycle (Figure 34). In this cycle, precipitation of anhydrite was enhanced in the subaerial or shallow-water sabkha conditions around the basin margins, which had the effect of extending the coastal shelf farther towards the centre of the basin. Taylor (1980) has calculated that the evaporating brine must have been recharged many hundreds of times by brackish seawater to account for the thickness of anhydrite precipitated during the first cycle. The halites, and magnesium and potassium salts of the second to fourth cycles are thickest in the centre of the Zechstein basin; most of the lateral thickness and facies variations in these cycles can be explained in terms of enhanced subsidence along the centre of the basin (Taylor, 1986).

Rhys (1974; 1975) allocated group status to the sediments of the first four Zechstein cycles in the southern North Sea, referring to these as the Z1, Z2, Z3 and Z4 groups respectively. Rhys (1974) recognised that a relatively thin and incomplete fifth cycle of Zechstein evaporites could be identified in some wells, but included these sediments within the Z4 Group. Rhys (1974; 1975) also introduced new formational names for the major lithostratigraphical units in each

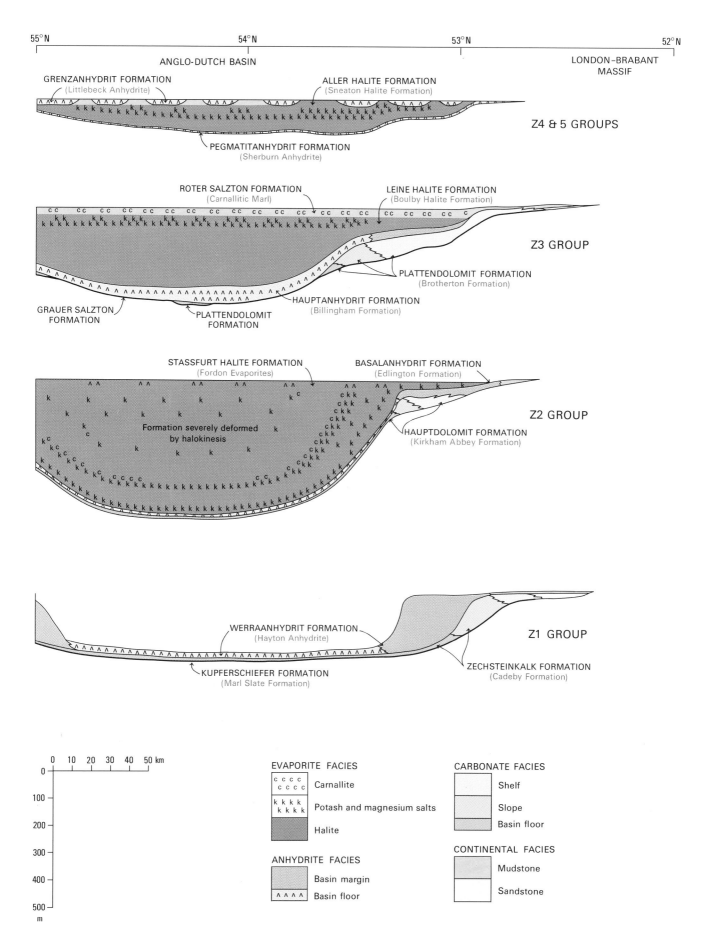

Figure 34 North–south sections showing successive cycles in the Upper Permian sediments of the southern North Sea. Offshore formations in black, onshore formations in red. The line of section is at 2°12'E.

of the Zechstein groups, based on German stratigraphical nomenclature. Subsequent research has resulted in revised subdivision of the first Zechstein cycle (Taylor, 1986), and the evaporites of the fifth cycle have been assigned to a new Z5 Group. Otherwise, the stratigraphical terminology proposed by Rhys (1974) is retained in this report. The stratigraphical relationships of the formations are summarised in Figure 34, in which the onshore nomenclature is after Smith et al. (1986).

Z1 GROUP

The Z1 Group of Rhys (1974) is divided into the Kupferschiefer, Zechsteinkalk and Werraanhydrit formations.

Kupferschiefer Formation

The basal sediments of the Upper Permian are dark grey or black, laminated, sapropelic, marine shales and carbonates. They contain chalcopyrite, sphalerite, galena and other sulphides, and display a distinctive gamma-ray signature in southern North Sea wells (Figure 35). In north-east England, there is evidence that the first of the Zechstein marine transgressions was sufficiently rapid and nonerosive to preserve unconsolidated ridges of Lower Permian dune sands up to 60 m in height (Smith, 1979). Glennie (1986c) has calculated that because the inundation by oceanic waters was into an area substantially below sea level, water depth rose by as much as 0.3 m per day during the initial stages of the transgression, so that all the dunes may have been submerged in less than a year. Sea level rose rapidly to as much as 250 m above the centre of the basin (Smith, 1979), establishing anoxic conditions across the whole of the offshore area. The resulting Kupferschiefer Formation offshore, and the Marl Slate Formation onshore, mantle the pre-existing topography, and form a continuous layer which is generally about 1 m thick and has a maximum thickness of 2.5 m.

Zechsteinkalk Formation

The carbonates of the Zechsteinkalk Formation are equivalent to the Cadeby (formerly Lower Magnesian Limestone) Formation of Yorkshire (Smith et al., 1986). Both formations overlap the basal Z1 marine beds southwards on to the London–Brabant Massif (Figure 34), indicating that they were deposited after a further rise in sea level (Taylor and Colter, 1975). Their thickness increases basinward to between 60 and 120 m across a 20 to 35 km-wide marginal carbonate platform (Figure 36), whereas they are less than 10 m thick across the floor of the early Zechstein basin. Because of its bathymetric range of deposition, the Zechsteinkalk Formation includes coastal-shelf, submarine-slope and basinal-marine carbonate facies. Its diverse fauna includes bivalves, brachiopods, crinoids, bryozoans and foraminifera.

Before the carbonate platform became established, a monotonous sequence of grey-brown, variably dolomitised, bioturbated, lime muds was deposited around the margins of the basin (Taylor and Colter, 1975). These pass gradationally upwards into shallower-water shelf sediments that include reef limestone from the outer shelf, and carbonate sands from the inner shelf. Most of the reefs were constructed by bryozoans, and are capped by stromatolites; algae were important producers and binders of sediment during the later stages of reef formation (Taylor, 1986). The reefs were fringed by sands composed of abraded skeletal debris; these trace shorewards into sheets of dolomitised oolites within a few kilometres of their reef source (Taylor and Colter, 1975; Taylor, 1986). On the inner margins of the shelf, there are sands containing fine or very fine spherical grains that Taylor and Colter (1975) interpreted as algal ooliths, pisoliths and oncoliths. These carbonate sands trace shorewards into dolomite-cemented quartz sands and conglomerates on the northern flank of the London–Brabant Massif.

Two transgressive–regressive subdivisions of the shallow-water facies have been identified in the Cadeby Formation of north-east England (Smith, 1981). The lower member contains small, isolated, bryozoan-algal patch reefs, whereas the overlying member is less fossiliferous and is characterised by large-scale cross-stratification. A massive shelf-edge reef, 100 m high and at least 35 km long, was constructed in County Durham during the early stages of the Z1 cycle. Many wells have penetrated through or adjacent to similar reefs off the coast of Norfolk, on the southern margin of the basin (Taylor and Colter, 1975).

The Zechsteinkalk Formation has its greatest thickness in the wedge of carbonate mud which was deposited as a submarine slope facies adjacent to, and basinwards of, the outer-shelf reefs (Figure 36). These slope carbonates include turbidites and slumped material, and were derived mainly from both winnowing of the shelf and finely comminuted skeletal reef debris (Taylor, 1986). The slope facies contains both pale, bioturbated, and dark, thinly bedded carbonate mudstones (Clark, 1986), and becomes progressively paler and less argillaceous upwards (Taylor, 1986).

The slope carbonates trace basinwards into dark, pyritic, carbonaceous dolomites and limestones which are very argillaceous at their base and top; these were deposited under anoxic conditions as a basin-floor facies. The upper argillaceous layer contains oncoliths, foraminifera and crinoids; it seems to have been deposited during a temporary fall in sea level at the centre of the basin to between 100 and 150 m depth, resulting in less-anoxic conditions (Taylor and Colter, 1975).

Werraanhydrit Formation

The base of the Werraanhydrit Formation is a sharp lithological break across the whole of the southern North Sea, probably caused by a sudden increase in the salinity of the Zechstein sea (Taylor and Colter, 1975). The formation incorporates those sediments which were assigned by Rhys (1974) to the Lower Werraanhydrit, Werra Halite and Upper Werraanhydrit formations; these are recognisable only around the margins of the basin, and should be regarded as members of one anhydrite-dominated formation (Taylor, 1980). Taylor (1980) speculated that the Upper Werraanhydrit Member was precipitated early in the Z2 cycle on the margins of the basin as the basal beds of the Z2 carbonates were accumulating on the basin floor. However, this solution to a difficult correlation problem was considered unlikely by Smith (1980) and Richter-Bernburg (1986), and all of the Werraanhydrit Formation is referred to the Z1 Group in this report.

The isopachs of the Werraanhydrit Formation (Figure 37) demonstrate the continuing influence of bathymetry on regional sedimentation rates during the remainder of the Z1 cycle. The anhydrites have their thickest development of 150 to 300 m about 25 km basinwards of the Z1 carbonate reefs, and then thin rapidly to between 12 and 25 m across the floor of the basin. Towards the palaeoshore, the formation abuts against the submarine slope of the underlying carbonates (Figure 34), and only its uppermost beds extend over the outer edge of the carbonate shelf to merge laterally into conti-

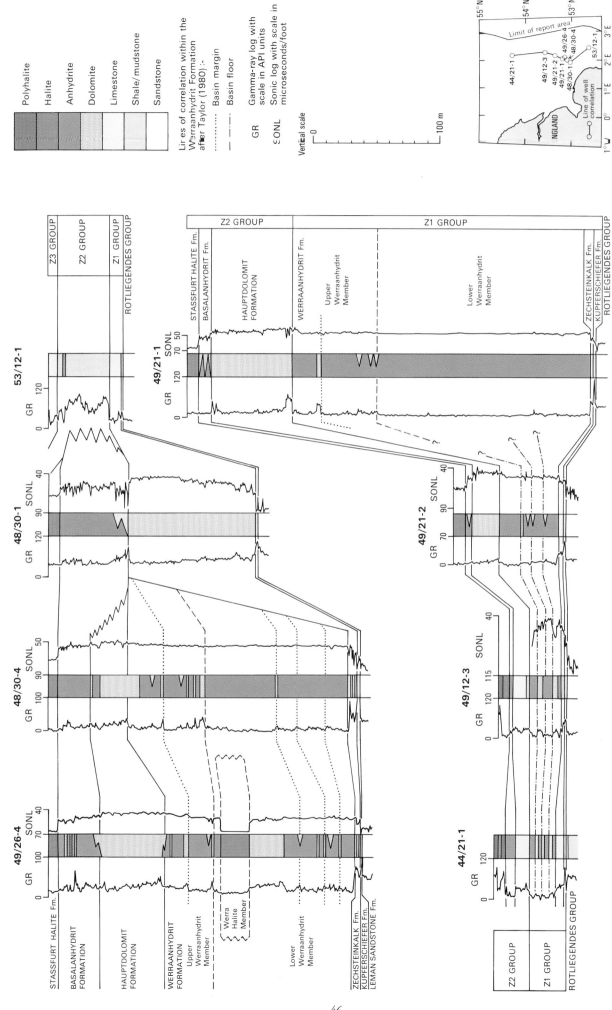

Figure 35 Correlation of carbonate and anhydrite formations in the first and second Zechstein cycles. Adapted from Taylor (1980).

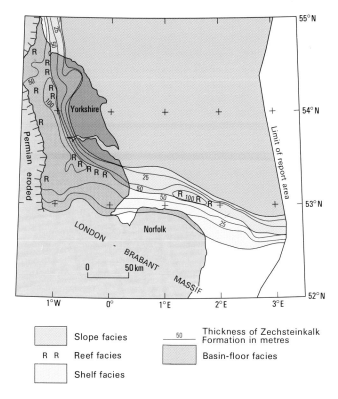

Figure 36 Distribution, thickness and facies of the Zechsteinkalk Formation in the Z1 Group, and equivalent formations in eastern England. Land information adapted from Smith (1989).

nental, grey and red, anhydritic shales (Taylor, 1980).

The basin-margin anhydrites have been described from a number of boreholes and mine workings in north-east England (Smith, 1974; Taylor, 1980). Here, and probably also offshore, about 80 per cent of the formation comprises felted, bluish white, microcrystalline to finely crystalline anhydrite in nodular or mosaic structure, with variable indications that bedding has been distorted by displacive growth, and perhaps by slumping. The remainder is composed of brown, microcrystalline dolomite forming scattered inclusions, laminations resembling algal mats, and beds less than a metre thick. Desiccation cracks occur towards the top of the formation.

Taylor (1980) has traced five relatively dolomitic or shaly intervals through the basin-margin anhydrites of wells in and adjacent to the Hewett and Leman gasfields. Gamma-ray peaks corresponding to these intervals subdivide the Werraanhydrit Formation into six anhydrite-dominated depositional units (Figure 35). The anhydrites have been interpreted by Taylor (1980) as basin-margin sabkha deposits which accumulated either subaerially or under shallow water. The dolomitic and shaly intervals represent breaks in the depositional sequence, probably caused by temporary lowering of water level. If so, each of the succeeding anhydrites represents a recharge of brine in the basin. The beds of halite encountered in well 49/26-4 (Rhys, 1974; Figure 35), and in only two other offshore wells, were deposited in isolated salt pans within the marginal sabkha (Taylor, 1980).

The basin-floor sediments of the Werraanhydrit Formation can be subdivided into four widespread and easily identified depositional sequences (Figure 35). In all four sequences, 1 to 2 m of nodular and mosaic, displacive anhydrite, resembling the basin-margin sabkha facies, grades upwards into 4 to 6 m of finely interlaminated anhydrite and dark, bituminous dolomite or limestone with some bands of anhydrite nodules (Taylor, 1980). The proportion of carbonate increases both upwards through each sequence and from the centre towards the south-western margin of the basin floor (Figure 37). Halite fills vugs, and forms discontinuous lenses between carbonate laminae near the base of the first sequence in some wells (Taylor, 1980); halite dissolution may have caused the disturbed bedding which occurs at this level in the centre of the basin.

The anhydrite at the base of each sequence was precipitated during supersaturation of brine above a critical concentration (Richter-Bernburg, 1986). This could have resulted from an increase in evaporation rates, or from a temporary fall in water level of 150 to 300 m, allowing shallow-water or emergent conditions to extend across the floor of the basin (Taylor, 1980). More likely, the supersaturation was caused by an increase in brine temperature due to climatic factors or a change in current flow within the basin (Richter-Bernburg, 1986). The overlying laminated sediments were deposited as the brines became progressively less concentrated; Taylor (1980) has estimated that the Werraanhydrit Formation includes about 77 000 laminae in the centre of the basin. The laminations are a fraction of a millimetre thick, inviting interpretation as annual varves, although Clark (1980a) has suggested that they may not be annual if each was deposited from the layer of turbid water that covered the basin floor after a major storm.

Z2 GROUP

The Z2 Group is subdivided into the Hauptdolomit, Basalanhydrit and Stassfurt Halite formations.

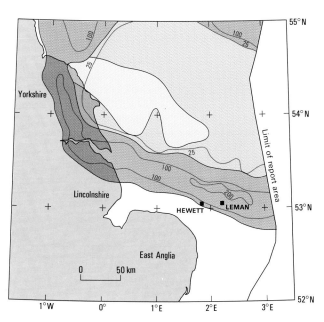

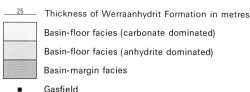

Figure 37 Distribution, thickness and facies of the Werraanhydrit Formation in the Z1 Group, and equivalent formations in eastern England. Land information adapted from Smith (1989).

Hauptdolomit Formation

The Hauptdolomit Formation was deposited during the second major influx of marine waters into the Zechstein basin. As during the Z1 transgression, intertidal, shallow-water and barrier environments were quickly established on a basin-margin carbonate shelf, and contemporary submarine-slope and basinal-marine carbonate facies were deposited from the shelf edge across the floor of the basin. Carbonate production exceeded subsidence nearshore, allowing the marginal shelf to expand basinwards over the submarine-slope facies in parts of north-east England (Smith, 1980).

The Hauptdolomit Formation has its maximum known thickness of between 40 and 75 m offshore. The equivalent Kirkham Abbey Formation of Yorkshire is up to 110 m thick along the outer margin of the carbonate shelf, and may be similarly thick close to the coast of north-east England (Figure 38). The carbonate shelf is between 10 and 30 km wide along the southern margin of the Zechstein basin. Contemporary deposits shorewards of the carbonate shelf (Figure 34) are the anhydrites and red and grey shales that comprise sabkha and continental facies on the northern flank of the London–Brabant Massif; these are incorporated for convenience within the Basalanhydrit Formation, and are less than 20 m thick. Basinwards, the carbonates thin rapidly off the southern shelf margin, but less rapidly off the western margin. The formation is between 9 and 20 m thick across the floor of the basin.

The fauna of the Hauptdolomit Formation is less diverse than that of the first-cycle carbonates, but it seems likely that the salinity was only slightly higher than that of normal sea water (Smith, 1980). The fauna at rock outcrop in north-east England is dominated by three species of bivalves, along with local concentrations of foraminifera, gastropods, ostracods and fish remains. There are no bryozoans, brachiopods, corals or crinoids; in the absence of these, there are no frame-built organic reefs in the Z2 carbonates (Taylor and Colter, 1975).

The shelf facies of the Hauptdolomit Formation is composed mainly of oolitic, pelletoidal, bioclastic and pisolitic grainstones (Taylor and Colter, 1975; Clark, 1986). The oolitic grainstones, which are cross-laminated in north-east England (Smith, 1980), were deposited within a system of sand barriers which became established along the shelf edge (Figure 38). Inshore of these barriers, pelletoidal carbonates accumulated in nearshore lagoons (Taylor and Colter, 1975) where some carbonate muds were also deposited (Clark, 1986). Pisoliths and algal sheets are common near the top of the shelf-edge sequence, where they accumulated as salinity began to rise and sabkha environments became established across the whole of the shelf area during the later stages of carbonate deposition.

The slope facies is composed mainly of dolomitised lime muds, which are commonly pelleted and burrowed (Taylor and Colter, 1975). Beds of finely laminated calcitic dolomite or dolomitic limestone that occur towards the base of the equivalent sediments in north-east England were deposited at or near the base of the slope (Smith, 1980). Lenses and beds of fine-grained or oolitic limestone occur within both the structureless and the laminated slope sediments. These have a similar lithology and fauna to the shelf carbonates, and represent allochthonous masses of sediment which have slumped over the shelf edge (Smith, 1980).

The basin-floor facies has been subdivided by Taylor (1986) into the Stinkschiefer and Stinkkalk members. Both members are composed of dark brown to black, bituminous, finely laminated carbonates, but there are more shale partings within the lower, Stinkschiefer Member. Coarsely crystalline limestone predominates at the centre, and microcrystalline dolomite around the margins of the basin floor. Anhydrite laminae occur up to 1 to 2 m above the base, and recur near the top of the basin-floor carbonates (Taylor, 1980), heralding a rapid transition into the overlying evaporites.

Basalanhydrit Formation

The Basalanhydrit Formation is approximately equivalent to the upper part of the Edlington (formerly Middle Marls) Formation of north-east England (Smith et al., 1986). The Edlington Formation is up to 55 m thick in Yorkshire, where it comprises a continental sequence of siliciclastic rocks in which the proportion of anhydrite and dolomite increases eastwards (Smith, 1980). Only the uppermost evaporite beds extend basinwards over the Z2 carbonate shelf in Yorkshire. In the southern North Sea, the Basalanhydrit Formation displays similar interdigitating and overlapping relationships with the shelf carbonates of the Hauptdolomit Formation (Figure 34). Basinwards, a much greater thickness (5 to 50 m) of anhydrite with thin beds of dolomite is maintained over the carbonate shelf before the Basalanhydrit Formation thins rapidly over the shelf edge (Figure 39). On the basin floor, the formation is thin or absent, or may be represented by the interlaminated carbonate and anhydrite at the top of the Stinkkalk Member.

The basal Z2 anhydrites of Yorkshire include both layered (possibly primary) and secondary nodular deposits which, along with the interbedded carbonates, accumulated in a peritidal, sabkha environment (Smith, 1980). Initially they were deposited shorewards of the carbonate shelf, but eventually extended over it. As salinity began to rise, a chain of gypsum-sand barriers may have superseded the carbonate-sand barriers of the outer shelf on the southern margin of the basin (Clark,

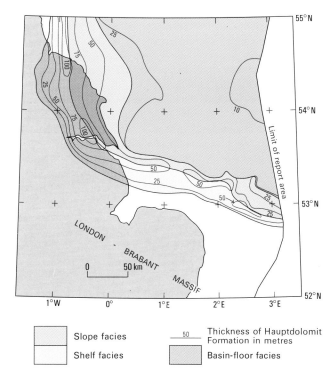

Figure 38 Distribution, thickness and facies of the Hauptdolomit Formation of the Z2 Group, and equivalent formations in eastern England. Land information adapted from Smith (1989).

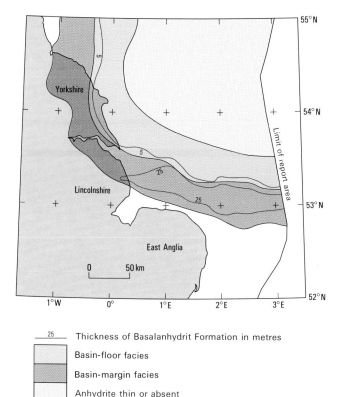

Figure 39 Distribution, thickness and facies of the Basalanhydrit Formation in the Z2 Group, and equivalent formations in eastern England. Land information adapted from Smith (1989).

1980a); this may account for the thicker and more widespread development of sabkha facies there than in Yorkshire.

Stassfurt Halite Formation

The Stassfurt Halite Formation comprises the complex sequence of evaporites which filled, or nearly filled, the Permian basin during the later stages of the second Zechstein cycle. More than 500 m of evaporites may have been precipitated from concentrated brines in the centre of the basin, but their primary depositional sequence and thickness are not easily determined as they have been severely deformed by halokinesis. Where they overlie the submarine slope and adjacent parts of the early Z2 basin floor, the evaporites are less severely affected by halokinesis and are between 100 and 250 m thick (Figure 40). Shorewards, the Stassfurt Halite Formation extends over the outer edge of the Z2 carbonate-anhydrite shelf on the southern margin of the basin (Figure 34), and is between 25 and 60 m thick along the former shelf edge.

Although dominated by halite, especially in the centre of the basin, the Stassfurt Halite Formation contains variable amounts of sulphate minerals (anhydrite, polyhalite and kieserite) and chlorides (sylvite and carnallite). Colter and Reed (1980) have used the relative proportions of these to establish that the sequence of evaporite precipitation was punctuated by two regional discontinuities which subdivide the equivalent Fordon Formation in Yorkshire into three evaporite subcycles. These discontinuities can be identified in many of the offshore wells. The two earliest evaporite cycles have a lenticular geometry in Yorkshire (Colter and Reed, 1980), thickening from the shelf over its submarine slope, but thinning eastwards across the floor of the basin. As with the underlying Z2 carbonates and anhydrites, this had the effect of extending the marginal shelf, and restricted sedimentation of the uppermost subcycle towards the centre of the basin.

On the outer margin of the basin floor, the lowest subcycle is between 15 and 50 m thick in offshore wells; it is mainly composed of halite, but as in eastern England (Colter and Reed, 1980), may contain thin bands and wisps of anhydrite towards the base. The middle subcycle is up to 275 m thick, and its basal 100 m has a distinctive gamma-ray signature in many wells along the southern margin of the basin, including well 49/24-4 (Figure 40). By analogy with onshore wells, anhydrite and polyhalite dominate at the base, whereas halite, polyhalite and kieserite predominate at the top of this subcycle. Much of the remainder of the middle subcycle is composed of halite; well logs suggest that there are only minor intercalations of anhydrite and polyhalite. Carnallite becomes an increasingly important component towards the top of the subcycle in a number of wells, and its intercalation with red and grey shales causes characteristically spiky gamma-ray profiles (Figure 40). The presence of these shales suggests virtually complete evaporation of parts of the Zechstein basin by the end of this middle subcycle. The upper subcycle is more than 100 m thick in the centre of the basin, and is characterised by an upward increase in halite at the expense of anhydrite; polyhalite is notably absent both offshore and in Yorkshire.

In the centre of the southern North Sea, the basal 80 to 120 m of the Stassfurt Halite Formation comprises a mixed sequence of anhydrite, polyhalite and halite, with a gamma-ray signature very similar to that of the basal beds of the middle subcycle around the basin margins. The lowest subcycle therefore appears to be either thin or absent in this region. The remainder of the Z2 evaporite sequence has been deformed by halokinesis in most wells, and the discontinuity between the middle and upper subcycles cannot be traced across the basin.

The cyclicity of the evaporites is difficult to demonstrate in the basin-marginal shelf sequences, where the basal 10 to 25 m of the formation in many wells is dominated by halite (Figure 40). The remainder of the formation is composed of interbedded polyhalite and halite, though carnallite has been recorded in some wells. The upper basinal evaporite subcycle appears to be absent from the shelf.

The Stassfurt Halite Formation also includes a discontinuous anhydrite layer, formerly assigned by Rhys (1974) to the Deckanhydrit Formation, which is between 1 and 25 m thick and occurs in about 70 per cent of wells. This anhydrite is diachronous, for it occurs above the middle subcycle of the Z2 evaporites on the shelf, and at the top of the overlying subcycle in the centre of the basin. Taylor and Colter (1975) deduced that the anhydrite is most likely a solution residue from the top of the Stassfurt Halite Formation, formed as less saline waters flooded the basin during the third major Zechstein marine transgression.

Z3 GROUP

This group consists of the Grauer Salzton, Plattendolomit, Hauptanhydrit, Leine Halite and Roter Salzton formations.

Grauer Salzton Formation

This formation comprises grey, illitic, unfossiliferous, marine shales which generate a prominent gamma-ray peak; they are between 1 and 3 m thick at the base of the Z3 Group in most offshore wells (Figure 41). Smith (1980) surmised that these

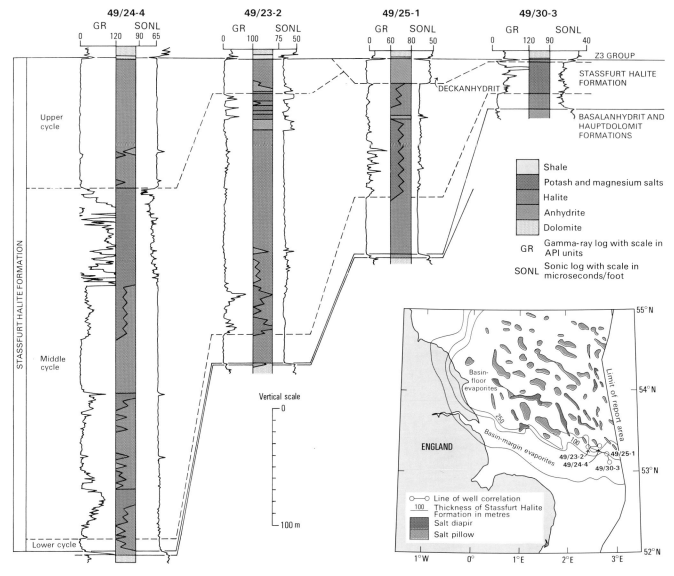

Figure 40 Well correlation of Z2 evaporite cycles on the southern margin of the Zechstein basin, with inset map depicting the distribution and generalised thickness of the Stassfurt Halite Formation, and equivalent formations in eastern England. Where deformed by halokinesis in the centre of the basin, the Stassfurt Halite is locally less than 50 m thick in zones of salt withdrawal, and more than 2500 m thick in some of the largest salt diapirs.

shales accumulated under shallow-water, highly saline conditions which prevailed across the whole basin during the early stages of marine transgression. The salinity may have been enhanced initially by dissolution of the Z2 halite, enabling shale deposition to continue until the sea was sufficiently diluted by the circulation of oceanic waters for carbonate production to begin. By that time, the transgression had already extended shale deposition over the Z2 carbonate-anhydrite shelf, for the Grauer Salzton Formation overlaps the Stassfurt Halite Formation and oversteps on to Z2 continental shales around the margins of the basin (Figure 34).

Plattendolomit Formation

The carbonates of the Plattendolomit Formation overlap the Grauer Salzton Formation and overstep on to Z2 continental deposits (Figure 34) across a zone which is between 5 and 30 km wide on the southern margin of the basin. Continuing basinal subsidence allowed shelf, submarine-slope and basin-floor environments to develop once more, though the bathymetric relief was much less pronounced than during previous transgressions. Water depths may not have exceeded a few tens of metres in the centre of the basin (Smith, 1980). As with the earlier carbonates, the Plattendolomit Formation of the southern North Sea and the equivalent Brotherton (formerly Upper Magnesian Limestone) Formation of Yorkshire thicken basinwards across a carbonate shelf, and have their maximum thicknesses of between 55 and 90 m along the outer shelf edge (Figures 34 and 42). Both formations thin rapidly across the submarine slope, and the Plattendolomit Formation is thin or absent in parts of the basin floor.

The shelf-carbonate facies is mainly composed of grey, microcrystalline, partly pelleted dolomite with thin, interbedded layers of grey shale. Porous, oolitic grainstones occur locally in shoal areas close to the shoreline of the Z3 transgression (Taylor, 1986). A restricted biota is entirely confined to the innermost 50 km of the shelf; it is dominated by two species of thin-shelled bivalves and the tubular calcareous alga *Calcinema permiana*. Smith (1980) has suggested that the salinity may have been slightly high, and a basinward decrease in water turbulence may have been one of the principal factors causing this fauna to disappear towards the outer shelf.

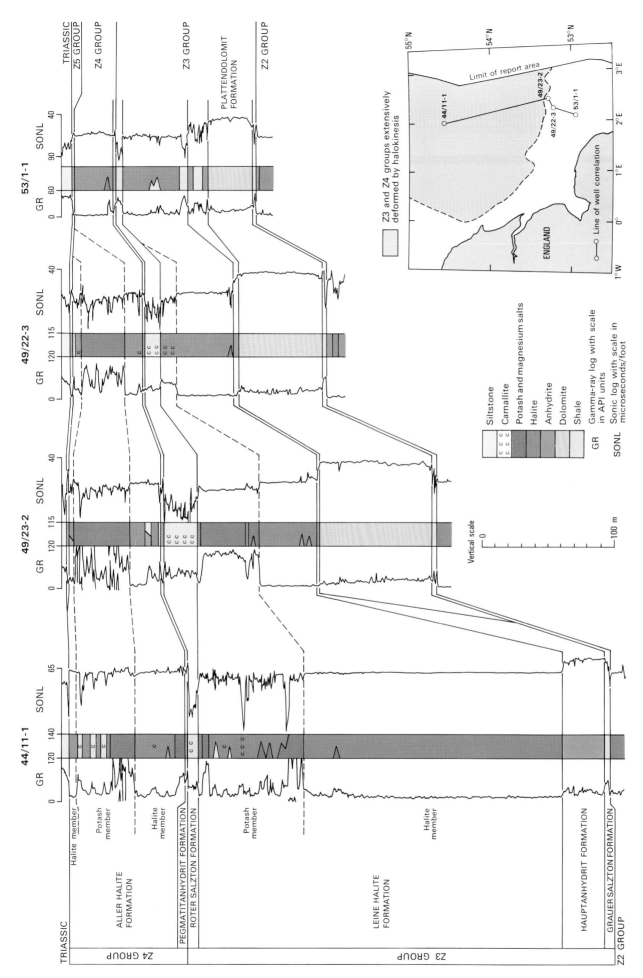

Figure 41 Well correlation of formations in the Z3, Z4 and Z5 groups. Adapted from Smith and Crosby (1979).

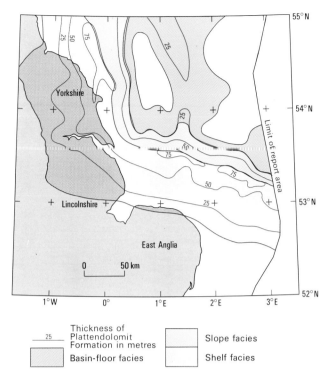

Figure 42 Distribution, thickness and facies of the Plattendolomit Formation in the Z3 Group, and equivalent formations in eastern England. Land information adapted from Smith (1989).

Where the algal tubes are present, they form layers up to a few centimetres thick consisting almost entirely of horizontal stems; these alternate with thicker layers of microdolomite containing scattered fragments of the alga (Taylor and Colter, 1975). Algal mats of stromatolitic dolomite and bands of nodular displacive anhydrite are widespread at the top of the shelf carbonates; they were deposited in minor sabkha cycles and indicate that carbonate production had built the shelf close to sea level once more (Taylor and Colter, 1975).

At the base of the Plattendolomit Formation there are a few metres of fine-grained, barren dolomite on the outer shelf (Smith, 1980). The remainder of the outer-shelf and the slope facies of the Plattendolomit Formation are composed of barren, dark brown, laminated microdolomite that contains abundant laths of replacement anhydrite (Taylor and Colter, 1975). Beds of nodular, dolomitic anhydrite occur within the carbonate in some wells. The proportion of anhydrite increases through the uppermost few metres of the slope carbonates, and the boundary between the Plattendolomit and the overlying anhydrite formation is transitional around the margins of the shelf.

Across most of the basin floor, the Plattendolomit Formation is represented by thin shales which are indistinguishable from the underlying Grauer Salzton Formation (Clark, 1986). Three wells in the area between 1° and 2°E have encountered a thicker development of anhydritic dolomite that was perhaps deposited on a shallow-water ridge extending from the Mid North Sea High across the floor of the basin.

Hauptanhydrit Formation

Isopachs of the Hauptanhydrit Formation reflect the bathymetry inherited from the preceding phase of carbonate deposition (Figure 43). Its anhydrites are between 25 and 70 m thick across areas of the basin floor, where the underlying Plattendolomit Formation is thin or absent. Both the Hauptanhydrit Formation and the equivalent Billingham Formation of north-east England thin rapidly shorewards over the carbonate slope, and both formations are generally less than 15 m thick across the Z3 carbonate shelf.

The Z3 anhydrites have been cored in many boreholes in north-east England, where the basin-margin evaporites accumulated in an exceptionally broad coastal sabkha (Smith, 1980). The formation is mostly composed of nodular or cumuloid anhydrite deposited in repeated minor sabkha cycles; these are interspersed with beds of laminated anhydrite which may have formed in shallow pools or during flooding. Unevenly laminated dolomite resembling the algal stromatolites in the underlying Z3 Plattendolomit Formation carbonates comprises 5 to 10 per cent of the evaporite sequence. Thin anhydrite breccias occur locally around the margins of the basin, indicating synsedimentary erosion and transport (Smith, 1980).

The lithology and depositional environment of the Hauptanhydrit Formation are less well known in the centre of the southern North Sea. The anhydrite is variably described as being amorphous or microcrystalline, and interpretation of well logs suggests that beds and laminae of dolomite decrease in abundance towards the top of the formation. Thin argillaceous laminae have been noted in a number of wells. Smith (1980) suggested that the basin-floor anhydrites could have accumulated either by precipitation from brine which was between 60 and 80 m deep in the centre of the basin, or by basinward extension of the sabkha facies across this area following a 60 to 80 m fall in water level.

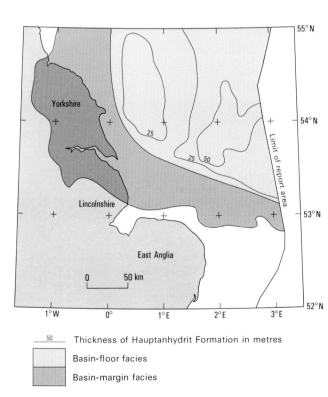

Figure 43 Distribution, thickness and facies of the Hauptanhydrit Formation in the Z3 Group, and the equivalent formation in eastern England. Land information adapted from Smith (1989).

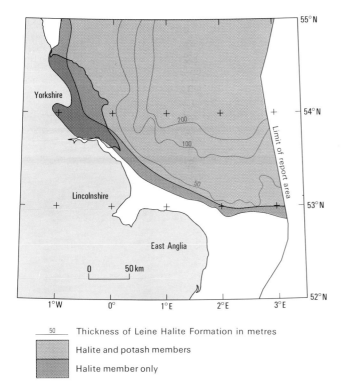

Figure 44 Distribution, thickness and lithology of the Leine Halite Formation in the Z3 Group, and the equivalent formation in eastern England. Land information adapted from Smith and Crosby (1979).

Leine Halite Formation

This formation thickens progressively basinwards to a maximum of about 300 m in the southern North Sea (Figure 44). Its base is sharply defined on well logs (Figure 41), but its boundary with the overlying Roter Salzton Formation is less clear in many wells in the centre of the basin. The Leine Halite and equivalent Boulby Halite Formation of north-east England can be readily subdivided (Figure 41) into a lower halite-dominated member and an upper potash-dominated member (Smith and Crosby, 1979).

The lower member thickens from 50 m on the Yorkshire coast to about 200 m in the centre of the southern North Sea. In cored boreholes in Yorkshire, the halite contains laminae and intergranular meshes of anhydrite with thin beds of mudstone. Mineral relationships and fabrics suggest that the basal 15 m of halite are diagenetically altered chemical precipitates from shallow water (Smith, 1980). Most of the remainder of the halite member was deposited in a marginal belt of salt flats which may have been more than 60 km wide.

The succeeding potash member is missing in a zone about 20 km wide around the margins of the Z3 evaporite basin; from this zone it thickens basinwards to about 100 m by apparently incorporating successively lower beds of the underlying halite by lateral transition (Smith, 1980). This implies a diachronous relationship between the two members. Sylvite is the principal potash-bearing mineral in north-east England, where the member is up to 16 m thick and forms commercial deposits. In this region, the member is composed of halite and secondary sylvite in a weakly layered mesh of carbonaceous mudstone or anhydrite, and there are beds of halitic or sylvitic mudstone and anhydrite (Smith, 1974). The distorted carbonaceous layers resemble algal mats (Smith, 1980), indicating continued deposition in basin-marginal salt flats. The potash member is overlain by up to 5 m of halite in Yorkshire, but this unit disappears towards the centre of the basin.

Where the Leine Halite Formation is thickest, polyhalite and carnallite are the principal potassium-bearing minerals (Taylor and Colter, 1975). Many wells in the centre of the basin, including 44/11-1, have penetrated a highly radioactive zone of concentrated potassium salts at the base of the potash member (Figure 41). The overlying sediments form a variable sequence of interbedded halite, potassium salts and mudstone; these tend to have a higher gamma-ray signature towards the top of the Leine Halite in beds which may be laterally equivalent to the lower part of the Carnallitic Marl of north-east England.

Roter Salzton Formation

The Roter Salzton and Carnallitic Marl formations respectively cap the Leine Halite offshore, and the Boulby Halite Formation in Yorkshire. They were assigned by Rhys (1974) and Smith et al. (1986) to the Z4 Group and equivalent Staintondale Group respectively. However, the overlying Pegmatitanhydrit Formation defines the transgression at the base of the fourth major Zechstein cycle (Taylor and Colter, 1975; Taylor, 1986) and is a more reliable log marker across the whole of the basin (Figure 41). Transgressive deposits define the bases of the other Zechstein groups in the North Sea (Rhys, 1974), and for this reason, the Roter Salzton Formation is assigned here to the Z3 Group.

The Carnallitic Marl Formation of Yorkshire was deposited on a basin plain in which the distal extremities of alluvial fans encroached over the evaporites around the margins of the basin (Smith, 1980). In these marginal areas, there is a rapid transition from the underlying halite formation to red mudstones. Intercalations of evaporites become increasingly abundant towards the centre of the basin, where the Roter Salzton Formation is virtually indistinguishable from the uppermost beds of the Leine Halite Formation in many wells, as it comprises red, saliferous mudstones. These have a maximum thickness of 35 m. In eastern England, the mudstones contain displacive and interstitial anhydrite, halite, carnallite and sylvite, suggesting temporary incursions of concentrated brines from the basin across its marginal plains (Smith, 1980). The proportions of these minerals increase offshore, and an extensive, highly saline playa lake may have persisted in the centre of the basin until the onset of the fourth major Zechstein transgression.

Z4 GROUP

The sediments of this group are divided into two; the Pegmatitanhydrit and Aller Halite formations.

Pegmatitanhydrit Formation

As with the earlier Zechstein transgressions, the basal sediments of the Z4 Group thicken basinwards across a marginal shelf on which the Pegmatitanhydrit Formation is up to 11 m thick, but thin rapidly across a submarine slope to form a continuous sheet 1 to 2 m thick across the floor of the basin. In eastern England, the equivalent Sherburn Anhydrite is underlain by a thin bed of barren dolomite or magnesite rock (Smith, 1974), but carbonates have not been recorded at the base of the Z4 shelf sequence in any offshore wells (Taylor, 1986).

The Pegmatitanhydrit Formation comprises a distinctive, grey anhydrite with coarse, secondary, halite crystals. It maintains a uniform lithology over a very wide area, and this, along with its even, parallel lamination and the absence of sabkha sedimentary structures, led Smith (1974) to surmise that a shallow, hypersaline, lagoonal sea extended over the basin during the Z4 transgression. There is no shelly fauna in the underlying carbonates onshore, implying that the saline brines were less diluted by circulating oceanic waters than during any of the previous Zechstein transgressions.

Aller Halite Formation

The Aller Halite Formation thickens progressively basinwards to a maximum of about 120 m in the southern North Sea (Figure 45). The formation appears to be anomalously thin over the Cleaver Bank High, perhaps due to leaching beneath the base-Cretaceous unconformity, which is relatively close above the Upper Permian in that region. Both the Aller Halite Formation and the equivalent Sneaton Halite Formation of north-east England can be readily subdivided into two halite-dominated members and one potash-dominated member (Figure 41).

The lower halite member increases in thickness from 19 m on the Yorkshire coast to about 35 m in the centre of the southern North Sea. In cored boreholes in Yorkshire, the halite contains laminae and a few thin beds of anhydrite, halitic anhydrite and anhydritic mudstone; sylvite is locally abundant at the top and base of the member (Smith and Crosby, 1979). In the centre of the basin, there are numerous intercalations of anhydrite and polyhalite near the base of the halites (Taylor and Colter, 1975), and the boundary between the Pegmatitanhydrit and Aller Halite formations is less clearly defined.

The succeeding potash member is up to 75 m thick in the southern North Sea, whereas equivalent potash and mudstone deposits have a combined maximum thickness of only 32 m on the Yorkshire coast (Smith and Crosby, 1979). In Yorkshire, the potash member contains variable proportions of halite and sylvite in a mesh of anhydrite and carbonaceous clay, and there are beds of halite, sylvinitic mudstone and anhydrite. The overlying mudstone is composed of silty halite in a matrix of brick-red, silty mudstone or siltstone. Most of the potash member in the centre of the southern North Sea appears to comprise interbedded sylvite, halite and perhaps polyhalite; gamma-ray radioactivity increases upwards in many wells, suggesting that sylvite and carnallite are intercalated within red mudstones at the top of the potash member.

On the east coast of England, the uppermost 5 m of the Z4 evaporites is dominated by halite, and there are only scattered films and local meshes of anhydrite and red, silty mudstone (Smith, 1974). In the centre of the basin, this member includes intercalations of silty mudstone, and has a maximum thickness of about 2 m.

Smith (1974; 1980) has surmised that the basal halite and potash members of the Aller Halite Formation accumulated on extensive salt flats of very low relief around the margins of the basin. They were precipitated from an increasingly saline playa lake which was subject to wide fluctuations of size and depth. The mudstone represents widespread encroachment of continental facies across the evaporite basin as the central

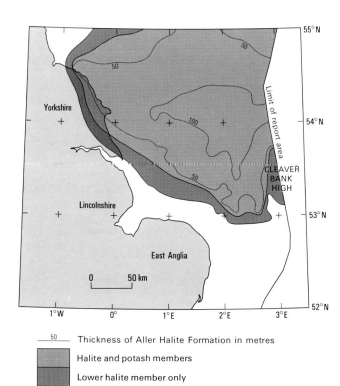

Figure 45 Distribution, thickness and lithology of the Aller Halite Formation in the Z4 Group, and the equivalent formation in eastern England. Land information adapted from Smith and Crosby (1979).

playa became temporarily reduced in size, and perhaps dried out completely for short periods. The upper halite member was deposited during a later expansion of the playa, probably caused by an influx of fresher water into the basin.

Z5 GROUP

Grenzanhydrit Formation

The fifth Zechstein transgression is represented in eastern England by a bed of weakly layered anhydrite (Smith, 1980), and although this extends across the southern North Sea as the Grenzanhydrit Formation, it has not been identified in many wells (Figures 34 and 41).

Taylor and Colter (1975) have used detailed well correlation to deduce that the Grenzanhydrit Formation may thicken basinwards to about 6 m. In some wells, the anhydrite splits to encompass a thin halite member so that the anhydrite beds are less than 1 m thick, close to the limit of downhole log resolution, such that they are difficult to identify. Smith (1980) has interpreted the minor fifth Zechstein cycle as representing a final, but short-lived expansion of the basin-centre saline playa. Continental facies had already been established beyond the limit of the Z5 transgression, and these spread rapidly across the whole of the basin around the end of Permian times.

8 Triassic

The configuration of the Variscan postorogenic foreland basin extending from north-east England across the southern North Sea into Germany and eastern Europe had been established during the Permian. This basin persisted through Triassic times, with the London–Brabant Massif, Pennine High and Mid North Sea High continuing to define its southern, western and northern margins. The Mid North Sea High had become a relatively low topographic feature, much reduced in size, and it is partly mantled by thin Triassic deposits. Upper Permian sediments in the basin are dominated by evaporites precipitated from concentrated brines, and marine carbonates and shales deposited during five incursions of less saline waters from the Boreal Ocean. The last marine incursion late in Permian times was followed by rapid contraction of the evaporite basin. This allowed continental and paralic facies, which had been restricted to its western and southern margins (Smith, 1989), to become established across the whole of the basin early in the Triassic. In the absence of fossils, the base of the Triassic across the southern North Sea is recognised at the abrupt transition to continental facies (Rhys, 1974).

In the offshore area, the Triassic sequence is dominated by reddish brown mudstones, although there are sandstones forming important gas reservoirs in the Lower Triassic, and thin beds of green mudstone, dolomite, anhydrite, sandstone, and widespread halite members in the Middle and Upper Triassic. The basal Triassic sediments of the Bacton Group (Figure 46) are up to 700 m thick, and were deposited in playa-lake, floodplain and fluvial environments (Fisher, 1986). The climate was undoubtedly arid, and Clemmensen (1979) has deduced that the North Sea area was then within a trade-wind zone in low northern latitudes. The dominant wind direction was from the east, resulting in generally low rainfall, hot summers and cool winters. Precipitation was principally in cloudbursts which initiated sheet floods that at first drained into a large, playa lake (Fisher, 1986). Fluvial sandstones make up most of the Lower Triassic sequence in eastern England, and rejuvenation of the London–Brabant

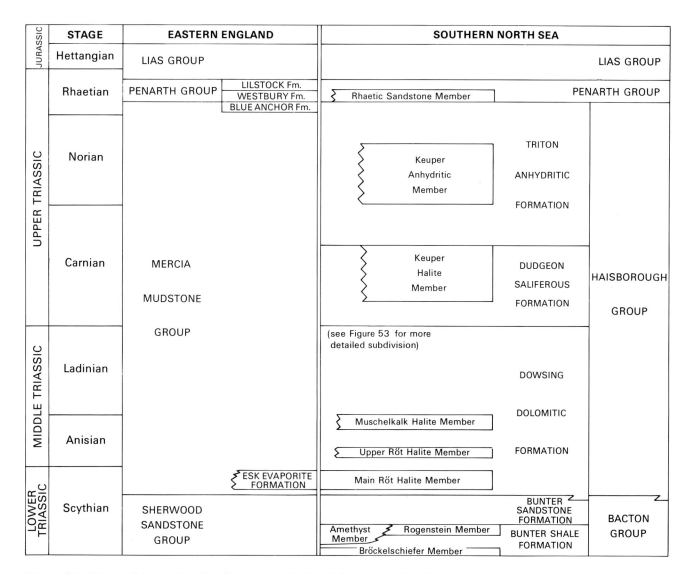

Figure 46 Triassic lithostratigraphy of eastern England and the southern North Sea.

Massif and Pennine High caused a progressive expansion of coarse-clastic deposition across the whole of the southern North Sea during Early Triassic times.

The remaining Triassic sediments (Figure 46), comprising the Haisborough Group (Rhys, 1974) and Penarth Group (Lott and Warrington, 1988), have their maximum thickness of 950 m at the southern end of the Sole Pit Trough. An abrupt change from sandstone to dominantly argillaceous lithologies occurs in all offshore wells; this basinwide sequence boundary is commonly referred to as the Hardegsen Unconformity. The peripheral source areas had been largely peneplaned by this time (Brennand, 1975), and had very low relief throughout the remainder of the Triassic Period. Most of the Haisborough Group was deposited in a shallow, hypersaline, perhaps quasi-marine environment, or on playa mudflats and distal floodplains (Fisher, 1986). Unlike the Bacton Group, there are substantial intercalations of evaporites and coastal sabkha deposits; these record four widespread marine incursions across the basin. In Germany, there is evidence that each of these incursions resulted from the temporary widening of a connection between the North European Triassic Basin and the Tethys Sea to the south (Ziegler, 1982). A more permanent connection was established late in the Triassic. During deposition of the Penarth Group, faunal evidence indicates that a shallow, warm, epicontinental sea with a reduced salinity of 25 to 30 parts per thousand, extended over most of north-west Europe (Hallam and El Shaarawy, 1982). Normal salinities and fully marine conditions were achieved in the basin early in the Jurassic.

The principal Triassic formations of north-east England are, like those of the Upper Permian, traceable through equivalent formations in the southern North Sea into The Netherlands and Germany with only minor thickness and lateral facies changes. Those thickness changes that do occur appear to be related to reactivation of Variscan basement faults from mid-Early Triassic times. These allowed the Sole Pit Trough to become the principal depocentre in the basin, and the Cleaver Bank High began to develop as a complementary area of reduced subsidence and thinner sediment accumulation towards the end of the Triassic. Contemporary halokinetic deformation of Upper Permian evaporites caused significant, but comparatively local, thickness variations of Middle and Upper Triassic formations adjacent to the Swarte Bank Hinge Zone (Walker and Cooper, 1987).

The Triassic sediments are buried beneath more than 1500 m of younger sediments in parts of the Anglo-Dutch Basin (Figure 3). Here, and elsewhere in the southern North Sea, their depth of burial and level of preservation have been affected by post-Triassic uplift of the Sole Pit Trough, Cleveland Basin and Cleaver Bank High, and by local diapirism of the underlying Permian halites. The uppermost Triassic sediments have been eroded at the southern end of the Sole Pit Trough and adjacent to the coast of north-east England, so that Middle Triassic beds subcrop beneath the Quaternary there. More than 1000 m of Triassic sediments were eroded from the crest of the Cleaver Bank High during the Late Jurassic.

There may be thin Triassic deposits south of the London–Brabant Massif in Kent (Warrington et al., 1980). Boreholes at Penhurst and Brabourne have penetrated unfossiliferous red mudstones and a few metres of sandy conglomerate between proven Carboniferous and Jurassic rocks. These sediments have long been regarded as Triassic in age (Warrington et al., 1980), but their lateral extent along the north-eastern margin of the Wessex–Channel Basin is unknown; if Triassic sediments do continue beneath the Dover Strait, they are likely to have a limited distribution.

The Triassic lithostratigraphy of the southern North Sea (Figure 46) was established by Rhys (1974). The mainly Lower Triassic Bacton Group is equivalent to the Sherwood Sandstone Group of England (Warrington et al., 1980). The overlying Haisborough Group, comprising Lower, Middle and Upper Triassic sediments, is equivalent to the Mercia Mudstone Group of England; it was subdivided by Rhys (1974) into three formations based on the relative abundance of dolomite, halite and anhydrite within the argillaceous sequence. Following recommendations by Lott and Warrington (1988), the top of the Haisborough Group has been redefined; sediments previously designated the Winterton Formation are now included within the Penarth Group, which encompasses most of the uppermost Triassic sediments of eastern England and the southern North Sea.

BACTON GROUP

Bunter Shale Formation

Most of the Bunter Shale Formation (Figure 46) comprises a monotonous sequence of red or red-brown silty mudstones, with only thin intercalations of siltstone, a few beds of dolomite, and traces of anhydrite. The formation is 315 m thick in its type well 49/21-2, and maintains a remarkably uniform thickness of between 350 and 400 m across the northern half of the Anglo-Dutch Basin (Figure 47). In its basal 20 to 70 m, there are more numerous beds of siltstone and local developments of sandstone within the Bröckelschiefer Member (Rhys, 1974). The Rogenstein Member (Rhys, 1974) at the top of the formation is distinguished by having a few thin beds containing ferruginous ooliths (Figure 48). In addition, fluvial sandstones of the Amethyst Member are interbedded towards the top of the formation near the English coast (Figure 47); these herald the

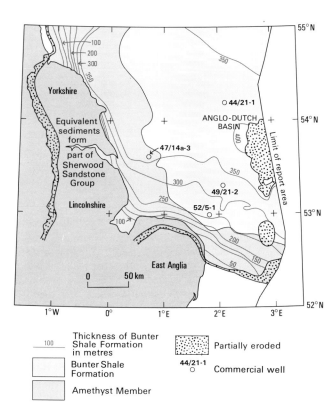

Figure 47 Distribution and thickness of the Bunter Shale Formation and the distribution of the Amethyst Member.

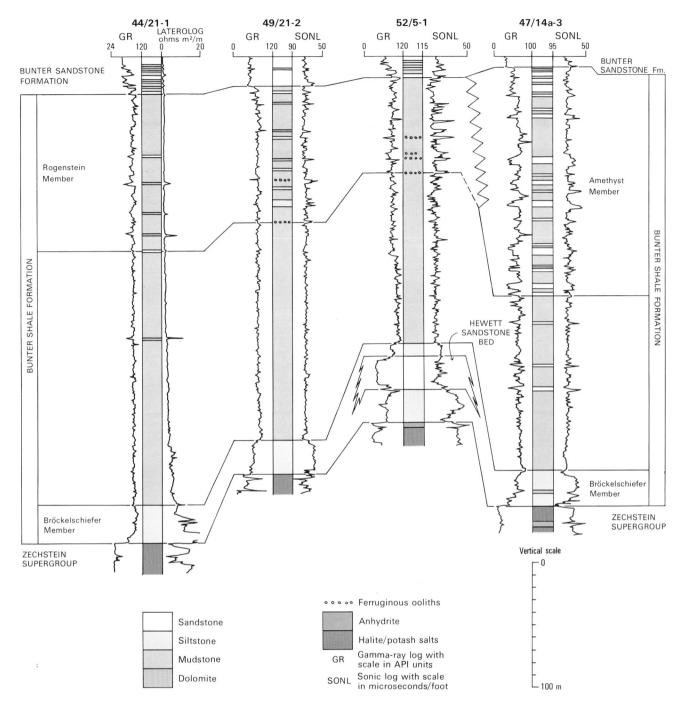

Figure 48 Correlation of the Bunter Shale Formation between four offshore wells. For locations see Figure 47.

more widespread influx of coarse-clastic detritus in the overlying Bunter Sandstone Formation.

The Bröckelschiefer Member has a distinctive, upward-increasing gamma-ray profile in many offshore wells, indicating upward fining (Figure 48); it also has a significantly higher sonic velocity than the overlying mudstones. The member was deposited as distal floodplains became established across the former Permian evaporite basin and were gradually superseded by a finer-grained lacustrine facies. Beyond the southern limit of the evaporite basin the base of the Bröckelschiefer Member is not easily resolved, as the uppermost Permian and basal Triassic sediments form a continuous sequence of red silty mudstones and siltstones. Within the basin there is a sharp lithological break separating the Bröckelschiefer Member from the underlying evaporites.

The Hewett Sandstone Bed occurs within the Bröckelschiefer Member in the Hewett Gasfield (Cumming and Wyndham, 1975) and along a narrow belt on the northern flank of the London–Brabant Massif (Figure 49). The well-sorted, medium- to coarse-grained, red-brown, quartzose sandstones, with a maximum thickness of 62 m, form the principal reservoir in the Hewett Gasfield. The sandstones have scattered pebbles and a few thin layers of conglomerate containing rounded clasts of sedimentary and metamorphic origin. Fisher (1986) has interpreted the Hewett Sandstone Bed as a localised influx of coarse-grained fluvial detritus derived from a short-lived Early Triassic phase of uplift and erosion of an adjacent part of the London–Brabant Massif.

Above the Bröckelschiefer Member, up to 370 m of red, silty, mainly lacustrine mudstones accumulated in the north ern half of the Anglo-Dutch Basin. Three or four beds contain abundant ferruginous ooliths towards the top of the Bunter Shale Formation (Figure 48), and the lowest defines the base of the Rogenstein Member (Rhys, 1974). Although the oolitic beds are widespread in the basin, they are generally less than 1 m thick. Despite this, they are easily identified on downhole logs, and provide the only basinwide correlative horizons within the Bunter Shale Formation. The first appearance of the ooliths is at 103 m below the top of the Bunter Shale Formation in its type well 49/21-2, but the Rogenstein Member is about 150 m thick in the centre of the basin.

Beds of sandstone up to 10 m thick become increasingly common within the lacustrine facies in the upper part of the Bunter Shale Formation, towards the western and southern margins of the Triassic basin. Onshore in eastern England, the equivalent sediments form part of the Sherwood Sandstone Group, which is composed almost entirely of fluvial sandstones (Warrington et al., 1980). The transitional facies containing interbedded sandstones and mudstones is up to 340 m thick immediately off Lincolnshire, and occurs in all wells around the Amethyst Gasfield; it is therefore here designated the Amethyst Member. The base of the Amethyst Member is defined by the first significant occurrence of sandstone, and its top at the generally sharp transition to the dominantly sandy sediments of the Bunter Sandstone Formation (Figure 48). The Amethyst Member extends offshore for up to 70 km from the English coast (Figure 47); its sediments represent an interdigitation of coarse-grained fluvial facies with distal-floodplain and lacustrine sediments around the fluctuating margins of a playa lake. Eastwards, the base of the Amethyst Member occurs at progressively higher stratigraphical levels, implying a gradual contraction of the playa lake during Early Triassic times. The Amethyst and Rogenstein members interdigitate at the offshore limit of the sandy sediments.

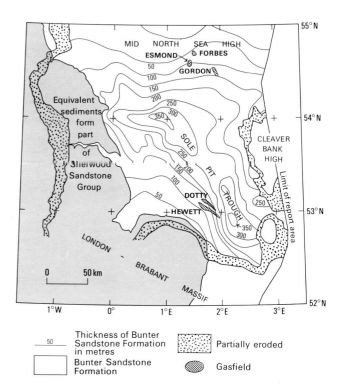

Figure 50 Distribution and thickness of the Bunter Sandstone Formation.

Bunter Sandstone Formation

As defined by Rhys (1974), the Bunter Sandstone Formation is an upward-coarsening sheet-sand complex of red, orange, and occasionally white sandstones which are mainly fine grained, but locally medium or coarse grained. The Bunter Sandstone Formation provides excellent gas reservoirs in the Forbes, Esmond, Gordon, Hewett and Dotty fields. Basinwards of the underlying Amethyst Member, log correlation indicates that the first influx of fluvial sands was more or less simultaneous across the southern North Sea area (Brennand, 1975). This further expansion of fluvial deposition was caused by increased tectonic activity in the source areas peripheral to the basin (Fisher, 1986). Contemporary movements on basement faults allowed the Sole Pit Trough to become the main depocentre, accumulating more than 350 m of fluvial sandstones (Figure 50). The Bunter Sandstone Formation is relatively thin over the crest of the Cleaver Bank High and on the southern flank of the Mid North Sea High, which were relatively distal from the source of the sands. The Cleaver Bank High may temporarily have been a zone of reduced subsidence in mid-Early Triassic times.

The Bunter Sandstone Formation is equivalent to the Main Bundsandstein Formation in the Dutch sector of the North Sea. In The Netherlands and in Germany, the sediments display a marked upward-fining cyclicity (NAM and RGD, 1980; Geiger and Hopping, 1968). Each of three regional cycles contains a basal sandstone unit which becomes thicker and more massive towards the south of the basin, and is overlain by claystones and siltstones that become thicker towards the basin centre. West of the Cleaver Bank High, this concept of cyclicity is not easily demonstrated in the UK sector, where the sediments are more dominantly sandy. There are only a few layers of red, green and grey, silty mudstone, mostly less than 2 m thick, and none can be correlated across the basin (Figure 51). Log properties and sedimentary structures indicate that the sands were deposited in fluvial channels by sheet floods (Fisher, 1986), and that they were

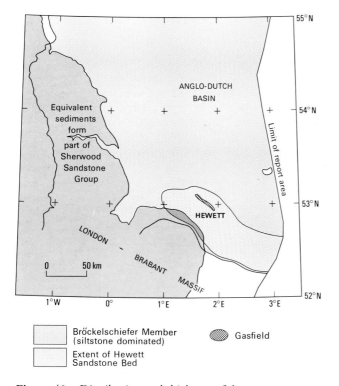

Figure 49 Distribution and thickness of the Bröckelschiefer Member of the Bunter Shale Formation.

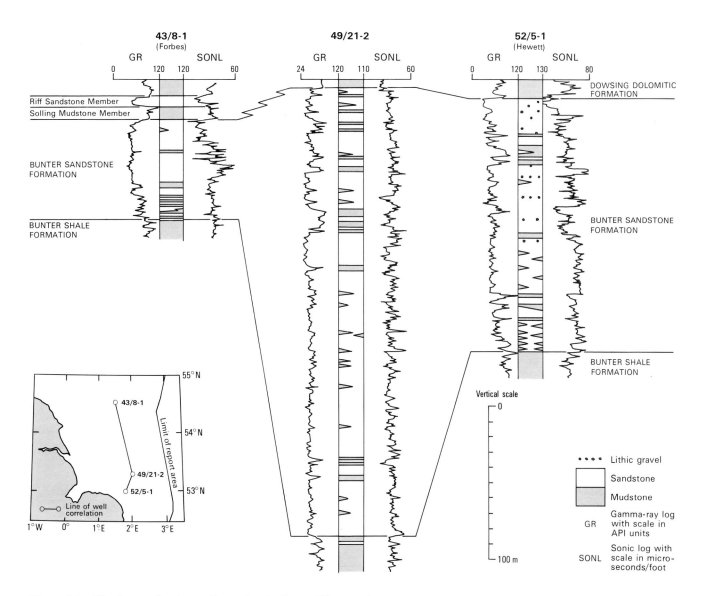

Figure 51 The Bunter Sandstone Formation in three offshore wells.

derived principally from the London–Brabant Massif and Pennine High to the south and west. The interbedded silty mudstones were deposited in ephemeral lakes. Bifani (1986) interpreted the Bunter Sandstone Formation of the Esmond, Forbes and Gordon gasfields as representing deposition by a number of coalescing alluvial fans dissected by fluvial channels in an arid to semiarid environment.

HAISBOROUGH GROUP

Dowsing Dolomitic Formation

The Dowsing Dolomitic Formation (Figure 46) has a maximum thickness of 420 m in the southern North Sea, and is more than 300 m thick over large areas of the Anglo-Dutch Basin and the Sole Pit Trough (Figure 52). Its upper formational boundary is less easily resolved in eastern England (Southworth, 1987), where it is generally less than 200 m thick. The formation is dominated by red, silty mudstones, but there are three intercalated halite members. These halites and their underlying sediments each record a temporary marine influx westwards across the basin, and each is overlain by a regressive sequence including sabkha deposits. The palaeoenvironmental interpretation and lithostratigraphy of the Dowsing Dolomitic Formation summarised in Figure 53 are based on Southworth (1987).

At the base of the Haisborough Group, the Solling Mudstone and Riff Sandstone members have been reported from only a few wells adjacent to Dutch waters north of the Cleaver Bank High. Here, these members have a combined thickness of 30 m, and are composed of red, silty, lacustrine or prodeltaic mudstones overlain by a relatively thin sequence of deltaic sandstones (Figure 54). The sediments record a limited retreat and temporary readvance of coarse-grained, fluvial deposition in the north-east of the UK sector.

Where it occurs, the Riff Sandstone Member becomes progressively more argillaceous basinwards (Southworth, 1987). Where it continues south-westwards beyond the limit of the Solling Mudstone Member, it is likely to be indistinguishable from, and has been incorporated within, the Bunter Sandstone Formation. The base of the Haisborough Group is therefore diachronous in the north-east of the UK sector.

Above the Riff Sandstone Member, deposition of the Lower Röt Mudstone Member extended argillaceous facies westwards across the Anglo-Dutch Basin and the East Midlands Shelf, but the mudstones did not encroach on to the northern flank of the London–Brabant Massif. Contemporaneous mudstones contain a marine fauna in Poland, and a brackish-marine fauna in Germany, but it is unclear as to whether truly marine conditions extended into the UK sector at this time. The red silty mudstones are about

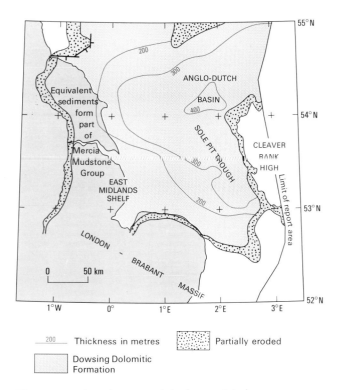

Figure 52 Distribution and thickness of the Dowsing Dolomitic Formation.

4 m thick in north-east England (Warrington et al., 1980), but are between 8 and 12 m thick in most of the offshore area. Successively younger argillaceous members of the Dowsing Dolomitic Formation overstep the Bunter sandstones to the south-west (Figure 53), recording the expansion of argillaceous sedimentation to peripheral areas during mid-Triassic times.

The overlying Röt Halite Member (Rhys, 1974), renamed the Main Röt Halite Member by Southworth (1987), is between 60 and 80 m thick in most of the northern half of the Anglo-Dutch Basin (Figure 55), but is more than 100 m thick near some Permian salt diapirs. Onshore, the halite is up to 40 m thick (Riddler, 1981) in the Esk Evaporite Formation of north-east England (Warrington et al., 1980). In the centre of the basin, the Main Röt Halite Member comprises five evaporite cycles (Southworth, 1987), with a thin but laterally continuous red mudstone at the base of each cycle deposited while marine influences persisted. Each mudstone is overlain by a thin layer of dolomite or anhydrite and capped by a massive halite bed. The halites record five intervals of marine regression from the area; they were precipitated in restricted lagoons during the first cycle, and in more widespread, shallow, evaporite basins during succeeding cycles.

Up to 23 m of interbedded anhydrites, dolomites and dolomitic mudstones, deposited as a contemporary sabkha facies around the landward margins of the halite basin (Figure 53), comprise the Main Röt Evaporite Member (Southworth, 1987). However, this member also includes a more widespread regressive sequence of sabkha deposits, generally less than 10 m thick, which prograded from the margins over the top of the Main Röt Halite Member.

Marine regression continued during deposition of the Intermediate Röt Mudstone Member (Southworth, 1987), which in England contains evidence of desiccation, and thickens north-eastwards across the offshore area from 10 m to more than 20 m. In eastern England, the member comprises cyclic alternations of laminated mudstones deposited in lagoons, and blocky mudstones that accumulated subaerially on playa mudflats.

The Upper Röt Halite, Upper Röt Evaporite and Upper Röt Mudstone members (Southworth, 1987) were deposited during the second major marine incursion and subsequent regression. Evaporite precipitation did not extend into England, and the Upper Röt Halite Member is restricted to the northern half of the Anglo-Dutch Basin (Figure 56), where it is between 5 and 11 m thick. The 3 to 6 m-thick sabkha deposits of the Upper Röt Evaporite Member were deposited both contemporaneously around the landward margins of the halite basin in a zone between 40 and 60 km wide, and as a regressive unit above the Upper Röt Halite Member (Figure 53). As regression continued, lagoonal and playa-mudflat deposition prograded once more from the margin across the whole of the southern North Sea. The resulting Upper Röt Mudstone Member is up to 80 m thick in the north-east of the Anglo-Dutch Basin (Figures 53 and 54).

Southworth (1987) recognised that beyond the landward limit of the sabkha deposits (Figure 53), the Intermediate and Upper Röt Mudstone members cannot be separated where lagoonal and playa-mudflat deposition were continuous between the first and third marine transgressions of the Dowsing Dolomitic Formation. Southworth (1987) referred these basin-marginal sediments to the Main Röt Mudstone Member, which is between 45 and 65 m thick on the East Midlands Shelf, and which overlaps the Main Röt Evaporite Member on to the Bunter Sandstone Formation along the northern flank of the London–Brabant Massif.

The Muschelkalk Dolomitic Member (Southworth, 1987) was deposited during the third major incursion of marine influences across the Triassic basin. Shallow, open-marine conditions existed in the eastern Netherlands and beyond (Ziegler, 1982), but the interbedded dolomites and dolomitic mudstones of this member were deposited mainly under quasi-marine conditions. During two intervals of relatively low sea level, playa-mudflat and lagoonal facies prograded temporarily across the basin-marginal area (Southworth, 1987). The Muschelkalk Dolomitic Member is between 40 and 55 m thick on the East Midlands Shelf, and up to 70 m thick in the Anglo-Dutch Basin.

The Muschelkalk Halite Member (Rhys, 1974) was deposited during the initial stages of marine regression, as quasi-marine conditions in the southern North Sea evolved into hypersaline waters. This member has a quite different areal distribution from the earlier halites of the Dowsing Dolomitic Formation (Figures 55, 56 and 57), for it extends much farther south towards the London–Brabant Massif. The Muschelkalk Halite Member is generally between 40 and 60 m thick offshore, but it is absent from eastern England, the East Midlands Shelf, and the north-west of the Anglo-Dutch Basin. There are numerous thin beds of red mudstone within the halite, but none of these are easily correlated through the offshore wells (Southworth, 1987). The halite beds are thinnest and least extensive at the base of the member (Figure 54), for initially the halites were precipitated in isolated lagoons, whereas thicker halite beds subsequently accumulated in more widespread evaporite basins (Southworth, 1987).

As with the earlier evaporites, the Muschelkalk Evaporite Member includes 15 to 24 m of thinly bedded anhydrites, dolomites and mudstones that accumulated around the landward margins of the halite basin, as well as regressive sabkha deposits that are between 8 and 15 m thick above the halite member (Southworth, 1987). As regression continued, the Muschelkalk Mudstone Member was deposited; it is composed of mainly red, but partly green, anhydritic mudstone, and is between 35 and 65 m thick. The equivalent sediments

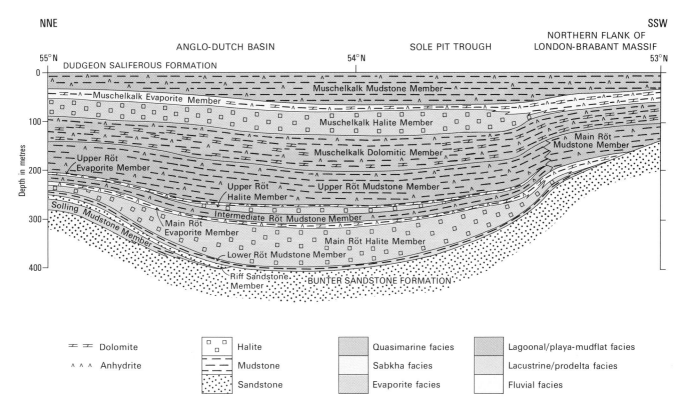

Figure 53 Schematic profile through the Dowsing Dolomitic Formation, showing the lateral relationships of the members proposed by Southworth (1987). The datum is the base of the Dudgeon Saliferous Formation.

in eastern England were deposited largely under subaerial conditions (Southworth, 1987), and the mudstone member represents an advance of subaerial, playa-mudflat facies across the southern North Sea, heralding a more prolonged retreat of marine influences from the area.

Dudgeon Saliferous Formation

The Dudgeon Saliferous Formation is thickest within the Sole Pit Trough, where it has its greatest thickness of 350 m at the southern end (Figure 58). The formation is less than 100 m thick both in the north-west of the Anglo-Dutch Basin and over the northern half of the East Midlands Shelf. Thickness changes are particularly abrupt across the Dowsing Fault Zone, indicating the local influence of contemporary fault movement on sediment accumulation.

The formation was deposited during a return to dominantly continental clastic facies across the North European Triassic Basin (Fisher, 1986). This resulted from a global eustatic fall in sea level combined with closure of the basin's south-eastern connection with the Tethys Sea (Ziegler, 1982). However, beds of halite that are intercalated towards the top of the Dudgeon Saliferous Formation record the fourth mid-Triassic incursion of marine influences across the basin, this time through a new connection with the Tethys Sea in southern Germany (Ziegler, 1982).

At its base, the Dudgeon Saliferous Formation is composed of a monotonous sequence up to 100 m thick of red-brown, sporadically green mudstones (Rhys, 1974). There are only a few thin layers of dolomite. The mudstones accumulated in subaerial, perhaps distal, floodplain environments (Southworth, 1987). Their boundary with the underlying mudstone member of the Dowsing Dolomitic Formation is represented in many offshore wells by an upward increase in sonic velocity and a slight increase in gamma radiation (Figure 54). The formational boundary is not easily resolved in some wells along the northern flank of the London–Brabant Massif, for subaerial floodplain deposition had already commenced around the landward margin of the playa mudflats.

Above the basal subaerial mudstone facies, there are numerous beds of halite interbedded with red mudstone in the centre of the offshore area. The lowest evaporite defines the base of the Keuper Halite Member, and Rhys (1974) has taken the top of the uppermost halite as the boundary between the Dudgeon Saliferous and Triton Anhydritic formations (Figure 59). Isopachs of the Keuper Halite Member (Figure 60) indicate that it is considerably thicker than the halite members in the underlying Dowsing Dolomitic Formation, yet it is almost entirely restricted to within, and east of, the Sole Pit Trough. It seems that connection with contemporary evaporite deposits in the Dutch sector was around the southern flank of the Cleaver Bank High.

The Keuper Halite Member records a return to dominantly subaqueous deposition in the centre of the southern North Sea, under conditions of fluctuating salinity. Beds of halite comprise up to 80 per cent of the member in the centre of the Sole Pit Trough, where they are between 1 and 20 m thick. Beds of anhydritic mudstone and anhydrite were deposited during the frequent intervals of reduced salinity in which hypersaline marine conditions extended far beyond the present limits of the halite. Evaporite precipitation took place during marine regression, and was confined to the most rapidly subsiding parts in the basin. Fisher (1986) noted that the thinnest halites are laterally impersistent, and suggested that precipitation took place in numerous, shallow, saline lakes, rather than the basinwide bodies of hypersaline water envisaged for the Röt and Muschelkalk evaporites.

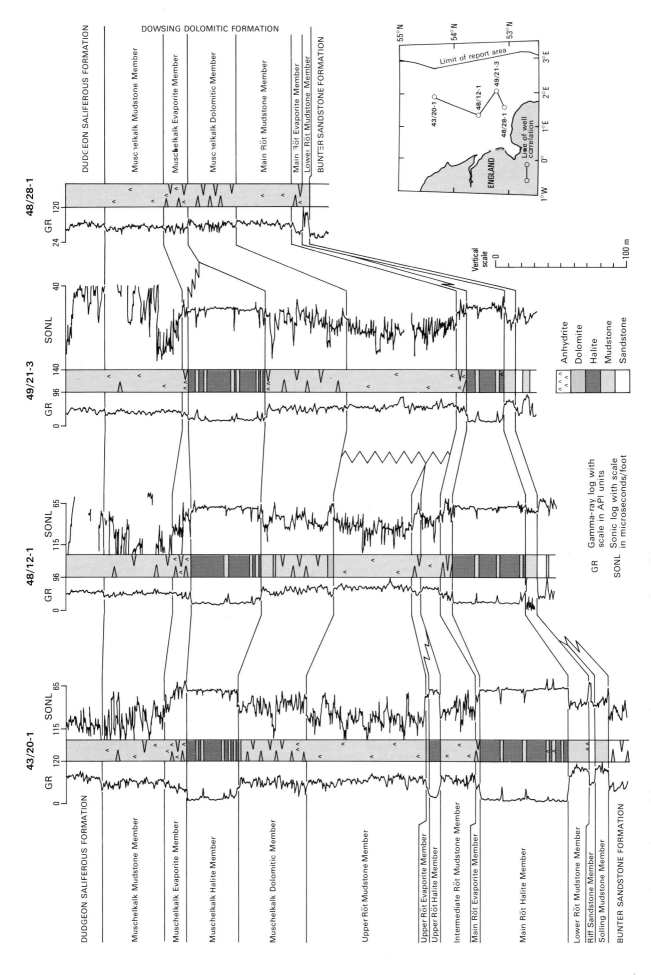

Figure 54 Correlation of the Dowsing Dolomitic Formation in four offshore wells.

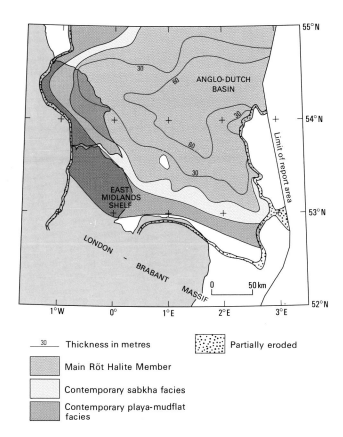

Figure 55 Distribution, thickness and facies of the Main Röt Halite Member.

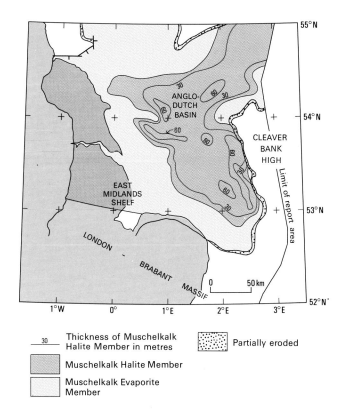

Figure 57 Distribution and thickness of the Muschelkalk Halite and Muschelkalk Evaporite members.

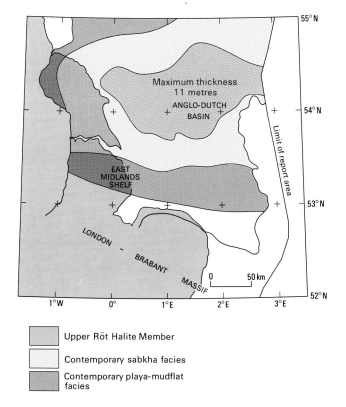

Figure 56 Distribution of the Upper Röt Halite Member and contemporary facies.

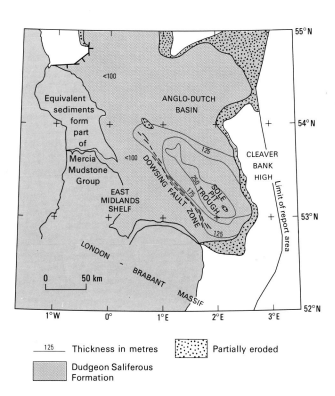

Figure 58 Distribution and thickness of the Dudgeon Saliferous Formation.

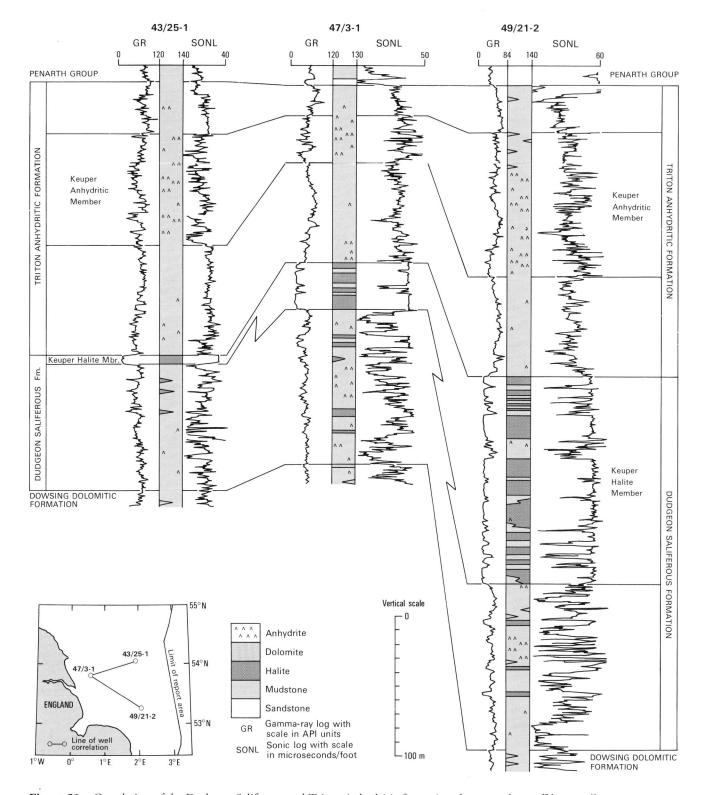

Figure 59 Correlation of the Dudgeon Saliferous and Triton Anhydritic formations between three offshore wells.

Triton Anhydritic Formation

The Triton Anhydritic Formation has its greatest thickness of 250 m at the southern end of the Sole Pit Trough (Figure 61). As with the underlying formation, its thickness changes abruptly across the Dowsing Fault Zone, demonstrating that sediment accumulation continued to be strongly affected by contemporary fault movements in this region.

The Triton Anhydritic Formation was deposited as continental conditions became established more continuously across the North European Triassic Basin. At its base in the Sole Pit Trough, the formation comprises a monotonous sequence of up to 100 m of red mudstones with only a few thin layers of anhydrite. These mudstones continue across most of the offshore area, but beyond the limits of the Keuper Halite Member they are not easily distinguished from similar mudstones in the underlying Dudgeon Saliferous Formation.

Higher in the Triton Anhydritic Formation, Rhys (1974) recognised that beds of anhydrite become sufficiently numerous to distinguish a Keuper Anhydritic Member. This member is characterised by a pronounced spikiness on the sonic logs, and slightly reduced gamma radiation in many wells (Figure 59). The Keuper Anhydritic Member is about 100 m thick in the Sole Pit Trough and, unlike the Keuper Halite, it continues on to the East Midlands Shelf and can be identified in many wells bordering the London–Brabant Massif. Taylor (1983) used stable isotopes to demonstrate that a continental brine regime prevailed during deposition of most of the mudstones in the equivalent Trent Formation of central England, although interbedded anhydrites indicate a possible minor contribution from marine-derived brines.

The mudstones at the top of the Triton Anhydritic Formation, above the anhydrite member, are predominantly grey-green in colour and are between 10 and 50 m thick. Most of the mudstones in the equivalent Blue Anchor Formation of central England were deposited in low-salinity lakes with indications of mixed marine and continental water sources (Taylor, 1983). As a quasi-marine facies, these mudstones signal the transition towards fully marine conditions that extended across the whole of the North European Basin during latest Triassic times.

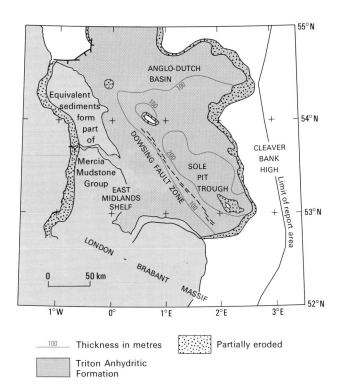

Figure 61 Distribution and thickness of the Triton Anhydritic Formation.

PENARTH GROUP

The Penarth Group (Figure 46), together with the basal beds of the succeeding Lias Group, comprise the uppermost Triassic sediments of eastern England (Warrington et al., 1980) and the southern North Sea (Lott and Warrington, 1988). Offshore, most of these latest Triassic sediments had been previously assigned by Rhys (1974) to the Winterton Formation, the base of which had been placed at a log marker within a relatively uniform sequence of grey-green mudstones. Lott and Warrington (1988) recognised that a more valid regional lithostratigraphical marker is provided by the transgressive sediments which were deposited as the Rhaetian seas extended across northern Europe into England. There is an abrupt change in lithology at the base of these deposits, reflected by a pronounced change in gamma-ray and sonic-log responses in most offshore wells (Figure 62). As so defined, the base of the Penarth Group can be correlated more easily across the whole of the offshore area, and occurs 7 m above the base of the Winterton Formation in its type well, 49/21-2 (Figure 62).

The Penarth Group is more than 25 m thick within the Sole Pit Trough, and has its greatest thickness of 75 m at its southern end (Figure 63). Here, the sediments are divisible into two units that correspond approximately with the Westbury and Lilstock formations of eastern England (Lott and Warrington, 1988).

The lower unit comprises very fine-grained silty sandstones and thin shales, although the proportion of sand increases towards the south-east, and well 49/21-2 penetrated 45 m of sandstone with only thin beds of grey-green and brown mud-

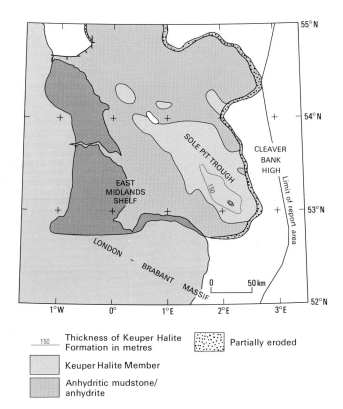

Figure 60 Distribution and thickness of the Keuper Halite Member.

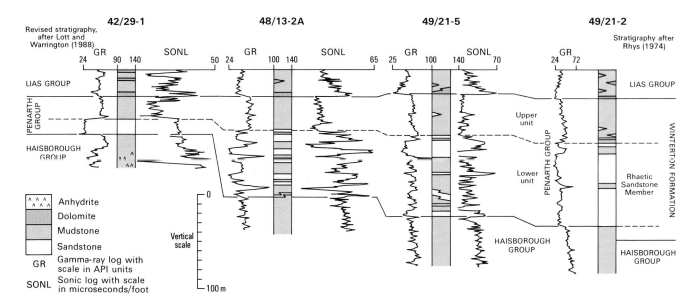

Figure 62 Correlation of the uppermost Triassic sediments between four offshore wells. Adapted from Lott and Warrington (1988). For locations see Figure 63.

stone. This sequence was designated the Rhaetic Sandstone Member by Rhys (1974), but is atypical. The gamma-ray records have a more serrated profile in most of the surrounding wells (Figure 62), indicating a greater proportion of shale beds. Lott and Warrington (1988) have used the gamma-ray profiles to deduce that the lower unit comprises a stacked series of upward-fining cycles. Transgressive sediments from the lower unit are less than 10 m thick where they have been cored in BGS borehole 81/44 off the Yorkshire coast (Lott and Warrington, 1988). These sediments comprise micaceous, pyritic sandstones in beds up to 0.3 m thick, alternating with dark grey to black, bituminous mudstones and shales. Pyritised shell debris and bioturbation are common, contrasting with their absence in the underlying sediments of the Haisborough Group.

The upper unit of the Penarth Group comprises relatively uniform, dark grey and brown, noncalcareous shales across the whole offshore area; there are only a few thin beds of fine-grained sandstone. Indeed the entire Penarth Group is composed of this argillaceous facies in the north-east of the Anglo-Dutch Basin, although, here, the lower boundary of the Penarth Group continues as a conspicuous log marker (Figure 59).

The marine shales and thinly bedded limestones of the Lias Group succeed the Penarth Group across the whole of the southern North Sea. Its basal limestone bed forms a prominent log marker in many wells (Figure 62), and defines the base of the Lias both in England (Warrington et al., 1980; Ivimey-Cook and Powell, 1991) and the offshore area (Rhys, 1974; Lott and Warrington, 1988). Warrington et al. (1980) and Ivimey-Cook and Powell (1991) used the appearance of the ammonite genus *Psiloceras* to define the base of the Jurassic at a level a few metres above this limestone bed in eastern England. The Triassic–Jurassic boundary is likely to fall within, but close above the base of the Lias Group across the southern North Sea also.

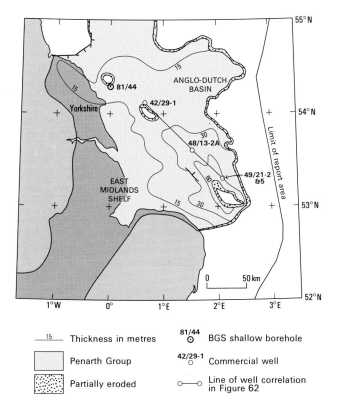

Figure 63 Distribution and thickness of the Penarth Group.

9 Jurassic

During Jurassic time, the report area formed part of a basin which extended eastwards from eastern Britain into The Netherlands and Germany, northwards via the Central Graben into the Norwegian–Danish Basin, and southwards into the Anglo-Paris Basin and Tethyan province (e.g. Ziegler, 1982). Such an extensive basin encompassed a wide range of depositional environments, which are reflected by considerable facies variation, particularly in the Middle and Upper Jurassic. Subsequent tectonism and erosion, primarily during the Late Jurassic to earliest Cretaceous, separated these basinal deposits into a series of isolated remnants. The classic successions of eastern England thus form the western margin of one Jurassic basin remnant, by far the greatest proportion of which lies offshore beneath the southern North Sea.

The Jurassic sediments preserved in the offshore area (Figure 64) are locally significantly thicker than their equivalents in eastern England. The maximum proven thickness of some 950 m compares with 600 m in the East Midlands, 700 m in the Cleveland Basin, more than 2000 m in the West Netherlands Basin, over 1400 m in the Central Graben and 3000 m in the Viking Graben. Commercial drilling in the southern North Sea area has however been restricted to positive structural features, and it is probable that the thickest Jurassic successions have yet to be drilled.

Jurassic strata crop out at sea bed or subcrop beneath Quaternary in two major areas of the southern North Sea (Figure 2). The most extensive is off the Yorkshire coast (Dingle, 1971; Lott, 1986; BGS California Solid Geology sheet) where strata proved by sea-bed sampling and shallow drilling range from Rhaetian to Kimmeridgian in age. Lithologically and biostratigraphically, they compare closely

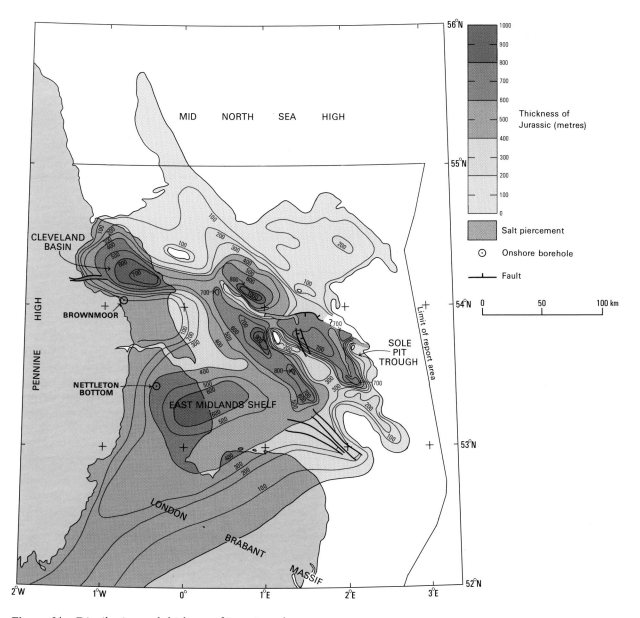

Figure 64 Distribution and thickness of Jurassic rocks.

with the succession onshore in the Cleveland Basin (e.g. Cox et al., 1987; Lott and Warrington, 1988).

The second extensive sea-bed outcrop is along the Sole Pit Inversion, where Jurassic strata ranging in age from Hettangian to Kimmeridgian are known (BGS Spurn Solid Geology sheet). The thickest Jurassic strata occur along the faulted western margin of the Sole Pit Inversion, and in the rim synclines associated with the major salt structures commonly found within this part of the basin (Figure 64). The Zechstein salt structures began moving in Middle Triassic times (Brunstrom and Walmsley, 1969), and continued to develop spasmodically throughout the Jurassic when local subsidence patterns, affected by salt withdrawal marginal to developing salt piercements, exerted a considerable influence on sediment thickness, and perhaps on local facies.

Despite the drilling of over 150 wells through the Jurassic succession of the southern North Sea, comparatively little has been published of its detailed stratigraphy. The poor hydrocarbon prospectivity of the Jurassic south of the Mid North Sea High has meant that few cores have been cut in the sequence. Nevertheless, it is possible to trace the main lithostratigraphical units recognised in eastern England into the offshore area using wireline logs and rock-cuttings descriptions.

The Jurassic succession offshore may be divided into a number of lithostratigraphical units which can be correlated throughout the basin, although this is made difficult by north to south facies changes. These changes are particularly marked in the Middle and Upper Jurassic, but more subtle variations affect the Lower Jurassic. Repeated oscillation of the shoreline during the Middle Jurassic was a response to sea-level changes that were partly of eustatic origin, and partly related to regional and local tectonics, including salt piercement. As a consequence, the facies belts moved backwards and forwards across the basin as the shoreline advanced and retreated. Transgression during the Bajocian, prompted by a eustatic sea-level rise, pushed the shoreline northwards, and the subsequent volcanic updoming of the Central Graben area was perhaps responsible for the southward retreat of the sea from the northern part of the area during the Bathonian. During the Callovian, the sea re-invaded the northern part of the area, and marine deposition continued throughout Oxfordian and Kimmeridgian times, re-establishing links with the Northern North Sea Basin via the Central Graben.

Rhys (1974) set up a preliminary, tripartite lithostratigraphical nomenclature for the Jurassic of the southern North Sea. A number of subsequent publications, including Kent (1975; 1980b), Selley (1976), and Brown (1986), have presented further summaries of the Jurassic succession of the area. At the base, Rhys (1974) recognised a Lias 'Group' of dark grey mudstones and interbedded argillaceous limestones. This underlies the West Sole Group that is dominated in the west by limestones, and in the north and east by clastic sequences. The upper division, the Humber Group, comprises mudstones, thin sandstones and limestones. These three groups were correlated approximately with the Lower, Middle and Upper Jurassic respectively, and a number of other lithologically distinct units within each group were informally recognised as correlatives of well-known formations in eastern England.

The Jurassic strata of the basin rest with apparent conformity on rocks of the Penarth Group both offshore and over most of the onshore outcrop. There is evidence that Jurassic strata overstep the Penarth Group on to the London–Brabant Massif and Market Weighton Block onshore, but no disconformity has yet been recognised from similar positive structures offshore. An examination of the distribution and thickness of the Penarth Group (Lott and Warrington, 1988) suggests that the patterns of differential subsidence which dominated the basin during the Jurassic were already in operation during the Triassic. The base of the Lias Group is marked by the development of a thin, but persistent, limestone unit.

The Jurassic succession of the southern North Sea is most completely preserved over the East Midlands Shelf. Log correlation of offshore wells on the shelf with cored sequences on its onshore portion, such as the Nettleton Bottom and Brownmoor boreholes (Figure 64), suggests that the lithologies of Jurassic sequences compare quite closely in the south of the basin. However, in the northern and eastern parts of the shelf, the Jurassic sequences thicken substantially into the Sole Pit Trough (Figure 64), and in part at least, are best correlated with the thicker successions of the Cleveland Basin.

The tripartite subdivision of the Jurassic succession offshore has been further refined by using the BGS database and more recently released well data. Re-examination of the sequence using these new data and boreholes from eastern England has allowed an improved lithostratigraphy to be developed for the offshore area (Figure 65).

LIAS GROUP

The Lias Group offshore as defined by Rhys (1974) includes Hettangian to early Toarcian strata, predominantly in a mudstone and argillaceous limestone facies; it is essentially equivalent to the Lower Jurassic. A laterally persistent basal limestone unit is probably equivalent, at least in part, to the Hydraulic Limestones of eastern England (Kent, 1980b). This basal limestone is one of a number of thin, extensive lithological units that can be recognised throughout the basin from their log responses. Typical Lias argillaceous facies are present also in the West Netherlands Basin and these can be traced via the Dutch Central Graben into the Norwegian–Danish Basin (Deegan and Scull, 1977). In Early Jurassic times, the Southern North Sea Basin was bounded to the south by the London–Brabant Massif; data from the offshore margin of this high are limited, but as yet no marginal clastic developments have been recognised. It is likely that, as onshore, the Lias Group thins towards, and onlaps, the massif (Donovan et al., 1979).

Recent BGS mapping of the Lower Jurassic in eastern England (Powell, 1984; Brandon et al., 1990) has shown how lithologically variable the Lower Jurassic sequence can be (Figure 65). In a borehole from the Cleveland Basin, Powell (1984) recognised five broad subdivisions, most of which can be identified offshore. The lowermost thick mudstone and siltstone unit of the Redcar Mudstone Formation (Hettangian to lower Pliensbachian) is separated from the variably thick mudstones and siltstones of the Toarcian Whitby Mudstone Formation by two thin units comprising more regressive, fine-grained sandstones, siltstones, and sandy oolitic ironstones. These regressive deposits are termed the Staithes Formation (lower Pliensbachian) and Cleveland Ironstone Formation (upper Pliensbachian). The topmost unit of the group, the Blea Wyke Sandstone Formation, is found only in the Cleveland Basin (Knox, 1984a), for its depositional limits were constrained by contemporaneous faulting (Milsom and Rawson, 1989). There is no evidence that this formation extends offshore, but the thick clastic successions assigned to the West Sole Group in both the Sole Pit Trough and the rim-synclinal areas may include an arenaceous, late Toarcian component.

The Lias Group sediments cropping out at the sea floor range from Hettangian to early Toarcian in age; sea-bed

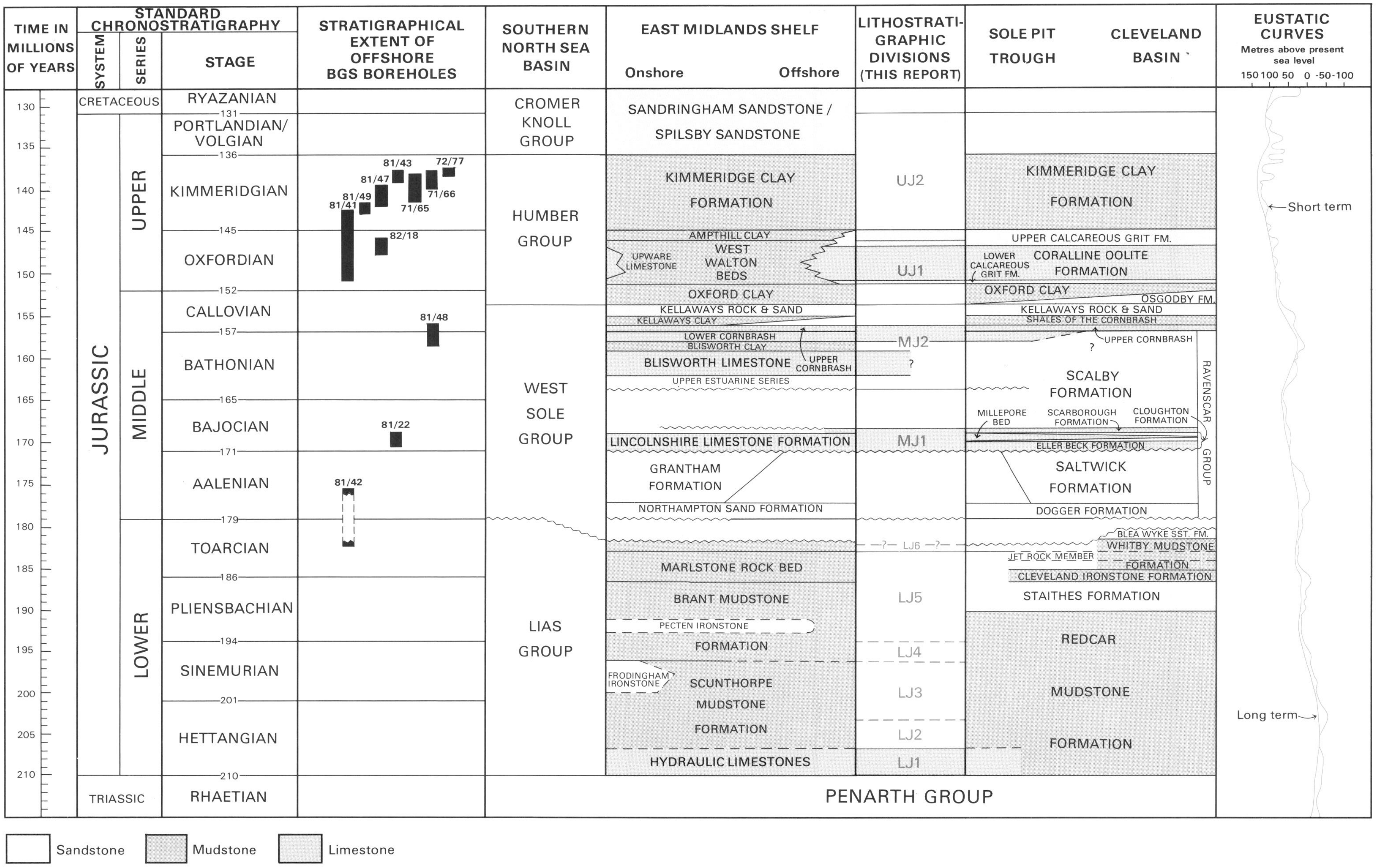

Figure 65 Jurassic lithostratigraphy of eastern England and the southern North Sea. Stratigraphies compiled from Hemingway and Knox (1973), Rhys (1974), Cope et al. (1980a; 1980b), Powell (1984) and Brandon et al. (1990). Eustatic curves after Haq et al. (1987).

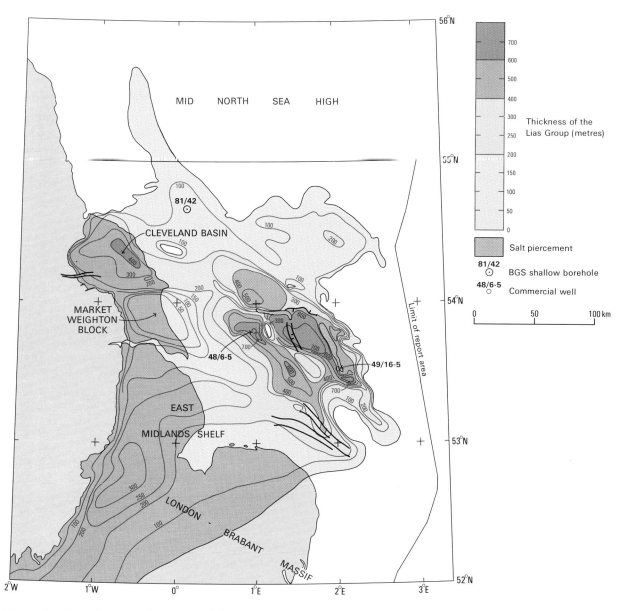

Figure 66 Distribution and thickness of the Lias Group.

samples have proved all stages of the Lower Jurassic other than the late Toarcian. BGS borehole 81/42 proved the topmost Lias Group strata off the Yorkshire coast to comprise dark grey, silty, calcareous mudstones with thin bituminous shales containing an abundant macrofauna of early Toarcian age (Lott, 1986). The absence of late Toarcian sediments can be related to a widespread phase of tectonism and associated erosion in pre-Aalenian times, both onshore and offshore (Hemingway, 1974). The boundary with the West Sole Group is therefore disconformable.

Lithological variations within the Lias Group are largely restricted to changes in carbonate and silt content. On sparker profiles, the Lower Jurassic sequence in the north of the basin is typically well bedded, with alternating strong (limestone) and weak (mudstone) reflectors. Seismic bedding is less clearly defined towards the base of the sequence, but the junction with the underlying Triassic succession is conformable. Localised facies developments, such as the Frodingham Ironstone (Sinemurian) that occurs marginal to the Market Weighton Block, have so far not been recorded in offshore wells, but a prominent seismic and lithological marker occurs in the Middle Lias where it forms a strong topographic ridge at the sea bed. This correlates with an upward-coarsening sequence of sandstones and oolitic ironstones, and is probably equivalent to the late Pliensbachian Marlstone Rock Bed in Lincolnshire, and the Cleveland Ironstone/Staithes formations of the Cleveland Basin (Ivimey-Cook and Powell, 1991). The upper boundary of the Lias Group is marked seismically by a change to the more chaotic reflectors of the West Sole Group, and is coincident with lithological changes readily observed on log responses.

Thickness variations within the Lias Group offshore are considerable (Figure 66). The greatest thickness so far proved in the UK sector is found in well 49/16-5, where 744 m of Lower Jurassic mudstones and argillaceous limestones were proved. Comparable thicknesses have been discovered around the West Sole Gasfield where 709 m were penetrated in well 48/6-5, but most of the area has less than 300 m of Lias Group sediments. These thickness variations are a reflection of differential subsidence patterns that had already been established in the basin during Early to Middle Triassic times.

The Lias Group can be divided on the basis of geophysical log characteristics into six lithostratigraphical units, termed LJ1 to LJ6. Each can be widely traced between wells drilled in the basin (Lott, 1986). The unit boundaries are defined by sharp changes in log response which reflect changes in lithology (Figures 67 and 68).

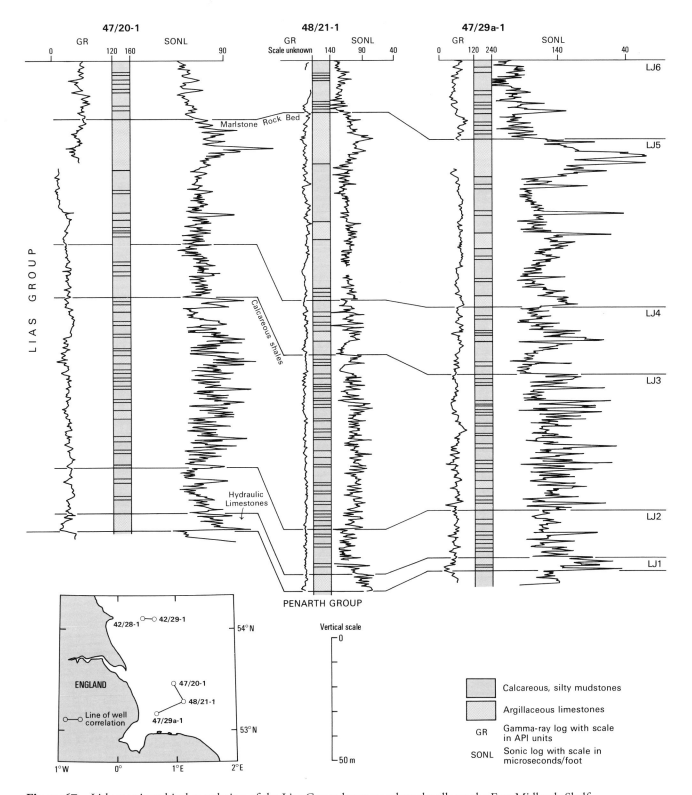

Figure 67 Lithostratigraphical correlation of the Lias Group between selected wells on the East Midlands Shelf.

The basal unit, LJ1, comprises hard, argillaceous limestones which immediately overlie lower-velocity shales of the Penarth Group. The geophysical borehole logs typically show a low gamma-ray response together with a high, spiky, sonic velocity response (Figures 67 and 68). The unit can be traced onshore into the East Midlands where it is probably equivalent to the Hydraulic Limestones, a sequence of alternating shales and shelly argillaceous limestones of early Hettangian age (Kent, 1980b). The unit represents the establishment of fully marine conditions within the basin in Early Jurassic times, following the Rhaetian transgression.

The base of unit LJ2 is defined by an upward increase in the gamma-ray response and a corresponding decrease in the sonic velocity (Figures 67 and 68). The unit comprises shales with a few thin beds of argillaceous limestone. There is an abrupt increase in the proportion of limestone beds in the overlying, comparatively thick unit LJ3; this is reflected in the spiky character of its sonic velocity profile (Figures 67 and 68), although the background gamma-ray reponse is not significantly different from that of unit LJ2. A number of minor cycles can be distinguished, showing upward transition from carbonate-poor to carbonate-rich beds. Over the East

Midlands Shelf, these cycles range up to 30 m in thickness, and are of considerable areal extent.

Unit LJ4 is a thin but extensive interval characterised by a comparatively high-gamma/low-sonic response, but including a medial low-gamma/high-sonic, carbonate-rich interval. The lower boundary of the unit is marked by an abrupt upward decrease in sonic velocity from the limestone-rich unit LJ3 (Figures 67 and 68), and equates with the Sinemurian–Pliensbachian boundary.

Unit LJ5 contains two important marker horizons of the Lower Jurassic, and over much of the area shows a 'waisted' gamma/sonic profile; this is particularly well developed on the East Midlands Shelf (Figure 67). In the lower part of the unit, a high-sonic/low gamma-ray interval probably correlates with the early Pliensbachian Pecten Ironstone of eastern England. In its upper part, the unit shows very high sonic velocity and low gamma-ray values resulting from oolitic, ferruginous or sandy beds which are in part equivalents of the Marlstone Rock Bed, Cleveland Ironstone, and Staithes formations of late Pliensbachian age (Powell, 1984). This interval is widely recognisable within the UK sector of the basin, and a correlation can also be made with a similar, regressive, late Pliensbachian sequence recognised in the Dutch sector (NAM and RGD, 1980; Michelsen, 1978).

The uppermost unit, LJ6, varies in thickness across the report area, either because of locally condensed deposition or as a consequence of pre-Aalenian tectonism and erosion. It is largely composed of calcareous silty mudstones of early Toarcian age, and characteristically shows a lower sonic velocity than the underlying unit. A thin, high sonic-velocity spike, probably equivalent to the organic-rich Jet Rock Member, occurs just above the base of the unit in borehole logs (Figures 67 and 68). No representative of the late Toarcian has so far been found offshore in the UK sector.

WEST SOLE GROUP

An abrupt change in sedimentation pattern followed widespread tectonism at the end of the Early Jurassic. The Middle Jurassic successions show much more lateral facies variation, resulting from transgressive and regressive events associated with eustatic sea-level changes and localised tectonic activity. During the late Aalenian and early Bajocian there developed a major volcanic domal feature centred on the Central Graben area of the North Sea (Leeder, 1983). The subsequent erosion and shedding of clastic sediments from this positive structure dominated facies developments in much of the North Sea area during the late Bajocian and Bathonian.

Limited well data appear to confirm that the London–Brabant Massif was a topographic barrier throughout Middle Jurassic times, with links developed only around its western and eastern margins. Like the Middle Jurassic sequences onshore, the West Sole Group thins southwards on to the massif (Figure 69). To the east of the Sole Pit Trough in the West Netherlands Basin, Middle Jurassic sequences have proved to be of brackish-water to shallow-marine facies (NAM and RGD, 1980). In the northern North Sea, the Middle Jurassic sequences are dominated by fluviodeltaic sequences of the Brent Group.

The considerable facies variations that have been identified from north to south within the Middle Jurassic of the report area, combined with their poorly fossiliferous nature (Kent, 1980a), make delimitation of stage boundaries within the group difficult. The facies units are nearshore, shallow-marine and fluviodeltaic in character, and have proved difficult

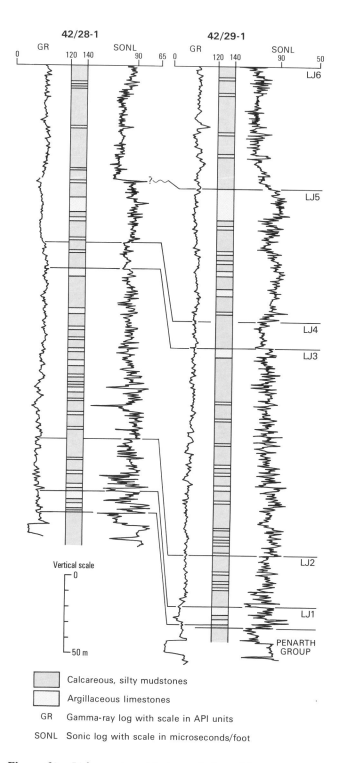

Figure 68 Lithostratigraphical correlation of the Lias Group between two wells off Flamborough Head. For locations see Figure 67.

to correlate biostratigraphically, even in the well-exposed successions of eastern England. Their correlation, both onshore and offshore, relies heavily on lithostratigraphical methods.

The West Sole Group comprises a series of interbedded marine, brackish-water and fluviodeltaic mudstones and sandstones with oolitic and bioclastic limestones; their distribution reflects the oscillatory nature of the shoreline across the basin. As a result of these primarily eustatically controlled shoreline movements, a distinct cyclic nature is commonly apparent in the sedimentary facies of the area.

The West Sole Group (Figure 65) only broadly corresponds to the Middle Jurassic as defined by Cope et al.

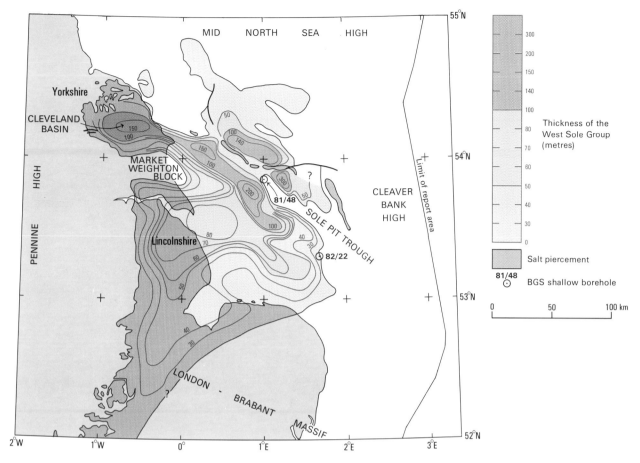

Figure 69 Distribution and thickness of the West Sole Group (Middle Jurassic).

(1980a), since it includes strata ranging from Aalenian to early Callovian in age (Rhys, 1974). The base of the group occurs at a well-defined lithological change in the majority of wells which penetrate the junction. However, as pointed out by Rhys (1974), the group may locally include beds at its base which are of latest Toarcian age, equivalent to the onshore Blea Wyke Sandstone Formation. In the absence of core, these sediments are difficult to distinguish from the overlying sandstone sequences of established Middle Jurassic age.

The Middle to Upper Jurassic boundary, taken at the base of the Oxfordian Stage (Cope et al., 1980a), lies within the Oxford Clay, a lithologically uniform mudstone sequence in much of the southern North Sea. In the absence of cored material, this boundary is difficult to identify precisely as it is based on the recognition of ammonite assemblages. A more readily identified boundary offshore is a lithological change close to the top of the Middle Jurassic, between the sandy Kellaways Rock and Sand and the overlying Oxford Clay. The thin but persistent Kellaways unit (Osgodby Formation) is traceable from Dorset to the Yorkshire coast without significant facies change, and has been proved in sea-bed samples off north-east Yorkshire. In addition, this unit forms a prominent geophysical marker in downhole logs, and is therefore taken here as the top of the West Sole Group. The preserved distribution of the West Sole Group as shown in Figure 69 is not as extensive as that of the Lias Group (Figure 66), and it has a lesser maximum proven thickness of some 300 m.

Rhys (1974) noted that it is possible in some wells to recognise Lincolnshire Limestone Formation and Cornbrash lithologies similar to those of eastern England, but otherwise made no subdivision of the Middle Jurassic. On the basis of well-log interpretations and cuttings, it is apparent that the group falls geographically into two divisions. Those wells drilled on the East Midlands Shelf have Middle Jurassic successions of a similar thickness and lithology to cored sequences drilled onshore in Lincolnshire and South Yorkshire, whereas wells drilled farther east and to the north have considerably expanded successions that show affinities with those of the Cleveland Basin (Figure 70). The type well selected by Rhys (1974) was 47/15-2 (Figure 70); it lies in an area transitional in facies between the East Midlands Shelf succession, dominated by interbedded, marine, oolitic limestones and calcareous sandstone units (e.g. well 47/13-1), and the thicker sandstone and mudstone dominated successions of the Sole Pit Trough and rim-synclinal areas (e.g. well 47/15-1, Figure 70). It seems likely that the latter zones of comparatively rapid subsidence acted as sumps which trapped much of the coarser-grained sediment shed from the Mid North Sea High (Figure 71), thereby delaying the general inundation of the East Midlands carbonate shelf ramp by detrital sediments until the Bathonian.

Correlation of the thick West Sole Group sediments of the Sole Pit Trough north-westwards into the Cleveland Basin, is hampered by lack of well data. The correlatives of the West Sole Group lie, at least in part, within the heterogeneous fluviodeltaic Ravenscar Group. Recent palynological studies have suggested that much, if not all, of the Ravenscar Group (with the possible exception of the youngest unit, the Scalby Formation) is of Bajocian age, and that Bathonian strata were probably not deposited in the Cleveland Basin (Woollam and Riding 1983; Riding and Wright, 1990). The northern limit of Bathonian deposition therefore probably lay close to the northern margin of the Sole Pit Trough. During the Bathonian, brackish-water sediments gradually prograded westwards from the trough into east Lincolnshire, eventually displacing the oolitic carbonate-ramp facies.

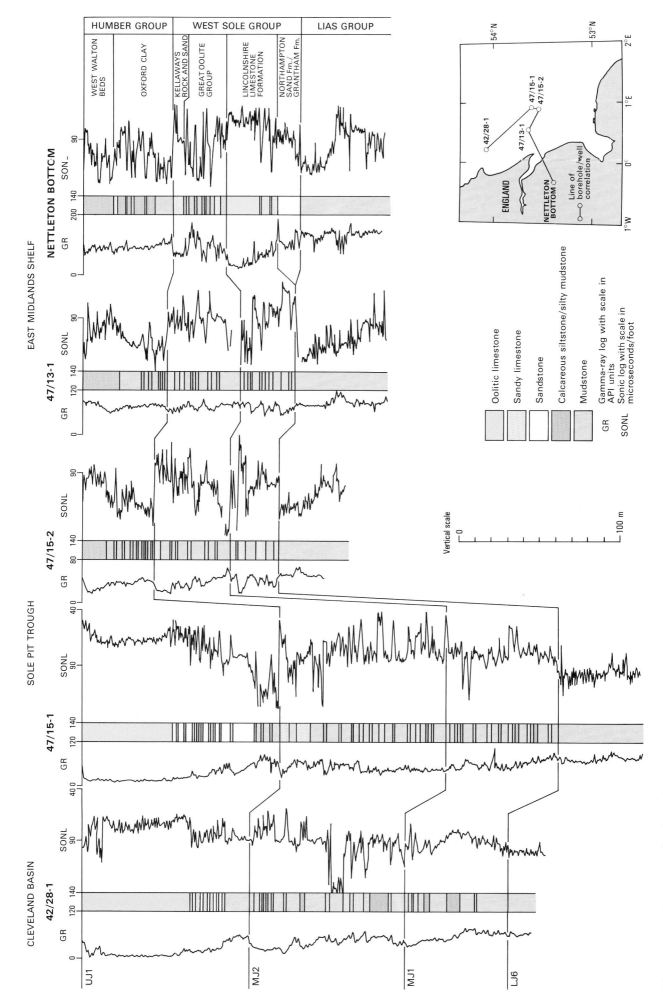

Figure 70 Lithostratigraphical correlation of the West Sole Group from selected wells.

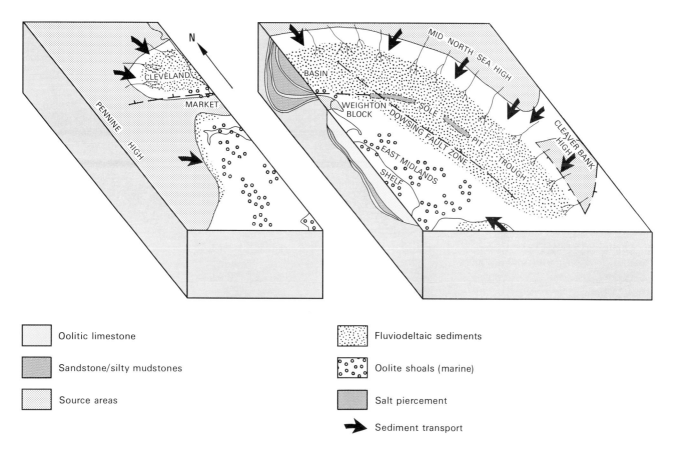

Figure 71 Model of the Bajocian depositional environment in the southern North Sea.

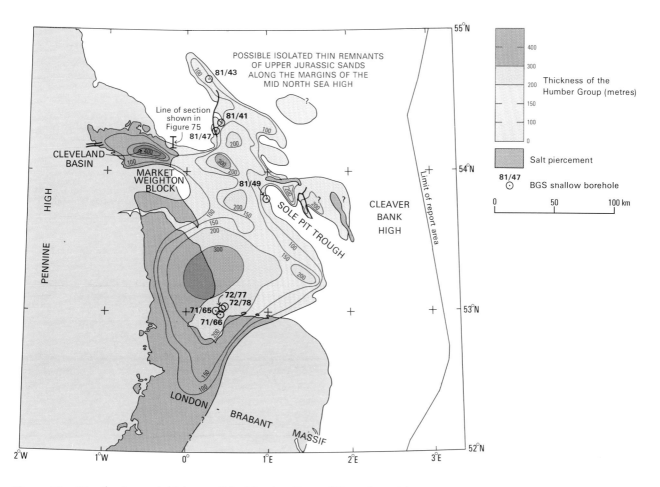

Figure 72 Distribution and thickness of the Humber Group (Upper Jurassic).

The rapid lateral facies changes in the West Sole Group preclude detailed correlations comparable with those made in the Lias Group. However, geophysical logs allow subdivision of the West Sole Group into two broad lithostratigraphical units, here termed MJ1 and MJ2 (Figure 70), for which offshore cores have been obtained only in BGS boreholes.

The base of unit MJ1 is characterised by a marked decrease in gamma-ray response and an increase in sonic velocity compared with the underlying Lias Group. The oolitic limestones that occur within the upper part of the unit in the south-west are gradually replaced eastwards by interbedded sandstones and mudstones. Log correlations with the Nettleton Bottom borehole suggest that these limestones are probably equivalent to the Lincolnshire Limestone Formation. There appears to be no offshore sequence equivalent to the ferruginous sandstones and ironstones of the Northampton Sand and the clays and sands of the Grantham Formation that underlie the Lincolnshire Limestones in eastern England (Figure 70). This suggests that these sediments are littoral developments that do not extend far into the present offshore area, except along the margins of the London–Brabant Massif. The unit shows a sharp top on log responses (Figure 70).

BGS borehole 82/22 cored about 6 m of sediments from the upper part of unit MJ1. The lithologies encountered range from pale grey, hard, sandy limestone to olive-green, muddy siltstone with thin sandstone stringers and lenses. The sequence yielded good palynomorph assemblages that suggest an early Bajocian age, equivalent to the Lincolnshire Limestone Formation (J B Riding, written communication, 1982).

Unit MJ2 shows an overall upward decrease in the gamma-ray response with an increase in sonic velocity (Figure 70); the top is defined by a marked high-velocity sonic spike. The unit is traceable from the East Midlands Shelf into the Sole Pit Trough. The lithologies are dominantly interbedded clays and sands which, by comparison with the Nettleton Bottom borehole sequence (Bradshaw and Penney, 1982), are probably equivalent to the Bathonian Upper Estuarine Series and Blisworth Clay. The unit is capped by the better-cemented calcareous units of the Callovian.

BGS borehole 81/48, on the western margin of the Sole Pit inversion, cored 0.5 m of greenish grey, laminated siltstones and fine-grained, carbonaceous sandstones that pass up into greenish grey, yellow and purple, mottled, waxy clay (7.06 m) topped by hard, grey, bioclastic limestone (1.55 m). The faunal association and bioclastic-grainstone lithology, overlying sediments of brackish-water, 'estuarine'-type facies, are typical of the Upper Estuarine Series/Cornbrash interval. The calcareous microfaunal assemblages present suggest a Callovian (Upper Cornbrash) age for the limestone intervals (I P Wilkinson, written communication, 1983).

HUMBER GROUP

Rhys (1974) subdivided the Humber Group of the offshore area into the Oxford Clay, Corallian and Kimmeridge Clay formations; its base equates with the top of the Kellaways Rock and Sand of eastern England. The Upper Jurassic successions of the southern North Sea, range from Callovian to Kimmeridgian in age; as yet no Portlandian/Volgian sediments have been proved, except within The Wash. The Humber Group is confined to a subcrop smaller in area than that of the West Sole Group (Figures 69 and 72), and has a maximum proven thickness of 300 m (Kent, 1980b). The group is much thicker on the East Midlands Shelf than are the earlier Jurassic groups.

The Humber Group comprises a series of marine, silty, fossiliferous, partly bituminous mudstones and calcareous mudstones, calcareous sandstones and oolitic limestones which can be closely correlated with the equivalent successions at outcrop in eastern England. Subdivision and correla-

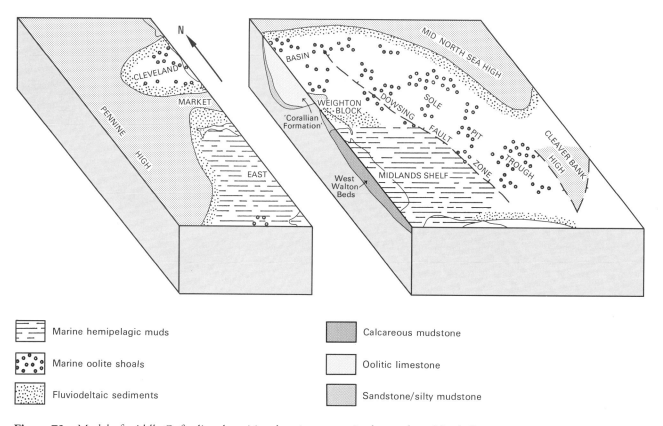

Figure 73 Model of middle Oxfordian depositional environments in the southern North Sea.

tion of the land sequence on the basis of its downhole log characteristics reveal the presence of marked north to south facies changes across the basin, particularly in the Oxfordian. In the Cleveland Basin, the sequence comprises Oxford Clay mudstone overlain by moderately thick, spicular sandstones and oolitic and bioclastic limestones of the Lower Calcareous Grit and Coralline Oolite formations; these are in turn overlain by the Ampthill Clay (Cox and Richardson, 1982). On the East Midlands Shelf, the Oxford Clay, West Walton Beds and Ampthill Clay consist predominantly of a mudstone/siltstone facies.

Offshore, the wells document this north to south transition in the Oxfordian, uninterrupted by the effects of the Market Weighton structure. The oolitic shoals and banks of the 'Corallian Formation' (Rhys, 1974), generally show a gradual transition southwards and westwards into lower-energy, muddy siltstones of the West Walton Beds (Figures 65 and 73). The oolite facies then reappears locally to the south, where the Upware Limestone fringes part of the London–Brabant Massif (Gallois and Cox, 1977).

These north–south facies transitions ceased to be effective during late Kimmeridgian times, although subsidence of the Cleveland Basin persisted, as indicated by thickening of individual beds within the Kimmeridge Clay (Cox et al., 1987). The Kimmeridge facies throughout the area consist typically of fossiliferous, dark grey, organic-rich mudstones, calcareous mudstones and cementstones.

The London–Brabant Massif probably formed a very low-lying barrier separating the southern North Sea area from the Anglo-Paris Basin during the Kimmeridgian, although evidence for its existence is sparse. The lithological uniformity of the Kimmeridge Clay from Wiltshire to Yorkshire is remarkable; even in sediments marginal to the supposed high, thin, fine-grained sandstones form only a very minor part of the sequence. The organic-rich mudstone facies dominated much of the North Sea Basin during the Kimmeridgian, excepting the West Netherlands Basin where there was a more marginal facies with lignitic sands and clays (NAM and RGD, 1980). The Humber Group may be divided into two units, UJ1 and UJ2, which can be traced throughout the offshore area.

Over the East Midlands Shelf, unit UJ1 shows a spiky, high gamma-ray profile decreasing in magnitude towards the top of the unit. The sonic log also shows a spiky profile representing interbedded, high-velocity limestone beds in a silty mudstone sequence (Figure 70). By reference to the Nettleton Bottom and Brownmoor boreholes, the unit correlates with the Oxford Clay and West Walton Beds. Changes in log profiles suggest that the lower part of the unit becomes more muddy eastwards, with fewer limestone interbeds.

At the eastern margin of the East Midlands Shelf there is a gradual upward transition into the spiculitic sands and oolitic limestone facies of the Corallian Formation; this is marked by a decrease in gamma-ray amplitude and an increase in sonic-velocity response. In adjacent parts of the Sole Pit Trough, and in the Cleveland Basin, it is possible to subdivide the equivalent sequence into a lower calcareous and arenaceous interval with spiky gamma-ray and sonic response (Lower Calcareous Grit Formation), and an upper, more massive interval of oolitic limestones (Coralline Oolite Formation) with low gamma-ray and uniformly high-velocity sonic profiles (Figures 65 and 70). The greatest thicknesses of these proved offshore, are 80 m of spiculitic sandstones in well 47/9-2, and 40 m of oolitic limestones in well 47/3-2 (Figure 74). The occurrence of these oolitic-facies sediments in the middle

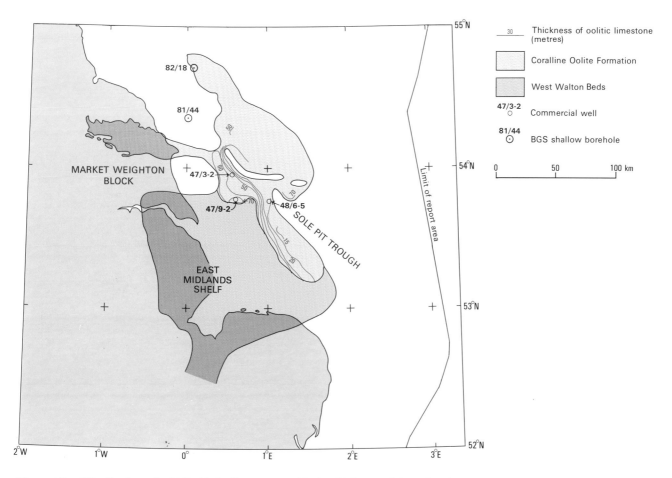

Figure 74 Distribution of middle Oxfordian strata, with the thickness of the oolitic limestones.

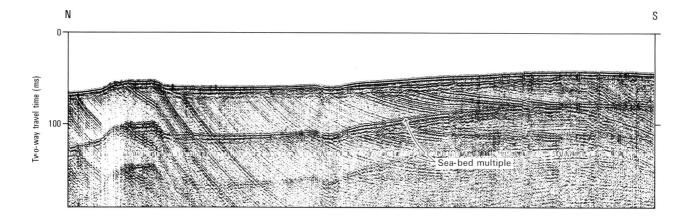

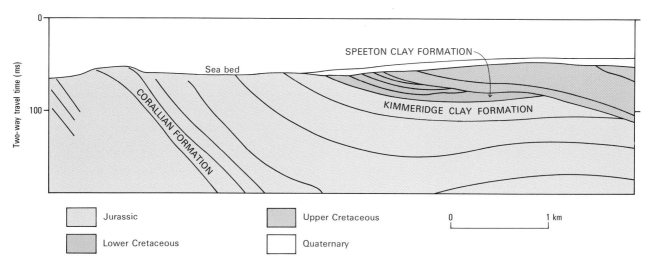

Figure 75 Sparker profile showing the unconformity between the Kimmeridge Clay and Speeton Clay formations to the north-west of Flamborough Head. See Figure 72 for location.

Oxfordian of the Cleveland Basin–Sole Pit Trough suggests that differential subsidence temporarily slowed there during the middle Oxfordian. At the same time, deeper-water shales were deposited continuously in central and western areas of the East Midlands Shelf, indicating a temporary shift in the depocentre.

Several shallow boreholes have been drilled by BGS over the 'Corallian Formation' outcrop in the northern part of the basin, and many short cores of the oolitic limestone facies have been sampled at sea bed (Lott, 1986). Borehole 81/41 cored thin sandstones and oolitic limestones of the Corallian Formation (equivalent to the Upper Calcareous Grit and Coralline Oolite formations of onshore usage). The pale grey to white, oolitic and bioclastic limestone, passes up into calcareous sandstone that varies from soft and friable to hard and cemented, and is bioturbated and fossiliferous, becoming less muddy towards the top. Borehole 82/18 cored 3.6 m of buff to grey, oolitic limestone (grainstone), with common bioclastic debris and sporadic, sutured, stylolitic beds.

Farther south, a 27 m-thick sequence of limestone was cored at sea bed near the West Sole Gasfield in well 48/6-5. The limestone is a buff to pale grey, hard, oolitic and bioclastic grainstone. On lithological grounds it may be equated with the Corallian Formation of Rhys (1974), or the Coralline Oolite Formation of Cope et al. (1980a).

Unit UJ2 includes the Kimmeridgian and younger Jurassic sediment, and comprises a rhythmic series of fissile, fossiliferous and bituminous mudstones with interbedded limestone/cementstone; these include representatives of the Ampthill and Kimmeridge Clay formations. The sequence typically shows a high gamma-ray response with a spiky sonic velocity profile, although no 'hot shales' with exceptionally high-gamma response have been observed. The lower boundary of the unit is not readily identifiable from log responses, except on the East Midlands Shelf (Penn et al., 1986).

BGS has drilled seven shallow cored boreholes into unit UJ2 (Figure 72), all of which proved sequences dominated by mudstone and shale. These can be correlated lithologically on a bed-by-bed basis with the successions of eastern England (Gallois and Cox, 1974; Wingfield et al., 1978; Cox et al., 1987). Of the seven boreholes, 81/47 and 81/49 show slightly expanded bed thicknesses compared to the same interval in eastern England; this is consistent with their siting closer to the axis of the Sole Pit Trough, once more the site of maximum deposition. The four boreholes off The Wash provide a complete section through the upper Kimmeridge Clay (Wingfield et al., 1978); their results emphasise how precise correlations with the onshore successions can be, and further demonstrate the lithological uniformity of the Kimmeridge Clay Formation over the basin.

The Jurassic–Cretaceous boundary in eastern England lies within the Spilsby Sandstone/Sandringham Sands (Casey and Gallois, 1973). This boundary, which is defined largely on palaeontological criteria, is not recognisable offshore because of the lack of core. For the purpose of this account therefore, the top of the Jurassic is taken as the top of the Kimmeridge Clay Formation. Over the East Midlands Shelf, this boundary is coincident with the marked lithological change at the base of the sands, but off the Yorkshire coast is placed at the lithologically less distinct junction between the more indurated Kimmeridge Clay Formation and the less indurated Speeton Clay Formation that overlies the Jurassic. On shallow-seismic profiles, the junction is commonly seen as a low-angle unconformity (Figure 75). The Jurassic–Cretaceous boundary has been cored in BGS borehole 72/77 in The Wash, and in borehole 81/43 (Figure 72) off the Yorkshire coast (Wingfield et al., 1978; Lott et al., 1986, 1989).

10 Cretaceous

Cretaceous rocks occur more widely than those of Jurassic age in the report area, although they are absent in the Cleveland Basin and Sole Pit Trough, where they were eroded during inversion at or near the end of the Cretaceous Period. Cretaceous sediments crop out (Figure 2) at the sea bed near Flamborough Head and in the Silver Pit, but west of the Sole Pit Trough they are overlain by thin Quaternary sediments. East of the trough, they are concealed beneath Tertiary and Quaternary sediments (Figure 3); they are most deeply buried beneath approximately 1000 m of these younger sediments adjacent to the median line. Above some Zechstein salt diapirs, Cretaceous rocks subcrop Quaternary sediments, but they are absent over other salt-piercement structures. South of the Thames Estuary the cover of Tertiary rocks is less complete, although thin Quaternary sediments commonly overlie the Cretaceous.

Both lithology and facies within the Cretaceous are varied. Lower Cretaceous marine deposits include the argillaceous Speeton and Gault clays, and the arenaceous Spilsby Sandstone, Sandringham Sands and Lower Greensand, whereas the Wealden Beds are nonmarine fluvial and lagoonal deposits, and the Carstone and the Red Chalk Formation are transgressive sequences. The overlying Chalk of the Upper Cretaceous is more varied than is widely assumed, with beds of marl, hardgrounds, and flint concretions.

The London–Brabant Massif separated the report area into two major sedimentary and faunal provinces of Cretaceous deposition, although its influence waned after the mid-Cretaceous. The northern province includes the Southern North Sea Basin, and covers the major part of the report area. The southern province includes parts of Kent, the eastern part of the English Channel and the Dover Strait; it is described in more detail in Hamblin et al. (1992). Figure 76 tabulates the formations that make up the Cretaceous succession in these two provinces.

At the beginning of the Cretaceous Period, the Southern North Sea Basin was occupied by a sea which extended eastwards through Germany into north-east Europe, and was linked to the Atlantic via the Viking Graben to the north (Figure 77a). The sea was cut off from the southern Tethyan Ocean by the landmasses of the Welsh and London–Brabant massifs. In west Norfolk, a marine incursion from the north briefly submerged this barrier at the start of the Cretaceous, but there were no major phases of marine deposition in the southern province early in the Cretaceous, when the London–Brabant Massif was an effective boundary between the two depositional basins (Jenkins and Murray, 1989). However, late Aptian to Albian transgression connected the northern and southern seas across the western part of the London–Brabant Massif (Figure 77b). This was the start of a major eustatic sea-level rise which eventually flooded the whole of the European area (Figure 77c) and was the greatest inundation of the continents since the Ordovician (Hancock and Kauffman, 1979). The seas reached a maximum of some 250 m above their present level (Figure 76) during the Turonian (Haq et al., 1987). A decrease in the supply of terrigenous material from the shrinking landmasses allowed the deposition of pure, calcareous chalk to be established widely on the continental shelves (Hancock, 1976).

Cretaceous sediments in the southern North Sea were deposited in many areas which had undergone uplift and widespread erosion during the Late Jurassic. Further intervals of local uplift and erosion during the Early Cretaceous resulted in a complex series of unconformities in the basal part of the Cretaceous sequence; these are commonly referred to as the late-Cimmerian Unconformity (e.g. Fyfe et al., 1981). On the offshore part of the East Midlands Shelf, Lower Cretaceous sediments rest with slight unconformity on the Late Jurassic Kimmeridge Clay Formation of the Humber Group (Figure 78). Eastwards from the Sole Pit Trough, northwards on to the Mid North Sea High, and towards the London–Brabant Massif to the south, the Cretaceous is strongly transgressive and rests on progressively older formations from Middle Jurassic to Triassic in age. On the London–Brabant Massif itself, Cretaceous rocks rest unconformably on both early and late Palaeozoic rocks. The Lower Cretaceous is absent in some areas east of the Sole Pit Trough where the Upper Cretaceous rests directly on Jurassic and Triassic formations.

In most of the area, the Lower Cretaceous is less than 100 m thick, but adjacent to the Dowsing Fault Zone as much as 900 m have been proved by drilling (Figure 79). Farther north, the Lower Cretaceous sediments south-east of Flamborough Head and the Speeton Cliffs are up to 400 m thick. Marine clays of Valanginian to Aptian age make up the thickest units within the Lower Cretaceous. The Upper Cretaceous is generally between 200 and 800 m thick, and is thickest to the east of the Sole Pit Trough at the median line (see Figure 82).

There are rapid lateral changes in the thickness of Cretaceous formations overlying or adjacent to many of the Zechstein salt structures, although on a regional scale the changes in thickness are more gradual. Late Jurassic uplift of the Sole Pit Trough had a major influence on Cretaceous depositional patterns; each Cretaceous stage appears to thin abruptly at the flanks of the Sole Pit Trough, and Cretaceous sediments are now absent over the inversion axis (Figure 78).

The Cretaceous of eastern England is largely well known from extensive cliff and quarry exposures and from the numerous water boreholes drilled into the Chalk aquifer. Yet even in well-exposed onshore areas, the boundary between the Jurassic and the Cretaceous is difficult to place, and this boundary is even more problematical offshore. Furthermore, few details of the Cretaceous sequence were recorded in the earliest offshore wells, and the use of different dating schemes by the companies involved (Hancock, 1986) has resulted in uncertainty regarding the distribution of the various stages of the Cretaceous. Most offshore hydrocarbon wells have been drilled on structural highs where the Cretaceous sequence is attenuated and incomplete.

Stratigraphical relationships in the southern North Sea are indicated in Figure 76. North of the London–Brabant Massif, Rhys (1974) equated the arenaceous and argillaceous sediments of the uppermost Jurassic and basal Cretaceous with the Spilsby Sandstone and Speeton Clay formations of eastern England. The Speeton Clay Formation is succeeded by a transitional development of characteristically red-coloured marls and chalky limestones of the Red Chalk

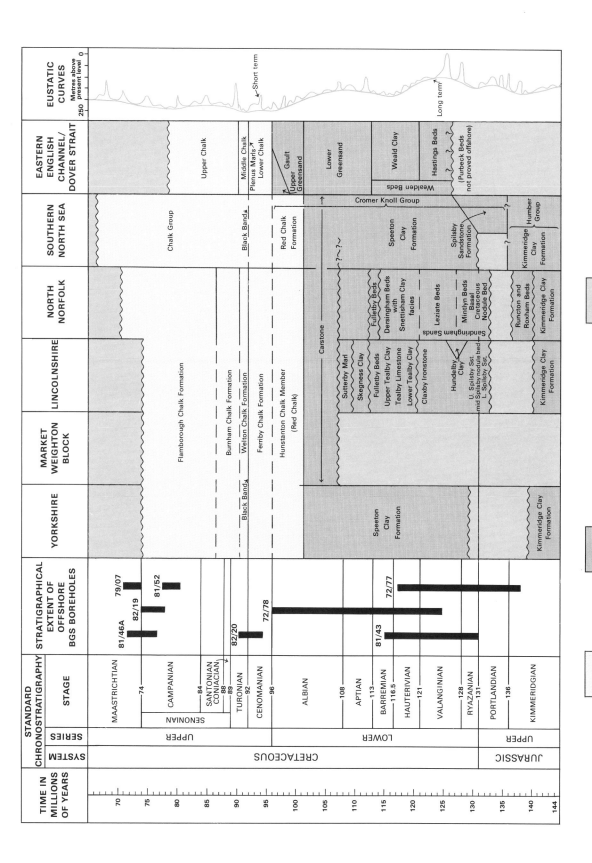

Figure 76 Stratigraphical correlation of Cretaceous rocks. After Rhys (1974), Rawson et al. (1978), Kent (1980b), and Haq et al. (1987).

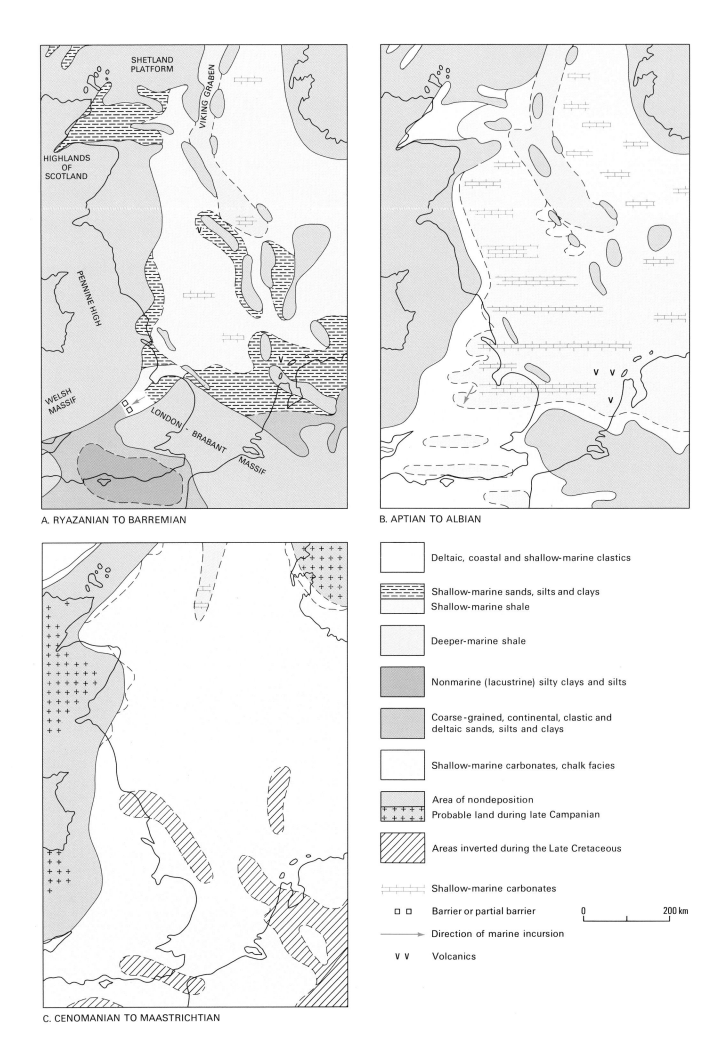

Figure 77 Generalised Cretaceous facies and palaeogeographies. Adapted from Anderton et al. (1979), Hancock (1975), Hart and Bigg (1981), Rayner (1981), Whittaker (1985) and Ziegler (1982).

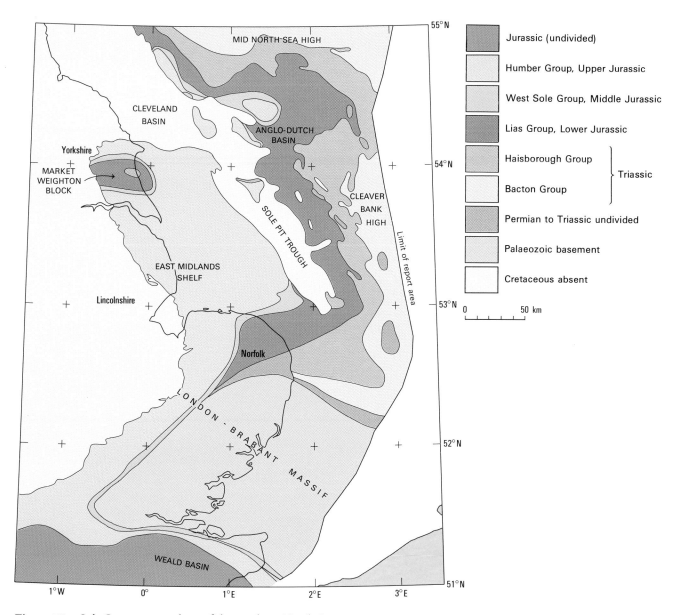

Figure 78 Sub-Cretaceous geology of the southern North Sea.

Formation. Rhys (1974) included all these formations within the Cromer Knoll Group.

South of the London–Brabant Massif, the Lower Cretaceous is divided into the upper part of the Purbeck Beds, the Wealden Beds, Lower Greensand, Gault and Upper Greensand. Onshore, these include a variety of formations which have local names, and all thin northwards on to the London–Brabant Massif (Gallois, 1965).

The Upper Cretaceous sediments of the southern North Sea were given group status as the Chalk Group by Rhys (1974). In southern England the classic division of the Chalk has been into Lower, Middle, and Upper Chalk, but their boundaries become ill-defined towards the London–Brabant Massif (Rawson et al., 1978). In the northern province these subdivisions are imprecisely known (Whittaker et al., 1985); here the Chalk was subdivided (Figure 76) into the Ferriby, Welton, Burnham and Flamborough Chalk formations by Wood and Smith (1978). To date, similar lithostratigraphical subdivisions have not been made in the offshore area.

CROMER KNOLL GROUP

Figure 80 shows the Lower Cretaceous sequence penetrated by well 48/22-2, which was adopted by Rhys (1974) as the type sequence for the Cromer Knoll Group in the UK sector of the southern North Sea.

Spilsby Sandstone Formation

The Spilsby Sandstone Formation was thought to mark the base of the English marine Cretaceous until Casey (1962), studying the ammonite succession, showed the lower part of the formation to be equivalent to the Purbeck Beds of southern England. It is therefore of Portlandian to Ryazanian age, and the Jurassic–Cretaceous boundary occurs within the formation. Onshore, a mid-Spilsby Sandstone Formation phosphatic nodule bed is taken as the base of the Cretaceous; this corresponds to the Cinder Bed (Purbeck) which marks the base in southern England (Kent, 1980b). In wireline logs from offshore wells, a prominent gamma-ray peak indicates the position of a possibly equivalent nodule bed, but this method lacks the biostratigraphical control needed to identify the Jurassic–Cretaceous boundary with precision. In BGS

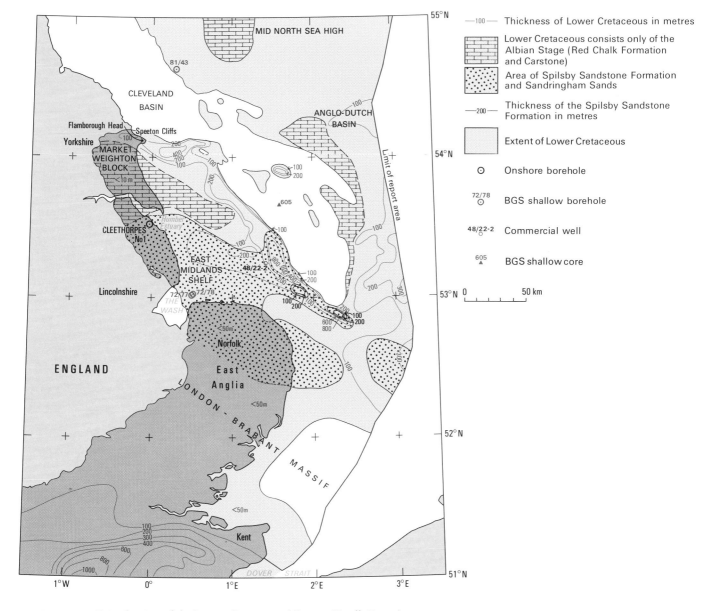

Figure 79 Distribution of the Lower Cretaceous (Cromer Knoll Group).

borehole 72/77 off The Wash (Figure 79), the basal Cretaceous Nodule Bed consists of 0.15 m of sand and clay with phosphatic nodules; it probably marks the boundary as well as being a very condensed deposit representing much of the Ryazanian Stage (Wingfield et al., 1978). In Norfolk, part of the Sandringham Sands is approximately equivalent to the Spilsby Sandstone Formation (Casey and Gallois, 1973).

The Spilsby Sandstone Formation occurs on the East Midlands Shelf and along the northern flank of the London–Brabant Massif (Figure 79). It has a maximum thickness of 200 m, and was deposited as a nearshore marine sand on the southern fringes of a sea which extended over most of the present North Sea during Late Jurassic and Early Cretaceous times. The sandstones are fine to coarse grained and generally uncemented, but have some hard, calcareous layers. They are a basin-marginal facies equivalent to the thick marine clays which were deposited within the Central and Viking grabens of the North Sea. The formation is partly glauconitic, includes thin beds of shale, and incorporates phosphatic nodules and calcareous concretions at certain levels, some of which are fossiliferous.

Isolated outliers of basal Cretaceous sediments occur over the inverted Sole Pit Trough, from which only one sample has been obtained. On the north-eastern flank of the trough, BGS core 605 penetrated 0.3 m of dark grey, fissile mudstone and grey-brown, slightly pyritic siltstone (Figure 79). Biostratigraphical analysis yielded a diverse, well-preserved dinoflagellate-cyst flora indicative of a late Ryazanian/Valanginian age. The mudstones and siltstones are deeper-water equivalents of the Spilsby Sandstone Formation, preserved from erosion in the hanging wall of a later Cretaceous fault.

Speeton Clay Formation

The Speeton Clay Formation in Yorkshire is the most complete development of marine Lower Cretaceous sediments in the UK (Neale, 1974), and ranges in age from Ryazanian to possibly late Albian. In Lincolnshire and Norfolk, the Speeton Clay Formation is represented by locally named formations consisting of sandstones, clays, limestones and ironstones (Kent, 1980b); these are basin-margin deposits, in contrast to those in Yorkshire and offshore which are a deeper-water facies. Detailed correlations have not been made between the onshore and offshore formations, but Figure 80 shows comparison of the sediments proved in the onshore

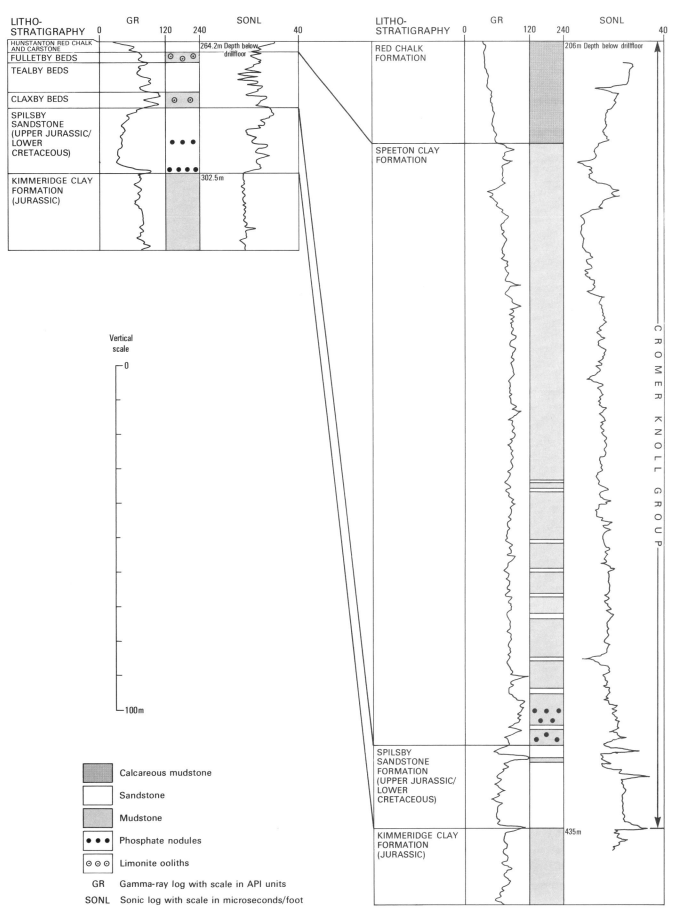

Figure 80 A comparison of the Lower Cretaceous in an onshore borehole, with the Cromer Knoll Group in the offshore type well. Cleethorpes No 1 adapted from Whittaker et al. (1985); well 48/22-2 adapted from Rhys (1974). See Figure 79 for locations.

Cleethorpes No. 1 borehole with those in well 48/22-2.

The Speeton Clay consists of variably calcareous mudstones and shales, with a range of brown, grey, green and grey-black colours. The formation is fossiliferous, has a few sandstone bands, and includes siltstone horizons and layers with abundant phosphatic nodules. In areas of its thickest development, where it is in excess of 600 m thick, mudstone is particularly dominant. BGS borehole 81/43, located 80 km north-east of Speeton Cliffs (Figure 79), cored 89.78 m of dark greenish grey, calcareous clay and brownish black, sporadically organic-rich and noncalcareous mudstones of Early Cretaceous age. These compare closely, both lithologically and sedimentologically, with the type section.

The clays commonly rest unconformably upon mudstones of the Late Jurassic Kimmeridge Clay Formation (Lott et al., 1986); the sparker profile across the Kimmeridge Clay–Speeton Clay boundary in Figure 75 shows a low-angle unconformity separating the two formations. The basal late Ryazanian beds form a poorly fossiliferous, carbonate-free, organic-rich facies similar to that of the Kimmeridge Clay. These grade upwards into carbonate-rich sediments which have a lower organic carbon content and are more fossiliferous (Lott et al., 1986; 1989). A similar facies transition in the northern and central North Sea areas has been taken as the boundary between the Kimmeridge Clay Formation and the Early Cretaceous Valhall Formation (Deegan and Scull, 1977). Riley and Tyson (1981) and Rawson and Riley (1982) attributed the facies change there to a major basin-flushing event which caused the relatively restricted and stagnating bottom waters of the North Sea Basin to be replaced by well-oxygenated waters, perhaps as a result of tectonic subsidence in the grabens. This was possibly associated with volcanic activity outside the North Sea, an idea supported by the presence of Ryazanian bentonites of volcanic origin in the Speeton Clay Formation of the southern North Sea (Knox and Fletcher, 1978; Lott et al., 1986).

Carstone and the Red Chalk Formation

The transgressions during the Aptian and the middle to late Albian more than doubled the area of the contemporary North Sea (Hancock, 1986). The Mid North Sea High, the Market Weighton Block and the London–Brabant Massif were submerged for the first time, resulting in the thin, transgressive deposits of the Carstone and the Red Chalk Formation resting unconformably on pre-Cretaceous sediments. Elsewhere there is a striking facies change between the Red Chalk Formation and the underlying clays of the Speeton Clay Formation. The Red Chalk is rich in species of planktonic foraminifera, which also occur in the uppermost part of the Gault Clay of southern England, into which the Red Chalk passes laterally across the London–Brabant Massif. This shows that the areas either side of the London–Brabant Massif were faunally united for the first time during the late Albian (Jenkins and Murray, 1989).

In Norfolk, Lincolnshire and on the Market Weighton Block, the Carstone is a fine- to coarse-grained or granular sandstone that grades upwards into the Red Chalk Formation. The Carstone has a maximum proven thickness on land of 12 m (Kent, 1980b) and although it has not been identified in offshore wells, similar sediments may occur near the coast between Flamborough Head and the Humber Estuary. Farther south, both the Carstone and the Red Chalk Formation have been proved at the entrance to The Wash. Borehole 72/78 (Figure 79) recorded 9.2 m of Carstone consisting of pink to deep red, silty clay with phosphatic nodules, shell debris, wood fragments and beds of yellowish green, oolitic (chamositic and limonitic), fine-grained sandstone.

The Red Chalk Formation is an impure limestone, varying in colour from pink to brick red, and containing rounded quartz grains and numerous fossils. Onshore, it is generally between 5 and 13 m thick, whereas offshore it usually ranges between 20 and 30 m, but is exceptionally more than 50 m thick. Its red colour may be due to ferruginous material derived from erosion of exposed Triassic marls on rising salt structures (Kent 1967), or alternatively to red mud washed from a low-lying, arid land rich in laterite (Kent, 1980b). Jeans (1973) has further suggested that diagenetic processes changed the original chalk lithology to form the Red Chalk. The Red Chalk Formation in borehole 72/78 consists of 3.2 m of deep pink to red or khaki-coloured limestone interbedded with red calcareous clay. The limestone includes a pebble bed containing phosphatic nodules, phosphatised pebbles and *Inoceramus* shell debris deposited on an intraformational erosion surface (Wingfield et al., 1978).

The Red Chalk Formation is mainly of Albian age, and onshore has been generally treated as a formation separate from the Chalk. However, more recent lithostratigraphical studies include it in the Ferriby Chalk Formation (Figure 76) as a partial equivalent of the Hunstanton Chalk Member (Wood and Smith, 1978; Gaunt et al., 1992). The Rhys (1974) offshore lithostratigraphical nomenclature included the Red Chalk Formation within the Cromer Knoll Group because a junction with the overlying Chalk Group is clearly indicated on gamma-ray and sonic logs. The geophysical log characteristics of the Red Chalk Formation show that this unit is mappable, and it is therefore retained as a separate formation (Crittenden, 1982a) with higher gamma-ray values and generally lower sonic velocities than those of the overlying Chalk Group (Figure 81).

THE LOWER CRETACEOUS OF THE SOUTHERN PROVINCE

South of the London–Brabant Massif, a complex sequence of marine and nonmarine depositional environments produced the varied deposits of the Purbeck Beds, Wealden Beds, Lower Greensand, Gault and Upper Greensand (Figure 76). In the report area, these formations are attenuated along the southern flank of the London–Brabant Massif (Figure 79), and are locally absent.

The Purbeck Beds are shelly limestones and mudstones with beds of gypsum and anhydrite; sands or brightly coloured clays become more common towards the top of the formation. They were deposited in shallow, saline lagoons which were occasionally flooded by freshwater. Like the Spilsby Sandstone Formation, the Purbeck Beds straddle the Jurassic–Cretaceous boundary, which is marked by a unit rich in oyster shells and known from its appearance as the 'Cinder Bed'. It was deposited during a short-lived incursion of marine waters from the northern basin. The Purbeck Beds have not been proved in the eastern part of the Dover Strait, although they are present in the Kent Coalfield, and are between 55 and 155 m thick in the Weald Basin (Hamblin et al., 1992).

The Wealden Beds are equivalent in age to the lower part of the Speeton Clay Formation, but are of nonmarine facies. In Kent they consist of sandy Hastings Beds beneath the muddier Weald Clay. The sandy sequences were deposited by braided rivers, whereas the clay sequences were laid down in shallow-water bays and lagoons (Allen, 1976; 1981). At the

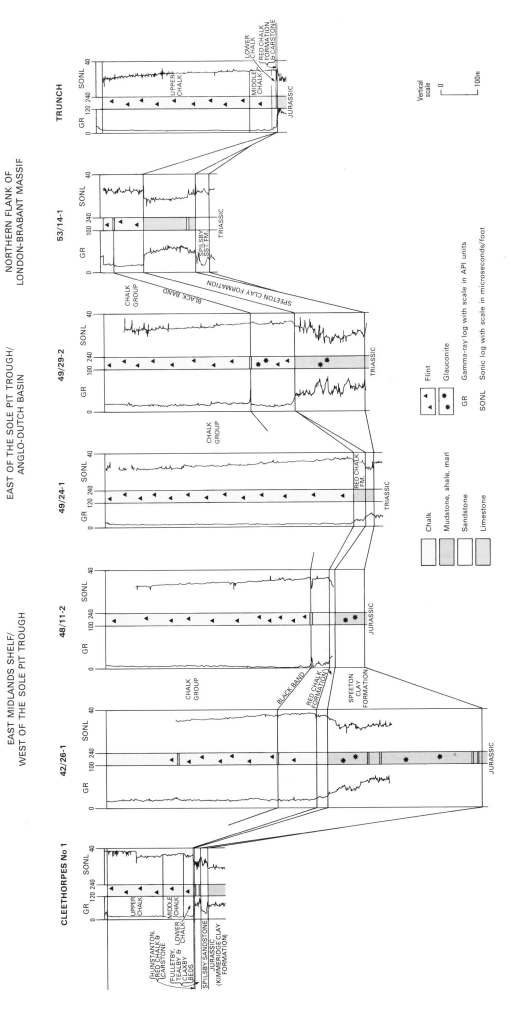

Figure 81 Correlation of Cretaceous rocks between selected wells across the southern North Sea and the adjacent land, using gamma-ray and sonic logs. Cleethorpes No 1 and Trunch adapted from Whittaker et al. (1985); well 49/24-1 adapted from Rhys (1974). See Figure 82 for locations.

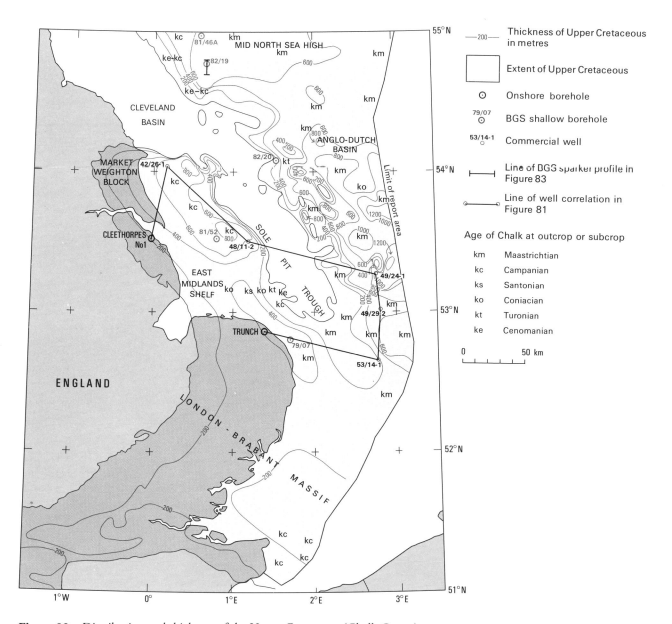

Figure 82 Distribution and thickness of the Upper Cretaceous (Chalk Group).

Kent coast the Wealden Beds are less than 100 m thick (Whittaker, 1985); it is likely that they extend offshore under the Dover Strait and thin northwards on to the London–Brabant Massif, as does the overlying Lower Greensand. The Lower Greensand is equivalent to the Aptian and early Albian sediments of the upper part of the Speeton Clay Formation. It is a lithologically variable series of muddy and sandy sediments which were deposited in shallow-marine and nearshore environments.

The Albian transgression caused deep-water marine conditions to extend over the London–Brabant Massif, resulting in the deposition of the Gault Clay in the southern province. The Gault consists of dark bluish grey to pale grey, soft, glauconitic, calcareous mudstones and silty mudstones rich in marine fossils. The basal part of the formation is silty or sandy, and phosphatic nodule beds occur at several levels, notably in the middle of the formation. The Gault has been proved offshore in boreholes drilled a few kilometres west of the limit of the report area for the Channel Tunnel investigations (Destombes and Shephard-Thorn, 1971). Locally the Gault becomes more sandy as it passes either upwards or laterally into the Upper Greensand which, as the basin became infilled, was probably deposited in a relatively shallow, nearshore environment.

CHALK GROUP

The two sedimentary and faunal provinces of the Early Cretaceous continued to influence Late Cretaceous deposition, even though the London–Brabant Massif had become submerged; the fossils present to the north have greater affinities with those of Europe than with those of southern England (Mortimore and Wood, 1983). On the massif, the successions are sedimentologically and faunally transitional (Mortimore, 1983) between harder, thin-bedded chalks in the northern province, and softer, more massive chalks with visible hardgrounds to the south. It is probable that the Chalk of the southern North Sea is similar to the onshore Chalk of the northern province, but as yet no detailed studies have been carried out.

In the southern North Sea, the base of the Chalk Group is recognised by an increase in gamma-ray response and an increase in sonic velocity (Figure 81). The Chalk Group has not been formally subdivided offshore, but subdivision onshore has resulted in local stratigraphical names. The nomenclature of Rawson et al. (1978), dividing the Chalk in southern England into Lower, Middle and Upper Chalk, is now accepted by many geologists to cover both the southern and northern provinces. Others prefer the lithostratigraphical subdivision of the northern province Chalk Group by Wood and Smith (1978) into the Ferriby, Welton, Burnham and Flamborough Chalk formations.

The thickest development of over 1200 m of Upper Cretaceous occurs east of the Sole Pit Trough (Figure 82), whereas the Chalk is missing from the inverted Sole Pit Trough and Cleveland Basin, where there has been significant uplift and erosion. However, Bulat and Stoker (1987), studying interval sonic velocities recorded on well logs, deduced that similar thicknesses of more than 1000 m of Upper Cretaceous sediments may also have been deposited in parts of these areas.

All stages of the Upper Cretaceous are represented over most of the southern North Sea. Their relative thicknesses are essentially the same as in eastern England, but the absolute thickness of the Chalk Group offshore is greater, partly because the younger stages have been largely removed by erosion onshore. The youngest Cretaceous stage, the Maastrichtian, provides 20 to 50 per cent of the total thickness of the Chalk Group in the North Sea (Hancock and Scholle, 1975).

East of the Sole Pit Trough, Maastrichtian Chalk underlies Tertiary or Quaternary sediments, except above some Permian salt diapirs where the youngest Cretaceous stages are missing due either to nondeposition or post-Cretaceous erosion. Adjacent to the south-eastern edge of the Sole Pit Trough, only Maastrichtian Chalk is preserved, and it rests directly on Jurassic or Triassic rocks. This suggests that the south-eastern area of the Sole Pit Trough may have remained as land until latest Cretaceous times, and may have been the source of the sandstone member of possible Turonian age intercalated within the Chalk Group in some wells adjacent to the London–Brabant Massif (Glennie and Boegner, 1981; Hancock, 1986; Van Hoorn, 1987). Over the London–Brabant Massif the age of the Chalk is not known, but Maastrichtian Chalk occurs on the northern flank. In the Dover Strait, Campanian Chalk crops out on the sea bed or is concealed beneath Paleocene and Quaternary deposits.

Westwards from the Sole Pit Trough, the Quaternary oversteps Chalk ranging in age from Cenomanian to Campanian, for Maastrichtian Chalk is absent. Sea-bed samples collected on the East Midlands Shelf contain rich, diverse and well-preserved Campanian microfaunal assemblages more similar to those of Norfolk than Yorkshire.

At the close of the Cretaceous and in earliest Tertiary times, virtually total marine submergence of the UK land area was replaced by almost complete emergence; only the deepest parts of the North Sea remained submerged from Maastrichtian into Danian times. In the southern North Sea area, marine sedimentation was continuous only to the northeast of the Sole Pit Trough, where wells have proved Danian Chalk in the Dutch sector.

Chalk is a very fine-grained, consistently pure, relatively soft, white limestone that consists of debris from planktonic algae. This matrix contains coarser calcite components such as calcispheres, foraminifera, and some fragmentary debris of larger invertebrates. In places the Chalk is intensely bioturbated, suggesting that the sea bed was in a thixotropic state, being permanently firm only at depths of a few tens of centimetres. It commonly consists of more than 98 per cent $CaCO_3$ (excluding flints), but there are small quantities of impurities throughout, chiefly quartz. These impurities are concentrated at certain horizons where beds of argillaceous marl represent an increase in terrigenous material. Discrete marl seams are a feature of both the Turonian and lower Campanian successions, but are particularly conspicuous in the Turonian of the northern province (Mortimore and Wood, 1983).

Hardgrounds consisting of nodules, pebbles, hardened chalk and numerous fossils represent a break in deposition in the shallower regions of the Chalk Sea. They may indicate subaerial exposure, or alternatively they were produced by mineralisation of current-swept sea beds, with minimal sedimentation (Kennedy and Garrison, 1975; Hancock, 1986). True hardgrounds are common only in the Upper Chalk. Much of the Chalk Group of the southern province comprises winnowed chalk with repeated hardgrounds, or chalk that has been redeposited as slumps, debris flows, or turbidites shortly after original accumulation (Hancock, 1976; 1986). North of the London–Brabant Massif, glauconitisation and phosphatisation are virtually absent, and hardgrounds occur only in local condensed sequences or within sedimentary channels (Mortimore and Wood, 1983).

The primary porosity of the Chalk is high, as much as 46 per cent, decreasing to 27 per cent where marls occur (Scholle, 1974). As a result of its very small pore sizes, the chalk has exceptionally low matrix permeability for a rock of such high porosity (Harper and Shaw, 1974). Compaction by deep burial and diagenetic recrystallisation have reduced its porosity in most of the Southern North Sea Basin, where the Chalk is harder and less massive than the 'earthy' Chalk of southern England. Indeed, diagenetic effects have provided most of the lateral and vertical lithological variation in the offshore sequence (Hancock and Scholle, 1975).

Flints commonly occur within the Chalk as irregularly shaped nodules and thin tabular sheets in layers which mostly follow bedding planes, though transgressive flints have been observed. The flint consists of silica in the form of microncsized, randomly arranged quartz crystals. The silica was probably derived from biogenic sources such as sponge spicules, although some small contribution from distant volcanic sources is possible (Anderton et al., 1979). In the southern province, the greater part of the Chalk succession is flint-bearing, but with more numerous flints in the upper portion. This contrasts with the onshore northern province where the Flamborough Chalk Formation is almost flint-free (Mortimore and Wood, 1983), as is the lowest part of the Ferriby Chalk Formation (Wood and Smith, 1978). Thick, tabular flints occur in the northern province where they characterise the Burnham Chalk Formation onshore, but they are absent south of the London–Brabant Massif.

Onshore, in both the northern and southern provinces, a belt of maximum flint development occurs at the top of the Turonian Chalk, always with marls above and below it. The belt can be traced to northern France, and is an approximately contemporaneous sedimentary phenomenon (Mortimore and Wood, 1983). This flint maximum has not yet been confirmed offshore.

The type well for the Chalk Group (Figure 81) in the southern North Sea is 49/24-1 (Rhys, 1974). In this and other wells, the Chalk shows little variation in its gamma-ray character, reflecting the uniform nature of its lithology across the basin. However the sonic log does show a characteristically gradual downward increase in velocity, the result of porosity loss caused by increasing depth of burial. The Chalk is typically white to greyish white, friable to moderately hard, microcrystalline, fossiliferous in parts, and there are traces of brown and grey-brown clay. Flints are common, but neither their colour nor their distribution have yet proved to be of use in subdividing the Chalk Group offshore.

In eastern England, the lithology below the Flamborough Chalk Formation remains remarkably constant, so that marl

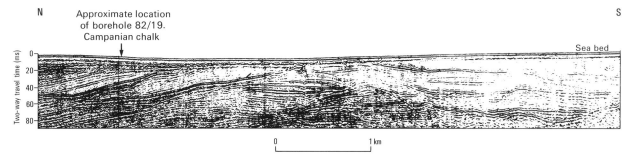

Figure 83 Sparker profile showing discontinuities in the Chalk. For location see Figure 82.

bands a few centimetres thick, and individual flint horizons, can be correlated from Flamborough Head into Lincolnshire (Wood and Smith, 1978; Gaunt et al., 1992). Barker et al. (1984) correlated borehole resistivity and gamma-ray logs with the lithostratigraphical subdivisions of Wood and Smith (1978), and Murray (1986) demonstrated that resistivity logs could be correlated from the northern to the southern province, providing evidence of the remarkable lateral continuity of marker horizons in the Middle and Lower Chalk. Offshore, such detailed regional correlations have not been attempted for the Chalk Group, although limited studies suggest that individual marl bands and flint horizons may be traced across the offshore area (Figure 81).

In southern England, a layer of dark greenish grey to black, laminated marl at the top of the Lower Chalk, characterised by the presence of the belemnite *Actinocamax plenus*, approximately marks the Cenomanian–Turonian boundary. A similar marl has been recognised in many wells in the southern North Sea by a marked gamma-ray and sonic response on the downhole logs (Figure 81). In southern England this layer is known as the Plenus Marl, and is up to 8.5 m in thickness; in eastern England and offshore, it is known as the Black Band. Palaeontological evidence suggests that the two do not correlate exactly, for the Black Band may be equivalent to the top part of the Plenus Marl (Jefferies, 1963; Hart and Bigg, 1981). The Black Band is associated with a complex of erosion surfaces, reworked pebbly beds and variegated, bioturbated marls (Hart and Bigg, 1981); it represents a brief but notable interval of clay deposition in the clear Chalk Sea. The clay is generally thought to have been deposited in a basin that had limited water circulation, resulting in oxygen-poor bottom waters (Schlanger and Jenkyns, 1976), for it is rich in planktonic foraminifera and radiolaria which could thrive in the oxygenated surface waters, but poor in benthonic fauna for which the bottom conditions were not suitable. This anoxic event occurred at the time of maximum transgression of the Chalk Sea, possibly coincident with a period of volcanic activity (Hart and Bigg, 1981).

The Black Band was cored in BGS borehole 82/20, 100 km east of Flamborough Head (Figure 82), where it is represented by 0.5 m of very carbonaceous, black shale with very hard, grey limestone characteristically containing phosphatic debris such as fish teeth. In the borehole, the Black Band separates Cenomanian Chalk containing abundant, greenish grey, soft marl from soft, white, Turonian Chalk rich in microfauna.

Synsedimentary structures within the Chalk, such as carbonate banks, slump beds, nodular chalks and hardgrounds, have been described in southern England (Kennedy and Garrison, 1975), and have also been identified beneath the English Channel (Larsonneur et al., 1975; Curry and Smith, 1975). A BGS sparker survey has revealed that similar structures occur beneath the southern North Sea; discontinuities in the Chalk section are imaged as a series of concave-up structures that have an apparent west–east axial trend and display cross-stratified internal seismic reflectors (Figure 83). They are very localised and were not detected on parallel seismic profiles only 7 km to the east or west. Nearby BGS borehole 82/19 (Figure 82) recovered 41.05 m of white, Campanian Chalk with angular flints and hard bands interpreted as hardgrounds; the microfauna is impoverished and broken, which may be indicative of reworking associated with the discontinuities.

Similar channel and scour structures identified beneath the English Channel have been attributed to erosion following shallowing of the Chalk Sea during regressional phases and local tectonic uplift (Quine, 1988). Mortimore and Pomerol (1987) related the occurrence of such unusual sedimentary features to underlying tectonic structures, providing evidence of tectonic movement during the Late Cretaceous in that region.

Onshore, and in the western part of the offshore area, there is a significant time gap between the oldest Cretaceous and the youngest Tertiary sediments. BGS borehole 81/46A (Figure 82) cored sand of Paleocene age resting unconformably at a sharp junction upon late Campanian to early Maastrichtian Chalk. The chalk is white to off-white, with pale grey clay laminations, burrows and large, grey flints. It is rich in microfauna, with both calcareous benthonic foraminifera and planktonic foraminifera. The uppermost part of the Chalk was either not deposited, or was removed by erosion prior to the deposition of the Paleocene sands. Even though this part of the report area was strongly positive through much of the Paleocene, there is no evidence that a shoreline existed near to the borehole site at the end of the Cretaceous, and it is probable that shallow-marine erosion, rather than subaerial erosion, accounts for the unconformity (Lott et al., 1983).

11 Tertiary

Tertiary sediments extend beneath much of the present-day southern North Sea (Figure 2), and their greatest thickness, about 2000 m, occurs at the centre of the basin to the east of the report area. The greatest preserved thickness of Tertiary within the report area is probably a little over 1000 m.

At the end of Cretaceous times, an extensive shallow sea covered much of north-west Europe, including most of the UK land area (Lovell, 1986). Sediments deposited at this time were dominated by carbonates, a pattern which continued in places into the earliest Tertiary (Danian). Although the British Isles became an emergent land area during the Danian, the North Sea area contained a semienclosed, epicontinental sea throughout the Tertiary. The geographical extent of this sea has varied with transgressive–regressive cycles, but it has occupied the same general area as the present North Sea. The palaeolatitude of the report area was approximately 43° to 47°N at the start of the Tertiary (Scotese et al., 1988), since when it has generally moved northwards to its present position at 51° to 55°N.

Early in Cretaceous times, crustal extension had been replaced by thermal cooling of the lithosphere as the principal mechanism for subsidence in the North Sea Basin (Badley et al., 1988). Sea-floor spreading began in the Norwegian–Greenland Sea during the mid-Paleocene Laramide phase of tectonic activity, preceded, and to a lesser extent accompanied by, extensive volcanism west of Scotland. This is evidenced in southern North Sea Paleocene to early Eocene marine sediments by the preservation of thin layers of airborne ash or tephra.

Paleocene uplift of the UK landmass brought an end to Danian carbonate deposition, and gave rise to the outbuilding of clastic, fan deposits eastwards into the Viking and Central grabens to the north of the report area (Stewart, 1987). These grabens formed an elongate and subsiding deep-water trough in the centre of the North Sea Basin (Figure 84). In the Central Graben, Palaeogene water depths have been variously interpreted as between 300 and 500 m (Lovell, 1986) or as much as 500 to 900 m (Parker, 1975).

In contrast, the southern North Sea area was subsiding more slowly, and Tertiary sedimentation was dominated by shallow (<200 m), marine-shelf and marginal-marine deposits. This meant that the Tertiary sequence records a succession of transgressive cycles which periodically inundated the area, largely as a result of eustatic sea-level changes. The sequence of marine sediments is therefore punctuated by regional unconformities and hiatuses. The movement of salt diapirs originating from underlying Permian evaporites probably also affected Tertiary sedimentation in the UK sector north of 53°N, where basal contours on the top of the Chalk show a strongly modified topography (Figure 85). This halokinesis is believed to have been most active during the Oligocene, contemporaneous with the main phase of Alpine orogeny (Van Hoorn, 1987).

Throughout the Tertiary, the waters of the southern North Sea were connected with those of the Norwegian Sea to the north (Figure 84). Other connections, either through the English Channel or through eastern Europe to Tethys, only existed during major transgressive episodes. For example, during the early Eocene a southern connection through the area of the modern Dover Strait clearly existed during Ypresian and Lutetian times, as indicated by the similarity of facies and fauna between the Southern North Sea and Hampshire–Dieppe basins. This connection allowed the migration of the Tethyan benthonic foraminiferid *Nummulites* into the Southern North Sea Basin (Blondeau, 1972). Species of *Nummulites* occur at several discrete stratigraphical levels in the sandy nearshore facies of the Belgian Tertiary, but are not recorded from eastern England. This may reflect current patterns in the area at such times, providing suitable shallow-water environments mostly on the eastern side of the basin.

Subsequent uplift along the Weald–Artois Axis severed this Dover Strait connection, and was responsible for the separation of the Hampshire–Dieppe Basin from the London Basin, the onshore part of the south-western North Sea Basin (Figure 84). The uplift began in late Lutetian times, and culminated during the Oligocene to Miocene (Lake and Karner, 1987). This southern connection was probably not restored until Pleistocene times.

King (1983) distinguished three depth-related benthonic foraminiferid biofacies within the North Sea Tertiary. One of these, the so-called '*Rhabdammina*' biofacies, indicates water depths in excess of 200 m and is dominated by noncalcare-

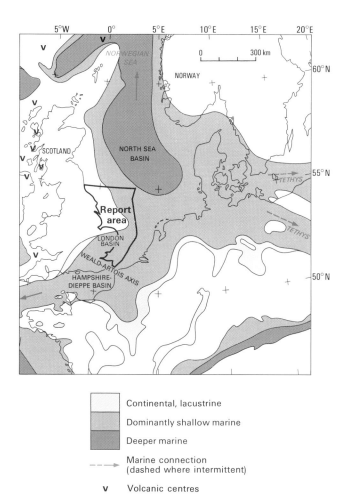

Figure 84 Generalised palaeogeography of north-west Europe during Paleocene to Eocene times. Modified after Ziegler (1982).

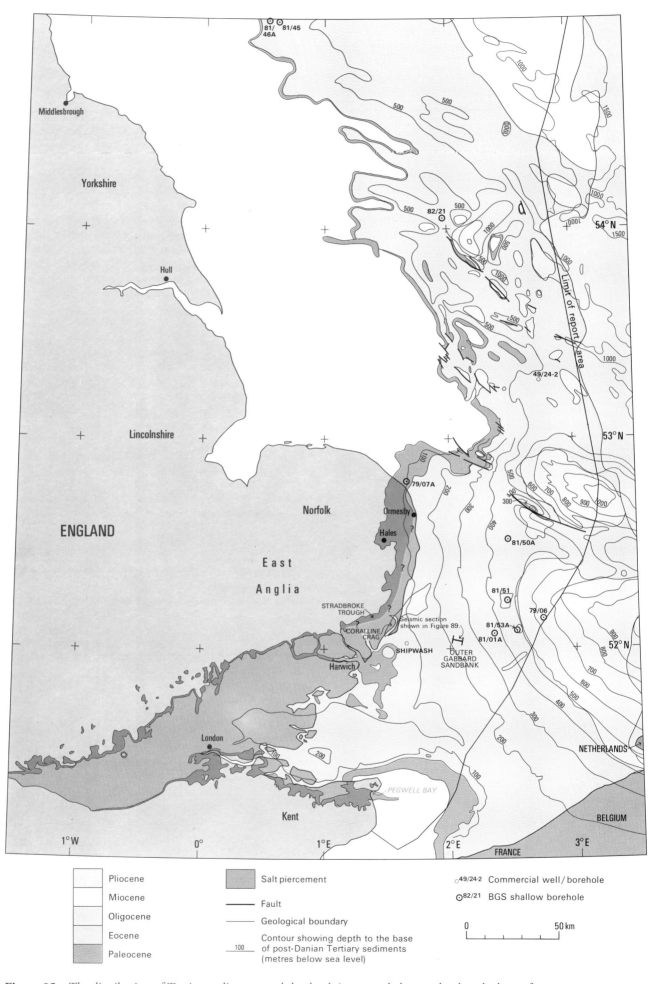

Figure 85 The distribution of Tertiary sediments, and the depth in metres below sea level to the base of post-Danian Tertiary sediments in the UK sector of the southern North Sea.

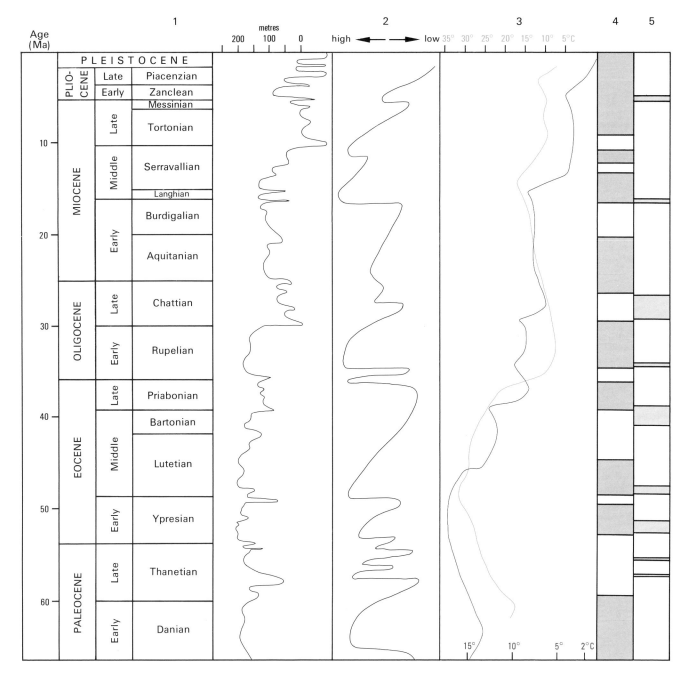

Figure 86 Tertiary stratigraphy, sea level and temperature.
1 Standard stage stratigraphy and eustatic sea-level curve. After Haq et al. (1987).
2 North Sea sea-level curve. After Kockel (1988a).
3 Isotopic palaeotemperatures. In blue from Tertiary benthic foraminifera after Savin (1977); in red from North Sea molluscs after Buchardt (1978).
4 Periods of connection (shaded) between the North Sea and the North Atlantic, as indicated by planktonic foraminifera. After King (1989) and C King (written communication, 1989).
5 Stratigraphical intervals with authigenic phosphorite concretions (shaded). After Balson (1990a).

ous, agglutinating foraminifera. The dominance of these foraminifera is believed (King, 1989) to reflect times of restricted water circulation, which led to slightly acid, reducing environments on the sea floor. Under such conditions, calcareous benthonic foraminifera cannot secrete $CaCO_3$ and are therefore absent from the sediments. Calcareous planktonic foraminiferal tests may also be absent due to postmortem dissolution within the sediment. These periods of restricted water circulation may have resulted from the separation of deeper water masses by topographic ridges or sills during eustatic lowstands, while surface water masses remained unaffected.

Planktonic foraminifera became most abundant during eustatic highstands which permitted the influx of North Atlantic water masses with their plankton. There is therefore a general inverse relationship between the abundance of non-calcareous agglutinating and planktonic foraminifera (King, 1989). Other possible evidence of intermittent connection with more oceanic waters comes from the occurrence of phosphorite concretions (Figure 86), which were formed only over brief intervals (Balson, 1987; 1990a).

A major eustatic sea-level fall in the late Oligocene has resulted in a regional unconformity. There are no marine deposits of this age or of the earliest Miocene over most of the southern North Sea. From mid-Miocene times onwards, warm-water microfaunas indicate a progressive shallowing of the central parts of the North Sea Basin (Ziegler, 1988), largely the result of greatly increased sedimentation rates (Bjørslev Nielsen et al., 1986), which outpaced subsidence. The deposition of deltaic sands, mostly from former Baltic rivers, westward-flowing north German rivers and the Rhine, continued throughout the Pliocene and into the Quaternary. The greatest thickness of these deltaic sediments occurs in the Dutch sector of the southern North Sea to the east. Within the UK sector, Neogene deposits are generally thin and consist of marginal-marine and shallow-shelf sediments.

Unlike areas of the North Sea farther to the north, where Paleocene sands have proved to be important hydrocarbon reservoirs, the Tertiary sequence in the report area has been nonprospective to date, which has resulted in relatively little attention. Hydrocarbon wells do not usually recover core from this part of the sequence, so that the log descriptions are generally based on examination of well cuttings. There have been few detailed palaeontological studies of these samples, and consequently the Tertiary stratigraphy depicted on commercial well logs is at best generalised. This is particularly true for the Neogene sequence, which cannot be differentiated from the overlying Quaternary deposits without detailed palaeontological analysis.

The BGS regional survey attempted to differentiate the Tertiary sequence at chronostratigraphical series level (BGS Solid Geology sheets). North of 53°N, where the Tertiary is buried beneath thick Quaternary deposits, interpretation is based mainly on shallow seismic-reflection profiles correlated to commercial well logs, and must be regarded as tentative, but south of 53°N the Quaternary cover is thinner, and cored boreholes and vibrocores (Knox, 1989) have increased control on the mapped boundaries. Detailed seismic stratigraphy studies of the Tertiary in Belgian waters include those of Henriet et al. (1989) and De Batist et al. (1989).

The offshore data have been integrated into regional lithostratigraphical and biostratigraphical schemes that are based mainly on studies of the land areas around the North Sea Basin, but with an increasing amount of data from the central and northern North Sea. This has permitted correlation with global or 'standard' stratigraphical schemes. A regional biostratigraphical scheme for the North Sea Tertiary was erected by King (1983), refined by Vinken (1988), and subsequently updated by King (1989).

PALAEOGENE

Palaeogene sediments are widespread both over the southern North Sea and on the adjacent land areas of eastern England, The Netherlands, Germany, Denmark, Belgium and northern France. The sequences, which are dominated by shallow-marine sediments, have been used as stratotypes for many Palaeogene 'stages', and thus form the conceptual basis for most of the currently used standard stages (e.g. Haq et al., 1987). However, the 'standard' planktonic foraminifera and nannoplankton stratigraphies originate from studies of deep-ocean sediments. In the Southern North Sea Basin, significant parts of the Palaeogene sequence are decalcified and yield neither foraminifera nor nannoplankton, which hampers correlation to the 'standard' stratigraphy. Dinoflagellate-cyst stratigraphy has proved useful in many parts of the sequence which otherwise would be difficult to date; a correlation between the dinoflagellate cyst stratigraphy and the geochronology of Haq et al. (1987) is shown in Figure 87. Data which offer additional means of correlation to the ocean record have also been obtained from magnetostratigraphy (Townsend and Hailwood, 1985; Cox et al., 1985; Aubry et al., 1986) and tephrostratigraphy (Knox, 1984b; Knox and Morton, 1983).

Paleocene

Paleocene sediments are found at the base of the Tertiary sequence throughout the area of Tertiary outcrop. In most places they unconformably overlie Late Cretaceous Chalk, although in the region of the Sole Pit Inversion, the Paleocene oversteps the Cretaceous locally onto Early Jurassic formations. At the south-eastern end of the Sole Pit Inversion, the upper surface of the Chalk forms a disturbed reflector on seismic records, the result of karstic topography (Jenyon, 1984). A narrow outcrop of Paleocene fringes the Tertiary basin (Figure 85), illustrating the erosional nature of the basin edge. In a few places between 53° 30'N and 54°N, the Paleocene sequence is penetrated by salt piercements which may extend into Eocene formations. Over most of the area, Paleocene deposits are between 40 and 80 m thick.

DANIAN

There are no unequivocal records of in-situ Danian sediments within the report area. Crittenden (1981; 1982b) described Danian planktonic foraminifera from chalky sediments in well 49/24-2 (Figure 85), but subsequently these have been deemed to have been reworked (King, 1983; Crittenden, 1986). As a result of mid-Tertiary inversion (van Wijhe, 1987), Danian sediments have a discontinuous outcrop in other parts of the Southern North Sea Basin. A small outlier in the south of The Netherlands consisting of friable calcarenitic limestones is separated from the rest of the Southern North Sea Basin, where chalky carbonates dominate (Letsch and Sissingh, 1983). Letsch and Sissingh (1983) indicate that greater than 50 m thickness of Danian sediments is present immediately adjacent to the report area in the Dutch sector at around 54°N, north-east of the Sole Pit Inversion axis. A more recent map (Kockel, 1988b) does not show the Danian outcrop extending into the UK sector, although it is likely that Danian sediments were formerly more widespread in the region (Knox, 1989).

THANETIAN

The remainder of the Paleocene succession is dominated by clastic sediments representing three Thanetian depositional cycles.

The first cycle consists of grey-green mudstones which are equivalent to both the Lista Formation in the central North Sea (Deegan and Scull, 1977) and the Thanet Formation of south-east England. In Norfolk, equivalent mudstones have been informally termed the Ormesby Clay by Knox et al. (1990). The first evidence of Tertiary pyroclastic sedimentation in the North Sea Basin occurs within this unit. This volcanic activity, Phase I of Knox and Morton (1983), is represented by a single tuff layer in a borehole at Ormesby in Norfolk (Cox et al., 1985), and by disseminated ash both in BGS borehole 79/07A (Morton, 1982; Knox and Morton, 1983) and in the Thanet Formation of Pegwell Bay in Kent (Knox, 1979).

The Thanet Formation of north Kent consists of up to 33.5 m of partially decalcified glauconitic sands and sandy

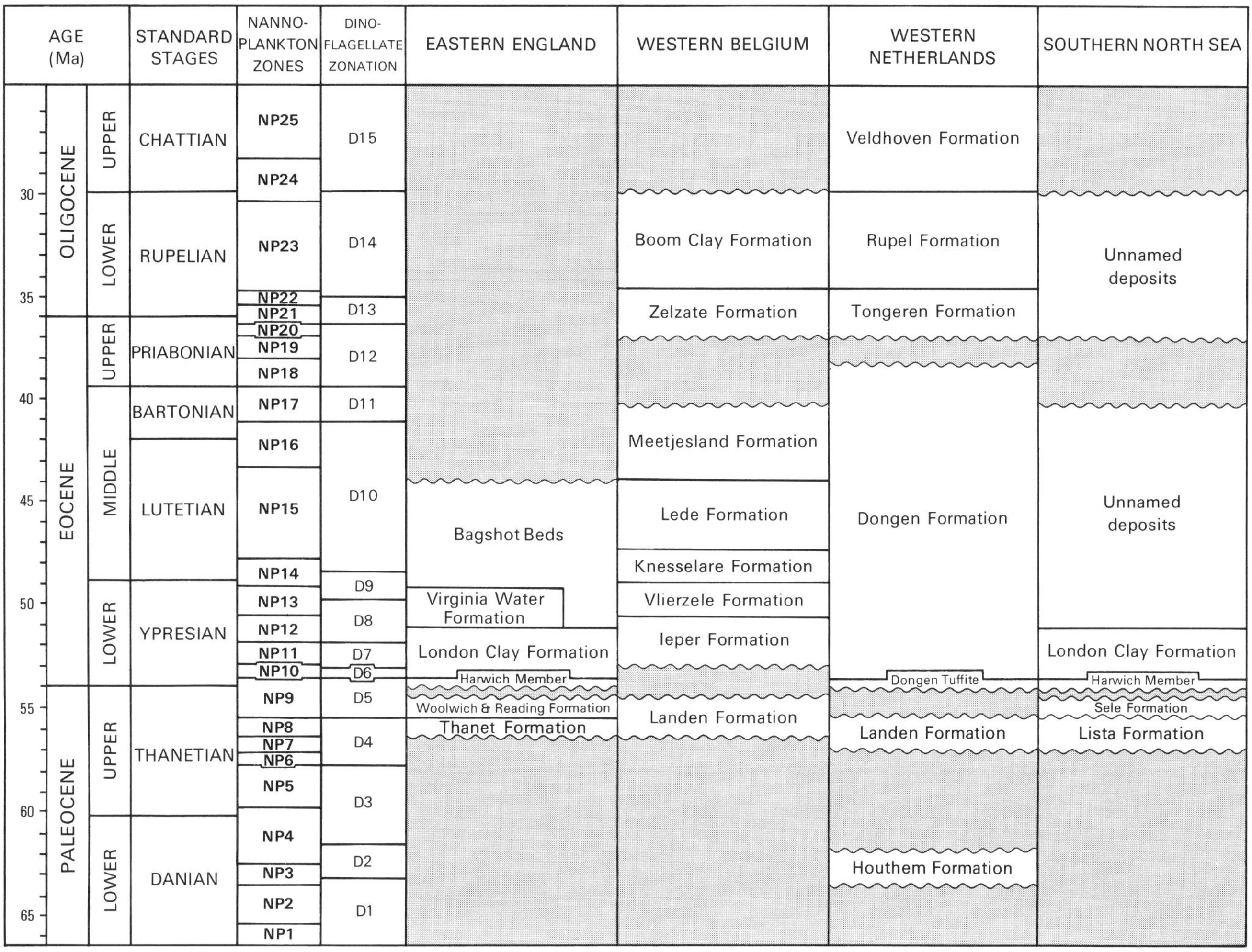

AGE (Ma)	STANDARD STAGES		NANNO-PLANKTON ZONES	DINO-FLAGELLATE ZONATION	EASTERN ENGLAND	WESTERN BELGIUM	WESTERN NETHERLANDS	SOUTHERN NORTH SEA
	OLIGOCENE	UPPER CHATTIAN	NP25	D15			Veldhoven Formation	
30			NP24					
		LOWER RUPELIAN	NP23	D14		Boom Clay Formation	Rupel Formation	Unnamed deposits
35			NP22					
			NP21	D13		Zelzate Formation	Tongeren Formation	
	EOCENE UPPER	PRIABONIAN	NP20					
			NP19	D12				
40			NP18					
		BARTONIAN	NP17	D11		Meetjesland Formation		
	MIDDLE		NP16					
45		LUTETIAN	NP15	D10	Bagshot Beds	Lede Formation	Dongen Formation	Unnamed deposits
			NP14			Knesselare Formation		
50			NP13	D9	Virginia Water Formation	Vlierzele Formation		
	LOWER	YPRESIAN	NP12	D8		Ieper Formation		
			NP11	D7	London Clay Formation			London Clay Formation
			NP10	D6	Harwich Member		Dongen Tuffite	Harwich Member
55			NP9	D5	Woolwich & Reading Formation	Landen Formation		Sele Formation
	PALEOCENE UPPER	THANETIAN	NP8	D4	Thanet Formation		Landen Formation	Lista Formation
			NP7					
			NP6					
			NP5	D3				
60			NP4					
	LOWER	DANIAN	NP3	D2			Houthem Formation	
65			NP2	D1				
			NP1					

Figure 87 Palaeogene stratigraphy in the southern North Sea region. Palaeogene standard stage stratigraphy, age, and correlation with nannoplankton zones, after Haq et al. (1987). Palaeogene dinoflagellate zonation and correlation after Costa and Manum (1988). Lithostratigraphies based on BGS Ostend Solid Geology sheet, NAM and RGD (1980) and C King (personal communication).

clays (Holmes, 1981). Offshore, only a few kilometres to the east, deposits of equivalent age are markedly thinner (13 m) and consist of glauconitic muds similar to the Ormesby Clay and the Lista Formation. They contain a dinoflagellate cyst assemblage indicating correlation with zone D4 of Costa and Manum (1988) (Figure 87).

The second cycle of grey, sandy, carbonaceous mudstone is equivalent to the Sele Formation of the central North Sea, and it rests with a sharp erosional contact on the underlying unit. This cycle contains pyroclastic ash layers attributed to a second phase of volcanic activity (Phase 2a) by Knox and Morton (1983). The lower part of the Sele Formation consists locally of sands (Knox, 1989) and is equivalent to the marine Woolwich Formation which interdigitates with the nonmarine Reading Formation of the London Basin. Onshore, these sediments are up to 30 m thick and consist of sublittoral glauconitic sands deposited under open-marine conditions in the east, with lagoonal silts and clays deposited behind a barrier-sand complex to the west (Ellison, 1983).

In the southernmost part of the offshore area, off the Kent coast, vibrocores recovered fine- to very fine-grained sands or muddy sands. The sediments are commonly decalcified, but locally contain abundant shell fragments. Dinoflagellate cysts from these sands indicate zone D5 of Costa and Manum (1988), and on seismic records their base may commonly be seen cutting down into the underlying Thanet Formation/Lista Formation (BGS Thames Estuary Solid Geology sheet; Henriet et al., 1989). Locally, the sands are lithified by a calcareous cement to form a hard, resistant sandstone that forms positive bathymetric features such as the Drill Stone off the Kent coast (BGS Thames Estuary Solid Geology sheet). Lithified sandstone dredged from Harwich Harbour (Thompson, 1911) may either be of equivalent age or alternatively derive from the lower part of the overlying Harwich Member.

Farther north, a cored borehole near the Shipwash sandbank recovered deposits similar to the onshore Reading Formation (Knox, 1989). In Norfolk, a greyish brown, silty or sandy mudstone overlies mudstones of the Ormesby Clay; this unit was informally termed the Hales Clay by Knox et al. (1990), and is equivalent in age to the upper part of the Sele Formation. At the northern margin of the report area, sediments equivalent to the Hales Clay were found directly overlying Late Cretaceous Chalk in BGS borehole 81/46A, where ten ash layers were recorded (Morton and Knox, 1990) compared with only four in Norfolk.

The third and uppermost Paleocene cycle observed in the southern North Sea consists of grey, sandy, carbonaceous mudstones with abundant volcanic ash layers. This unit is equivalent to the Harwich Member of the London Clay Formation onshore (King, 1981), and records the main phase of pyroclastic deposition in the area (Phase 2b of Knox and

Morton, 1983). Sixty-three ash layers were recorded in borehole 81/46A (Morton and Knox, 1990), and over 300 layers belonging to this phase have been identified in the Norwegian sector (Malm et al., 1984). The total ash thickness in this unit in the North Sea is approximately 2 m, which Knox and Morton (1988) calculate as indicating a total erupted volume of 6000 km³ of magma. The regional thickness trends of the ash layers indicate that the source lay to the north or north-west of the North Sea, and the similarity of geochemical characteristics (Morton and Knox, 1990) indicates a single source area.

The Harwich Member, both onshore and offshore, is largely decalcified and contains only a very limited calcareous microfauna. However, the dinoflagellate assemblage indicates assignation to the Hyperacanthum Zone (Costa and Downie, 1976), equivalent to zone D5 of Costa and Manum (1988), the top of which is considered by Costa et al. (1978) to represent the Paleocene–Eocene boundary and to be equivalent to the NP9/NP10 nannoplankton zonal boundary (Figure 87). In contrast, Knox (1984b; 1990) used indirect evidence to place pyroclastic phase 2b within nannoplankton zone NP10, and therefore considered the Harwich Member to be Eocene. The former interpretation has been used on BGS offshore maps and is followed here.

In the Thames Estuary region east of Kent, a cross-bedded, glauconitic, sandy unit, the Oldhaven Beds, has been sampled in vibrocores at the southern edge of the Paleocene outcrop. This unit also contains a *Hyperacanthum* dinoflagellate assemblage, and is believed to be a shallow-water facies, probably equivalent to the lower part of the Harwich Member (BGS Thames Estuary Solid Geology sheet). Volcanic grains have been found in the Oldhaven Beds onshore (Knox, 1983).

Eocene

Eocene sediments are found overlying Paleocene formations over almost all of the area of Tertiary outcrop (Figure 85). Over much of the report area, the top of the Eocene sequence is generally truncated, and is directly overlain by Quaternary deposits. Where most fully developed, the Eocene sequence in the area is over 800 m thick.

Ypresian

The Ypresian transgression probably represents the most widespread marine incursion into the North Sea area during the Tertiary. The climate at this time may also have been the warmest experienced during the Tertiary (Buchardt, 1978), and corresponds (Figure 86) to a period of global warm temperatures (Savin, 1977). Early Eocene deposition in the southern North Sea was dominated by mud; even at the western edge of the outcrop there are no signs of marginal facies. It is therefore likely that the palaeocoastline lay far to the west of the preserved outcrop, and that the nearshore facies have since been removed, although they are preserved onshore in the eastern London Basin. The onshore term London Clay Formation has been used in the UK sector because of both the consistency of facies between onshore and offshore and the previous widespread usage of this name in the North Sea (e.g. Davis and Elliott, 1957).

Studies of the London Clay Formation flora show this formation to contain the highest proportion of potentially tropical taxa of any British Tertiary sediments (Collinson, 1983). There has been much debate regarding the palaeoenvironmental significance of the London Clay flora (Daley, 1972), but certainly the climate was warm, and a muddy shoreline that supported a mangrove swamp extended inland along a few large rivers bordered by dense forest with abundant lianas (Collinson, 1983). The macrofauna includes crocodile and turtle remains that were borne out to sea by the rivers. The exact position of the coastline is unknown, but it is possible that much of the plant debris was rafted many kilometres from the river mouths.

In BGS borehole 81/46A at the extreme northern edge of the report area, the London Clay Formation is approximately 76 m thick. It consists of clays and silty clays believed to have been deposited in a more offshore setting than in the type area onshore (Lott et al., 1983). Deposition rates at this location and elsewhere appear to have been markedly lower than farther south during early Ypresian times, for the 16 m-thick D6–7 interval (Figure 88) is significantly thinner than that in the London Basin.

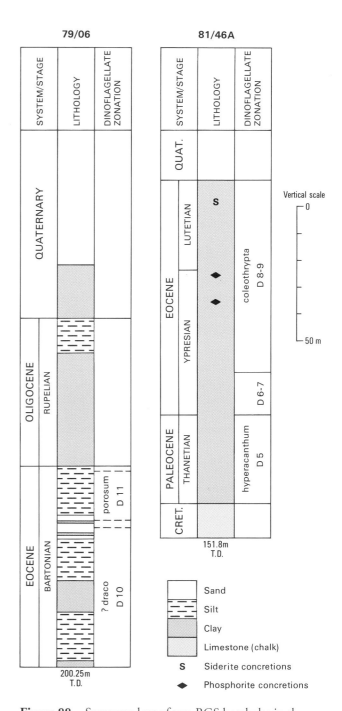

Figure 88 Summary logs of two BGS boreholes in the southern North Sea which recovered Palaeogene sediments. For locations see Figure 85.

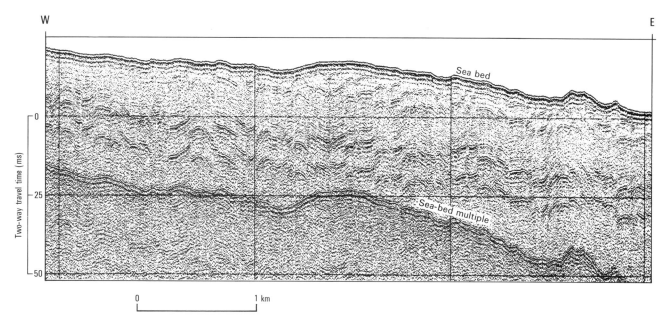

Figure 89 BGS high-resolution boomer record showing small-scale faulting within the London Clay Formation. For location see Figure 85.

In the Thames Estuary region, the London Clay Formation is approximately 150 m thick (including the late Paleocene Harwich Member), and consists of clayey silts, silty clays and silts within which a number of transgressive/regressive cycles can be recognised (King, 1981). Palaeodepths probably reached a maximum of more than 100 m (King, 1981; 1984), with shallower-water, tidally influenced sediments prominent at the top of the formation.

In the southernmost part of the area, where the London Clay Formation is at or near the sea bed, high-resolution seismic records reveal a complex pattern of small-scale faulting with throws of 2 m or less (Figure 89). The faults affect neither the base nor the top of the formation, and have been interpreted as soft-sediment deformation features induced by compaction during Ypresian times (De Batist et al., 1989; Henriet et al., 1991). Similar features are seen within fine-grained Ypresian sediments in the easternmost parts of the Hampshire–Dieppe Basin, but they appear to die out westwards as the facies becomes sandier.

Conformably overlying the London Clay Formation in south-east England are the late Ypresian Lower Bagshot Sands, the Virginia Water Formation of King (1981; 1984). These consist of very fine- to fine-grained, well-sorted sands with alternations of cross-stratified sands, wavy-bedded silty sands and clays, and flaser-bedded sands. They are believed to have been deposited as a nearshore, tidally influenced sand body (King, 1984), and include facies originating in subtidal channels (Goldring et al., 1978).

LUTETIAN

The Lutetian sediments in south-eastern England are the Middle and Upper Bagshot Sands, which consist of fine-grained sands, silts and pebble beds (Lake et al., 1986). Offshore they have not been assigned a name, but Lutetian sediments have been sampled in the southernmost part of the report area south of 52°N, where they crop out at, or close to, the sea bed. They are also known from BGS boreholes 81/45, 81/46A and 81/50A (Figures 85 and 88), where they consist of silty clays.

BARTONIAN

Bartonian sediments are not preserved in eastern England, but are recorded in offshore boreholes. In BGS borehole 79/06 (Figures 85 and 88), the Bartonian sediments consist of very fine- to fine-grained, glauconitic, silty sands which have yielded a dinoflagellate cyst assemblage equivalent to zone D11 of Costa and Manum (1988) (Figure 87). The radiolarian and diatom assemblage of this interval compares closely with that in the Zomergem Member of the Meetjesland Formation of Belgium (King, 1990). Bartonian sediments in BGS borehole 82/21 (Figure 85) consist of micaceous silts and silty sandy clays, and are also assignable to dinoflagellate zone D11 (Costa and Manum, 1988). Bartonian sediments also crop out at or near sea bed in the Belgian sector, where they consist of alternating layers of clay and glauconitic, micaceous, medium- to fine-grained sand.

PRIABONIAN

Priabonian sediments are generally absent from the report area. The Zelzate and Tongeren formations of Belgium and The Netherlands are of late Priabonian to early Rupelian age (Figure 87), and a seismic unit believed to be equivalent to these formations has a surface outcrop which extends north-westward from the Belgian coast. However, this unit has been sampled only within 25 km of the Belgian coast (BGS Ostend Solid Geology sheet), and has been included with the Eocene in Figure 85. In borehole 79/06, sands containing fragments of *Nummulites* cf *orbignyi* at 130 m depth (Figure 88) may correlate with the Priabonian part of the Zelzate Formation (King, 1990).

Oligocene

Oligocene sediments have a patchy distribution, and usually rest with slight angular unconformity on the underlying Eocene formations (Figure 85). In Belgium and in the south-west Netherlands, the earliest sediments of Rupelian age occur within formations of sandy sediments that straddle the

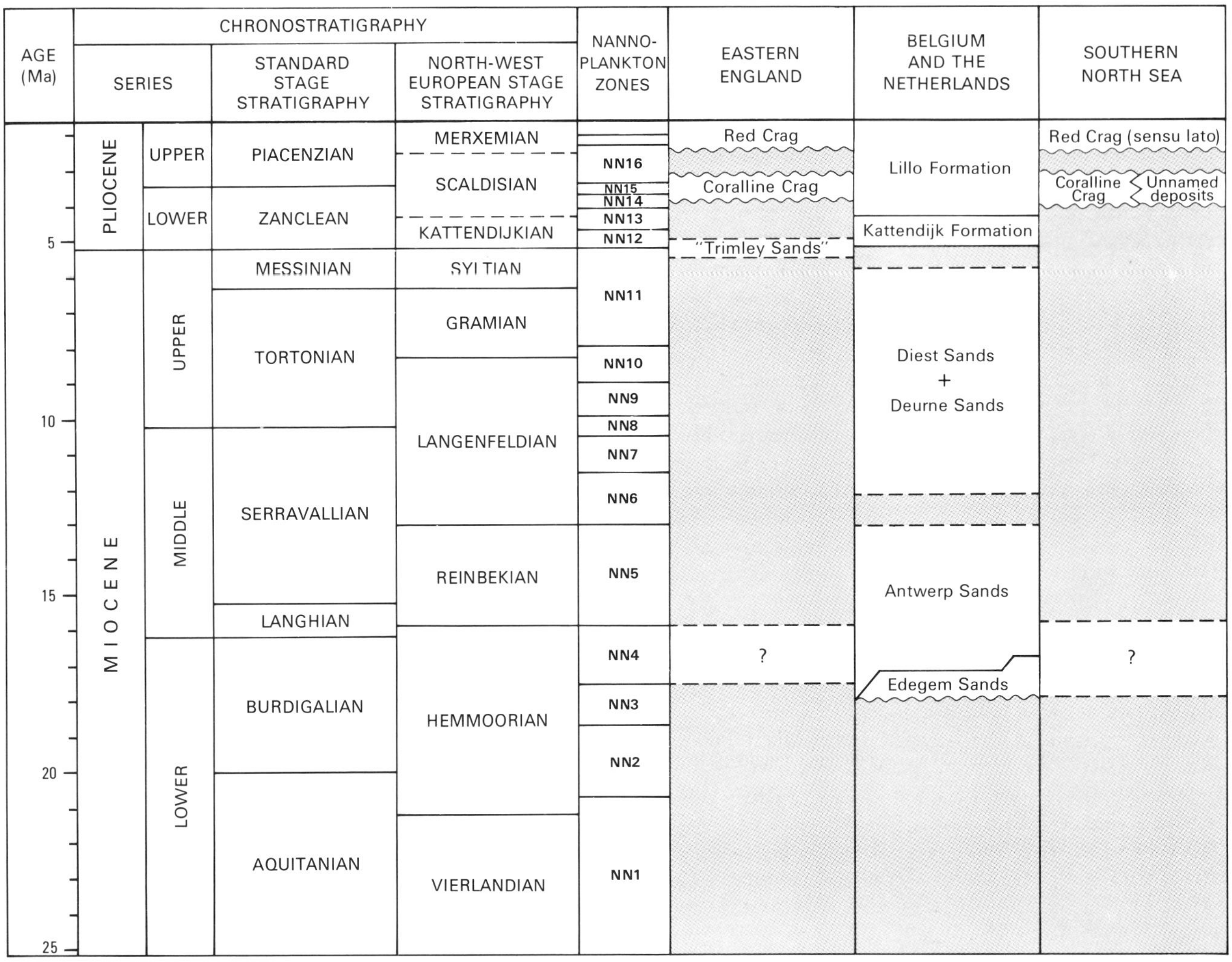

AGE (Ma)	CHRONOSTRATIGRAPHY			NANNO-PLANKTON ZONES	EASTERN ENGLAND	BELGIUM AND THE NETHERLANDS	SOUTHERN NORTH SEA
	SERIES	STANDARD STAGE STRATIGRAPHY	NORTH-WEST EUROPEAN STAGE STRATIGRAPHY				
	PLIOCENE UPPER	PIACENZIAN	MERXEMIAN	NN16	Red Crag	Lillo Formation	Red Crag (sensu lato)
			SCALDISIAN	NN15 / NN14	Coralline Crag		Coralline Crag / Unnamed deposits
5	PLIOCENE LOWER	ZANCLEAN	KATTENDIJKIAN	NN13 / NN12	"Trimley Sands"	Kattendijk Formation	
	MIOCENE UPPER	MESSINIAN	SYLTIAN	NN11			
		TORTONIAN	GRAMIAN			Diest Sands + Deurne Sands	
10			LANGENFELDIAN	NN10			
				NN9			
				NN8			
	MIOCENE MIDDLE	SERRAVALLIAN		NN7			
				NN6			
15			REINBEKIAN	NN5		Antwerp Sands	
		LANGHIAN					
	MIOCENE LOWER	BURDIGALIAN	HEMMOORIAN	NN4	?	Edegem Sands	?
				NN3			
20				NN2			
		AQUITANIAN	VIERLANDIAN	NN1			
25							

Figure 90 Neogene stratigraphy in the southern North Sea region. Neogene standard stage stratigraphy, ages and correlation with nannoplankton zones after Haq et al. (1987). North-west European stage stratigraphy after Hinsch (1986). Lithostratigraphies after Balson (1990b), Janssen (1984), and Jenkins and Houghton (1987).

Eocene–Oligocene boundary. The outcrop of these formations is at or near the sea bed close to the UK territorial boundary, but they have been included with the Eocene in Figure 85. Farther north in borehole 79/06 (Figure 88), 55 m of stiff to very stiff clays and silty clays were recovered; their calcareous microfauna (Hughes, 1981) indicates correlation with both the Boom Clay Formation of Belgium and the Rupel Formation of The Netherlands. The apparent absence of earliest Rupelian sediments in this locality (Hughes, 1981; Willems, 1989) indicates a period of erosion, a relatively short distance from areas where the sequence is more completely preserved in the Belgian sector (BGS Ostend Solid Geology sheet). In the extreme north of the report area, a slightly thicker sequence (c.110 m) is found (BGS Silver Well Solid Geology sheet).

A major eustatic sea-level fall at the end of Rupelian times (Haq et al., 1987) resulted in widespread erosion of late Rupelian sediments, which show a well-preserved erosional topography on seismic profiles (Henriet et al., 1989). As a result of this widespread regression, Chattian sediments appear to be absent from the report area, and are restricted to regions farther east, closer to the central axis of the basin.

NEOGENE

The standard stage stratigraphy for the Neogene is based largely on sequences in the Mediterranean region, and has not found wide application in northern Europe. Figure 90 therefore shows commonly used northern European stages with their likely correlation to the standard stage stratigraphy of Haq et al. (1987).

Miocene

Miocene deposits are of very restricted occurrence within the report area, for their outcrop lies mainly to the east and north, associated with the axial part of the North Sea Basin. The edge of the Miocene outcrop enters the extreme north-east of this area (Figure 85), where the sequence is believed to be around 30 m thick.

Although the early to middle Miocene transgression was probably one of the most extensive during the Tertiary, it appears to have left relatively little evidence in the south-western North Sea Basin. The Miocene sediments of the north-eastern corner of the report area were probably deposited during this transgressive episode. Fossil teeth of the mid-Miocene giant shark *Carcharocles megalodon* (formerly *Carcharodon megalodon*) occur in phosphatic lag gravels at the base of

Pliocene formations in eastern England, indicating the former presence of mid-Miocene sediments there (Balson, 1990b).

In Belgium, the Miocene is dominated by condensed sequences such as the extremely glauconite-rich Antwerp and Deurne Sands. These units are separated by stratigraphical hiatuses, and both have well-developed, basal, pebble-lag conglomerates.

On the mainland of eastern England, a deposit of possible Miocene age, the Lenham Beds of Kent, consists of ferruginous sands and gravels preserved within solution pipes in the surface of the Late Cretaceous Chalk. The sands are decalcified, and age determinations have traditionally relied on the identification of molluscan shell moulds, believed to indicate time equivalence with the Diest Sands of Belgium. Re-examination of the Lenham Beds molluscan fauna (Janssen in Balson, 1990b) has indicated that they show greater affinities to Atlantic, rather than North Sea, faunas. The Lenham Beds were therefore deposited as part of a transgression from the west through the English Channel region, whilst the Diest Sands were deposited within the geographically separate North Sea Basin.

Pliocene

A period of regression during the late Miocene as a result of a major eustatic sea-level fall (Haq et al., 1987) preceded a renewed transgression in the latest Miocene/earliest Pliocene. Pliocene deposits, like those of the Miocene, have an extensive outcrop to the east of the report area, but there are only a few, small, discontinuous outcrops within the area, particularly in the south (Figure 85), where they overlie Eocene formations with conspicuous unconformity. The upper surface of these Pliocene outliers is truncated; the thickest sequence observed is 65 m within an incised trough (Balson, 1989b).

The position of the Plio-Pleistocene boundary is controversial, particularly in north-west Europe. Firstly there is dispute over the location of the boundary in geochronological terms; traditionally the boundary in this region has been taken at 2.3 Ma, but the internationally accepted boundary is now placed at 1.64 Ma, just above the Olduvai subchron (Aguirre and Pasini, 1985). Secondly, there is the problem of location of this boundary within the sedimentary sequence. In the report area, difficulties are caused by the presence of nearshore, coarse, clastic facies which lack the diagnostic microfaunas and floras necessary for accurate age assignation. On seismic records, rapid lateral facies changes mean that the nature of reflectors may change over short lateral distances, and it seems likely that some deposits presently mapped as Pleistocene may prove to be of Pliocene age. The Pliocene outcrops shown in Figure 85 are those of deposits of unequivocal Pliocene age. A deposit in eastern England known as the Red Crag Formation, and sediments of equivalent age offshore, are not shown, for although part of the formation is likely to be Pliocene, its upper stratigraphical limit is poorly defined.

Evidence for sediments of latest Miocene/earliest Pliocene age comes from the phosphatic lag deposits at the base of the Coralline Crag and Red Crag of eastern England. These deposits contain large sandstone cobbles which originated as phosphorite concretions in a deposit of muddy sands informally termed the 'Trimley Sands' (Balson, 1990b). Deposits of equivalent age have not yet been identified offshore, but the presence of a fully marine fauna, including whales, in the onshore sediments, indicates that much of the present southern North Sea would have been inundated at the time.

In eastern England, the Coralline Crag is a mid-Pliocene formation (Balson, 1990c) of skeletal carbonate sands and silty sands, with an elongate outcrop near the coast which continues some 14 km offshore (Figure 85). Onshore, it has been divided into a number of sedimentary units. A lowermost unit of silty sands approximately 7.5 m thick has a basal lag deposit rich in phosphatic nodules (Balson, 1980). This unit is unconformably overlain by coarser-grained sediments rich in the skeletal debris of molluscs and bryozoans from which aragonitic grains have generally been removed by dissolution, and the sediment lithified into a porous limestone.

In the offshore portion of this outcrop, vibrocores near the coast have recovered aragonite-leached sediments similar to these uppermost facies, indicating that the leaching is not a recent subaerial phenomenon. Seismic records show a thickness of approximately 25 m, similar to that proven onshore. Where the sediments have been lithified, a strong seismic reflector is seen at the top of the formation. Dipping internal seismic reflectors may represent boundaries between sets of crossbeds. The underlying unlithified sediments are similar in seismic character to the overlying Plio-Pleistocene sediments, so that the extent of the unlithified part of the Coralline Crag is not well defined. Farther to the east, sediments of an approximately similar age to the Coralline Crag have been obtained in BGS boreholes and vibrocores (Figure 85).

In BGS borehole 81/51, 16.3 m of lower Pliocene, glauconitic, muddy sand was recovered from an outcrop where sediments of this age are believed to be approximately 20 m thick. The carbonate content of this sand is much lower than the Coralline Crag, and macrofossils much rarer. The extent of this outcrop is poorly known, but is likely to be small and not connected to the Coralline Crag outcrop to the west.

To the south-south-east, BGS borehole 81/53A penetrated approximately 45 m of shelly, glauconitic sand which yielded foraminifera (C King, written communication, 1991) indicating correlation with zone NSB14 of King (1989), which includes the Coralline Crag. Seismic records show that this borehole penetrated an infilled valley cut into Palaeogene sediments. About 20 km to the south of borehole 81/53A (Figure 85) is another small isolated body of Pliocene sediment infilling a north-easterly orientated valley eroded into Palaeogene formations and containing over 65 m of sediment (Balson, 1989b). Vibrocores into the topmost part of the infill have recovered very glauconitic muddy sands with a shelly fauna of molluscs, bryozoans, echinoids and *Terebratula* (Balson, 1989b). The sediments are thus very similar to the lower parts of the Coralline Crag of eastern England although richer in glauconite. This correlation is supported by the foraminiferal evidence.

Where it can be determined, these infilled valleys have a broadly east-north-easterly alignment similar to that of the Coralline Crag outcrop interpreted by Balson (1983) as being the primary depositional form of a nearshore tidal sandbank facies. It may therefore be speculated that the infilled valley forms, like the onshore Stradbroke Trough (Funnell, 1972), might similarly relate to regional tidal-current patterns.

In the vicinity of these valleys, the sea-bed sediments commonly contain Pliocene fossils including the robust umbonal parts of *Terebratula maxima* Charlesworth. Lagaaij (in Houbolt, 1968) recorded typical Pliocene bryozoans in the sediments of the Outer Gabbard sandbank. The sea-bed sediments nearby also contain abundant phosphorite pebbles (Balson, 1989a) which probably originate from the erosion of basal lag deposits of previously existing Pliocene formations. A BGS vibrocore to the east of the Outer Gabbard sandbank recovered Pliocene glauconitic sands from a deposit too thin to be seen on seismic records. Taken together, this evidence suggests that many thin lenses of Pliocene sediment may exist along the southern feather edge of the Pliocene outcrop, the remnants of a former sheet of glauconitic, fine-grained sands.

The Red Crag of eastern England is a formation of marine, shelly sands and gravels which unconformably overlies the London Clay Formation over most of its outcrop, but oversteps onto the Chalk at its western limit. The Red Crag consists of cross-bedded, shallow-marine, subtidal, shell sands overlain by sands and muds interpreted by Dixon (1979) as intertidal sand-flat deposits. The age of the Red Crag is controversial and has been traditionally placed with the Plio-Pleistocene boundary either at its base or within the formation. More recently, the Red Crag Formation has been placed within the late Pliocene, with an age between 3.2 and 2.4 Ma based on the presence of the planktonic foraminiferid *Neogloboquadrina atlantica* (Funnell, 1987; 1988). Unfortunately, the lithostratigraphical definition of the Red Crag is problematic, so that some sediments traditionally termed 'Red Crag' may be of Pleistocene age, at least using the definition of the Plio-Pleistocene boundary at about 2.3 Ma.

Offshore, sediments termed Red Crag Formation (Cameron et al., 1984; BGS Flemish Bight Solid Geology sheet) are found within a broad trough resting unconformably on Palaeogene formations. BGS boreholes have proved a thickness of 41 m of glauconitic muddy sand with a benthic foraminiferal fauna similar to that of the Red Crag of eastern England (Cameron et al., 1984). *N. atlantica* has also been found in these sediments (Funnell, 1987), indicating a late Pliocene age. At the base of the Red Crag Formation in BGS borehole 81/50A (Figure 85), phosphatic mudstone pebbles were found similar to those at the base of the Red Crag onshore (Balson, 1980). Seismic records indicate a maximum thickness of 70 m for this formation, which has a lenticular geometry and is unconformably overlain by the early Pleistocene Westkapelle Ground Formation (Cameron et al., 1989a).

12 Pleistocene

It has long been recognised that much of the floor of the North Sea was shaped during the last glaciation (Eisma et al., 1979), and it used to be assumed that earlier Pleistocene sedimentation had also been controlled by alternating glacial and interglacial conditions, with low and high sea levels respectively (Jansen et al., 1979). Marine, interglacial, Pleistocene sediments preserved in East Anglia are less than 80 m thick and elsewhere in Britain the generally thinner Pleistocene succession is overwhelmingly of glacigenic sediments. Thick, marine sequences were therefore assigned to the Neogene (Rhys, 1974) rather than to the Pleistocene.

V N D Caston (1977; 1979) first demonstrated the great thickness of Pleistocene deposits throughout the North Sea Basin, relating it to tectonic subsidence. Subsequent mapping in the southern North Sea has shown the majority of the deposits to be early to middle Pleistocene, shallow-water, deltaic sediments (Balson and Cameron, 1985). These represent the offshore continuation of a major delta system previously identified in The Netherlands (Zagwijn and Doppert, 1978; Zagwijn, 1979, 1989). A thin capping of middle to late Pleistocene glacigenic and nondeltaic marine sediments (Cameron et al., 1989b) completes the resemblance to the Dutch Pleistocene succession (Zagwijn, 1985).

Mapping has not yet proceeded far enough towards The Netherlands' coast to allow precise correlation of the offshore succession with established onshore Dutch Pleistocene formations. Neither is the offshore succession yet fully correlated with the Pleistocene deposits of eastern England, owing partly to a lack of mappable onshore equivalents. The offshore stratigraphy is essentially seismostratigraphical (Balson and Cameron, 1985; Laban et al., 1984), whereas that onshore is lithostratigraphical and biostratigraphical. In general, a separate suite of formation names has therefore been adopted offshore (Figure 91).

The base of the Pleistocene is taken here as the beginning of the Praetiglian Stage, dated at about 2.3 Ma (Zagwijn, 1989; Gibbard et al., 1991); this date is slightly younger than the transition between the Matuyama and Gauss magnetic polarity chronozones at 2.45 Ma (Harland et al., 1982). Although based on a regional bioclimatic concept (Zagwijn, 1974) which is not globally applicable (Aguirre and Pasini, 1985), this definition of the base of the Pleistocene is nevertheless useful in the Southern North Sea Basin. It corresponds approximately to the base of the Westkapelle Ground Formation (Figure 91), which is traceable on seismic profiles across the Southern Bight above an angular unconformity (Balson and Cameron, 1985). This unconformity is exposed in East Anglia as a marine planation surface (West, 1980).

The North Sea Basin is a Palaeozoic to Holocene multistage rift zone within the north-west European craton (Thorne and Watts, 1989). Low to very low rates of subsidence during Oligocene and Miocene times, together with mid- to late-Miocene uplift (Kooi et al., 1989), were followed by rapid subsidence during the Pliocene and Pleistocene. Isopachs of the Pleistocene sediments (Figure 92) show that the principal axis of subsidence in the southern part of the basin trends north-north-westwards, broadly parallel to the UK coastline from the Firth of Forth to The Wash. This trough of maximum subsidence lies to the west of a corresponding Tertiary one in the Dutch sector. Uplift of the basin flanks in East Anglia (Zalasiewicz and Mathers, 1988) has produced local regression and several unconformities; if this marginal uplift was simultaneous with increased deepening of the basin centre, it could be attributed to Alpine compression superimposed on the long-term tectonic subsidence (Kooi et al., 1989).

In calculating rates of basin subsidence, the Brunhes–Matuyama polarity transition is a useful datum. North of 55°N in the central North Sea, the transition is found in interglacial, shallow-marine sediments 270 m below sea level (Skinner and Gregory, 1983; Stoker et al., 1985); palaeontological indications are that the water depths at the time did not exceed 15 m. Hence, assuming a sea level similar to today's, subsidence there has been some 255 m over 730 000 years, giving a rate of about 0.35 m per thousand years. Because this determination lies near the axis of greatest subsidence, the average rate of maximum subsidence since that time is not likely to have exceeded 0.4 m per thousand years, a figure that compares well with a maximum rate of 0.5 m per thousand years estimated by Clarke (1973).

Between 53°50'N and 54°30'N, anticlines and synclines are developed in the Pleistocene deposits as a result of salt movement. These halokinetic structures are commonly related to suites of faults, such as the Dogger Fault Zone (Figure 13), which controlled sedimentation locally until at least early middle Pleistocene times. Faults displacing late Elsterian deposits have been noted by Balson and Cameron (1985). There is geomorphological evidence from the Outer Silver Pit that salt movement is probably still continuing; if so, there is no reason to suppose that fault movement has ceased.

A series of geological profiles across the report area is shown in Figure 93, and a schematic summary representation of the Pleistocene succession is presented in Figure 94. The succession can be divided into eight elements (A to H) which fall into two unequal divisions of differing geometries.

The lower, regressive, almost wholly deltaic division is made up of elements A and B, and comprises 80 per cent or more of the total. Its sediments are thick and extensive, and were deposited under relatively stable climatic conditions (Long et al., 1988). The division displays gradational and predictable lithofacies changes manifested in a limited variety of acoustic signatures whose arrangement indicates basinward delta progradation (Cameron et al., 1987).

The upper, fragmented, transgressive/regressive, nondeltaic division consists of elements C to H, which are either thin, tabular and somewhat discontinuous, or thick, semilenticular and discrete. The nondeltaic division was deposited under a wide range of climatic conditions (summarised in Nilsson, 1983), resulting in varied lithologies and acoustic signatures. The two episodes of downcutting which form the lower boundaries to the semilenticular elements C and H occurred during the Elsterian and Weichselian glaciations respectively.

DELTAIC DIVISION

This division consists of elements A and B, and within the report area element A has been subdivided into eight seismostratigraphical formations (Figures 91 and 94) that are types of alloformation (Owen, 1987). Of these, the Smith's Knoll

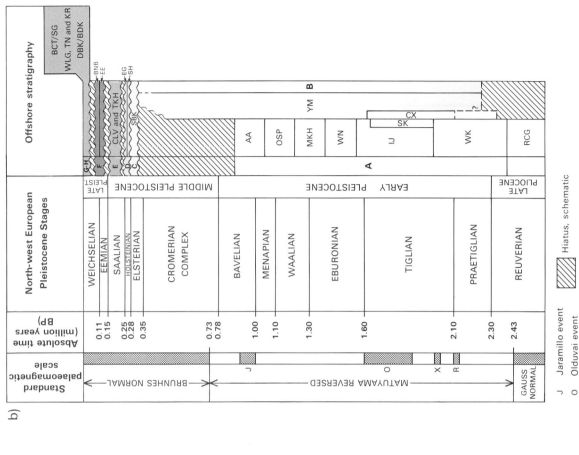

Figure 91 (a) Pleistocene formations in the southern North Sea referred to in the text. From BGS/RGD Quaternary Geology sheets. (b) Pollen-zone divisions of the Pleistocene, with palaeomagnetic and absolute timescales (Zagwijn, 1979; 1985; 1989), and suggested positions of UK sector offshore formations within this scheme, based on BGS/RGD data. Early Pleistocene formations are separated by hiatuses of unknown duration; the lines are drawn with reference to the base of the succeeding formation.

102

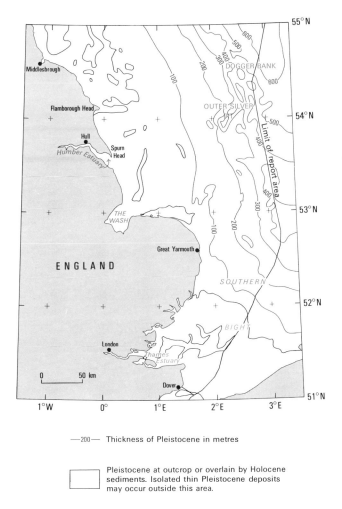

Figure 92 Generalised thickness of Pleistocene sediment in the southern North Sea.

Formation and IJmuiden Ground Formation are laterally equivalent, and the Crane Formation is anomalous in that it does not belong to the same sedimentary system. As noted in Cameron et al. (1987), the geometry of these formations (except the Crane Formation) tends to be lenticular where fully developed in the south, but wedge-shaped in the north (Figure 94), a variation ascribed to differential subsidence. Consequently, towards the north-east of the report area, seismic recognition of individual formations becomes increasingly difficult, particularly where internal and bounding reflectors are nearly parallel (BGS Silver Well Quaternary Geology sheet). The formations of element A (except the Crane Formation) all pass into the lower part of element B, which consists solely of the Yarmouth Roads Formation, a basinward-thickening wedge (Figures 93 and 94). At the western margin of the basin, the combination of slow subsidence and low sedimentation rates, except in the Southern Bight, has resulted in the formations being thin, near-tabular and separated by almost-parallel discontinuities.

From their geometries, elements A and B are interpreted as the marine and largely nonmarine parts respectively of a northward-advancing delta complex made up of two amalgamated deltas. A small western delta contiguous with East Anglia received sediment from Britain, while a much larger eastern delta extended from the Low Countries and received sediment from the European mainland (Cameron et al., 1987).

Several authors (Laban et al., 1984; Cameron et al., 1984, 1989a; Balson and Cameron, 1985; Long et al., 1988) consider that the formations of the deltaic division developed during high sea-level stands, and that the reflectors separating them represent basinwide unconformities caused by eustatic falls of sea level. Thus, the formations and their bounding reflectors have been linked respectively with the thermal maxima and minima interpreted from the pollen assemblages of early to middle Pleistocene sediments in The Netherlands (Zagwijn, 1985).

However, mapping in parts of the UK sector (BGS Indefatigable and Silver Well Quaternary Geology sheets) has suggested that the reflectors bounding the marine formations of element A, although persistent over long distances, are not developed basinwide, and are disconformities rather than unconformities (Figure 93). Furthermore, the local unconformity in the Southern Bight separating elements A and B (Figure 94) may be due to marginal tectonic uplift (Balson and Humphreys, 1986) during the long history of deposition of sediments laterally equivalent to elements A and B (Zalasiewicz and Mathers, 1988). Any coincidence of formation boundaries with the interpreted thermal minima may therefore be fortuitous.

Elsewhere, the somewhat discontinuous seismic reflector separating elements A and B is coincident with a gradual lithofacies change from fully marine to intertidal, supratidal and terrestrial sediments. This facies change is accompanied by the disappearance of marine dinoflagellate cysts and calcareous microfauna and by increased abundance of pollen from dune and saltmarsh plants, as well as from trees (Cameron et al., 1984). Not only is this facies change demonstrably diachronous (Figures 93 and 94), but the simple architecture of the deltaic division is in contrast to the intricate architecture displayed by regressive/transgressive sequences (Colman and Mixon, 1988; Christie-Blick, 1991). Both diachroneity and architecture can be adequately explained by a model of overall regression through delta-top advance, in which the formation boundary reflectors probably represent subtle lithological contrasts due to shifting depocentres and to spasmodic subsidence rates. However, Cameron et al. (1989a) account for these attributes of the deltaic division by invoking a model related to eustacy in which the limits of successive transgressions were displaced progressively basinwards, resulting in net regression.

Element A

It is convenient to treat the seven deltaic-marine formations of element A (Figure 94) together with the nondeltaic Crane Formation. Up to four acoustic facies and subfacies are consistently developed within each of the seven deltaic formations. These acoustic facies correspond to systematic lithological variations (Cameron et al., 1987) representing three principal depositional environments: prodelta, delta front and delta top (Figure 95). Since these formations lie in a sheaf-like fashion against each other (Figures 93 and 94), at any point in a vertical profile through several of these formations there should be a predictable upward transition through the acoustic facies from prodelta through delta front to delta top; this contention is borne out by available borehole data.

Palaeomagnetic measurements made on borehole cores from these formations give predominantly reversed magnetic polarity, which is inferred to belong to the largely early Pleistocene Matuyama epoch (Figure 91). Several normal polarity measurements (Cameron et al., 1989a) may correlate with known normal events in The Netherlands' sequence (Figure 91), but as neither the number nor the duration of these are fully resolved, they can be only tentatively used for dating (Cameron et al., 1984).

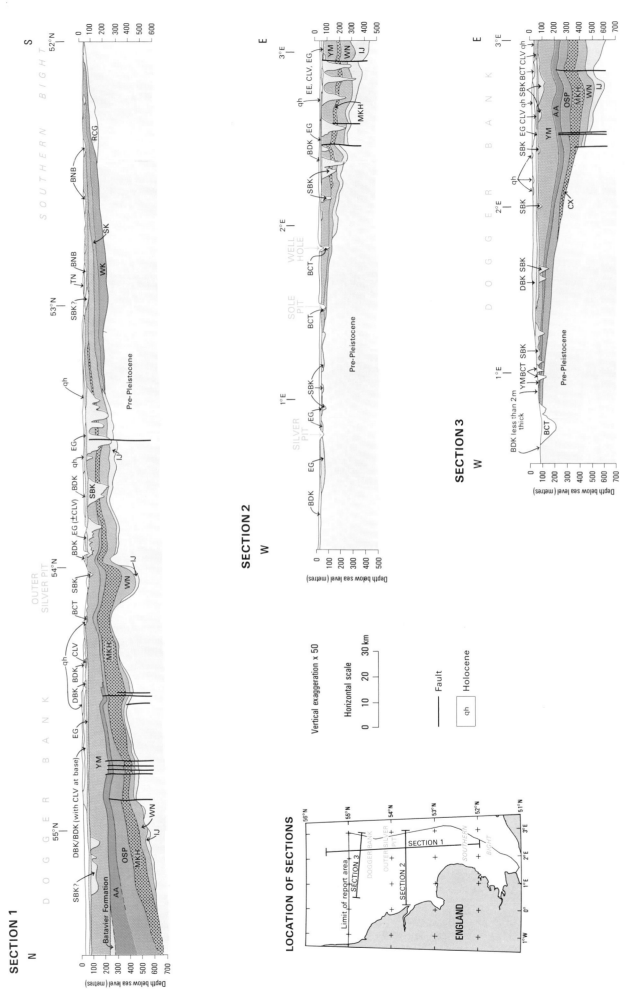

Figure 93 Cross-sections showing the dispositions of the Pleistocene formations of the southern North Sea. For key to abbreviations see Figure 91.

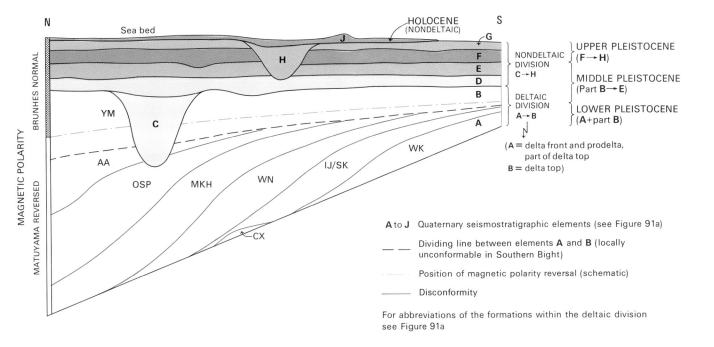

The reflectors separating the formations of element A are isochronous lines; hence the base of element B is diachronous. In the north, the shape of the isochronous line depicting the Brunhes-Matuyama polarity transition dips in the same manner as the formation boundaries.

Figure 94 Schematic illustration of the relationships of the stratigraphical elements and early Pleistocene formations in the southern North Sea.

The Westkapelle Ground Formation (Figure 96) is the lowest formation of element A, and has recently been recognised as Praetiglian and Tiglian in age, possibly extending down into the very late Pliocene (BGS Silver Well Quaternary Geology sheet). It crops out only in shallow water off the north-east coast of East Anglia (BGS East Anglia and Spurn Quaternary Geology sheets) where its upper levels have been partly removed by erosion. In the report area, it is mainly of prodelta acoustic facies, attaining a thickness of 20 to 50 m in the Southern Bight. This facies has been sampled in BGS boreholes (Figure 96), in which it consists of silty clays with partings of fine-grained, glauconitic, bioturbated sands passing gradually upwards into predominantly mud-free sands. Foraminiferal assemblages corroborate these lithological indications of decreasing water depth and increased energy levels (Cameron et al., 1984; 1989a). Two normal polarity zones within the formation were thought by Cameron et al. (1984) to be possibly the 'X' and Réunion events (Figure 91), and pollen spectra from borehole 81/50A are reminiscent of Thurnian (probably Tiglian — Gibbard et al., 1991) assemblages in East Anglia (Cameron et al., 1984).

The comparatively condensed nature of the Pleistocene sequence in the Southern Bight means that the delta-front facies of the Westkapelle Ground Formation is characterised by a nonsigmoid reflector configuration (Figure 95). The deposits are sandier than those of the prodelta facies, and were derived from Britain (Cameron et al., 1987). Figure 96 shows that the delta-front facies is probably represented in a borehole at Ormesby (East Anglia), where shelly, bioturbated, subtidal sands and muds are found in the lower part of the Pleistocene succession (Harland et al., 1991). In the same borehole, the delta-top facies is an overlying, intertidal, mud-dominated unit. Along the East Anglian coast, the Westkapelle Ground Formation has the same base as, and is equivalent to, the upper part of the Red Crag Formation (*sensu* Zalasiewicz and Mathers, 1988), a relationship founded on lithofacies and geometric relationships (BGS East Anglia Quaternary Geology sheet, section 1). Confusion has however been engendered by the dual use offshore of the term 'Red Crag Formation' for both an underlying Pliocene seismostratigraphical unit up to 70 m thick with a Waltonian-type foraminiferal assemblage (Cameron et al., 1984; 1989a) and for a seismostratigraphical unit which is laterally equivalent to part of the Westkapelle Ground Formation (BGS East Anglia Quaternary Geology sheet, section 4).

Further supply of sediment from Britain produced the overlying Smith's Knoll Formation, which by definition (Cameron et al., 1989a) is restricted to a narrow zone east of the delta-front facies of the Westkapelle Ground Formation (Figures 96 and 97). The formation is 20 to 55 m thick and has a subtabular geometry, with a wedge-like termination against the coeval IJmuiden Ground Formation over a 5 km-wide zone at its southern end, where contrasting lithologies are inferred. At its northern end, the transitional zone is up to 15 km wide, implying little, or very gradual change of lithology (Cameron et al., 1989a). To the north of 52° 50'N, the acoustic contrast between the Smith's Knoll and IJmuiden Ground formations becomes indistinct, and the seismically well-layered sediments have been assigned to the latter formation (Figure 97).

The Smith's Knoll Formation is therefore entirely characterised by the delta-front acoustic facies, and comprises up to three overlapping sets of divergent, inclined, seismic reflectors (Cameron et al., 1989a). Samples of this facies from BGS boreholes (Figure 97) consist of muddy, fine-grained, glauconitic, locally micaceous sand with minor intercalations of silty clay and of pebbly or shelly sand. Dinoflagellate-cyst and pollen assemblages resemble those of the Antian Stage (approximately mid-Tiglian — Gibbard et al., 1991) of East Anglia (Cameron et al., 1984). This observation is noteworthy since on facies and geometric grounds the Smith's Knoll

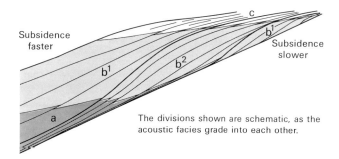

Relationship of acoustic facies to lithofacies

Acoustic facies	Reflector configuration	Environment	Dominant lithofacies
a	Parallel, continuous, even	Prodelta	Mud
b	b¹: Subfacies: divergent, even ("nonsigmoid") b²: Subfacies: divergent, sigmoidal ("sigmoid")	Delta front	Sand
c	Parallel to subparallel, discontinuous, shingled to even	Delta top (marine)	Sand or rhythmites of mud/sand

Figure 95 Acoustic facies and subfacies within seven of the eight formations of element A. Modified from Cameron et al. (1987).

Formation should be equivalent to much of the Norwich Crag Formation, which includes sediments of Antian age (BGS East Anglia Quaternary Geology sheet; Zalasiewicz and Mathers, 1988).

It is surmised that the delta-top facies, which prior to erosion would have lain mostly to the west of the delta-front facies (Figure 97), is probably represented on land by the 'Pastonian' and 'Pre-Pastonian' (late Tiglian — Gibbard et al., 1991) fully marine to intertidal, regressive sediments (West, 1980). These are assigned by some authorities to the Norwich Crag Formation (BGS East Anglia Quaternary Geology sheet) and by others to the lowest part of the Cromer Forest Bed Formation (*sensu* Zalasiewicz and Mathers, 1988).

During deposition of the Smith's Knoll Formation, contemporaneous sediment influx from the European mainland eventually overwhelmed the British supply, partly overstepping the Smith's Knoll Formation and causing shallowing of Southern Bight waters by extending the delta plain of The Netherlands (Zagwijn, 1989). This influx created the IJmuiden Ground Formation, a lenticular body up to 190 m thick and notable for the strong westward development of its sigmoid delta-front subfacies (Figures 95 and 97). It was built up as a series of smaller overlapping lenticular units (Cameron et al., 1989a). Interbedded sands and clays in the upper third of the delta-front facies in Dutch borehole S1-63 (Figure 97) have been dated as late Tiglian (BGS Flemish Bight Quaternary Geology sheet; Cameron et al., 1989a). Time-equivalent sediments belonging to the prodelta facies consist of bioturbated clay, silt and fine sand in the lower part of the Pleistocene succession of borehole 89/05; there is an upper zone of normal palaeomagnetic polarity in these sediments that may represent the Olduvai event (Figure 91).

While the Southern Bight was being filled rapidly with deltaic sediments, sediment supply in the UK sector north of 54°N was so restricted that only very thin, distal, prodelta clays were being deposited in the deepest parts of the basin.

Shorewards, there were bars of constantly reworked material, of which the Crane Formation is the only preserved example (Figure 97). Shaped in cross-section like a slightly asymmetric half-lens up to 25 m thick, it is characterised by internal seismic reflectors generally inclined to the south-east. The partial overstepping of the Crane Formation by up to 30 m of the youngest parts of the IJmuiden Ground Formation (Figure 97) indicates that the former has a minimum age of late Tiglian. A mollusc fauna similar to that of Baventian and Pre-Pastonian (mid- and late Tiglian — Gibbard et al., 1991) 'crags' of East Anglia (Norwich Crag Formation) was recovered in cuttings from well 44/21-2 (T Meijer, written communication, 1991). The samples include many bored, worn and weathered shells, with well-rounded, coarse, sand grains; all evidence of slow accretion and reworking in a high-energy environment. Thus the Crane Formation probably represents a long time-span, so its lowest parts may well equate with much of the Tiglian and Praetiglian Westkapelle Ground Formation.

Further overstepping of the Crane Formation from the south and east by prodelta sediments of the Winterton Shoal Formation was a harbinger of major delta expansion into the UK sector. This expansion is recorded by the sigmoid delta-front subfacies of the Winterton Shoal Formation (Figure 98), which attains a thickness of some 150 m within troughs adjacent to salt diapirs. Three BGS boreholes have sampled the formation towards its thin, western edge. Boreholes 82/21 and 89/05 penetrated bioturbated, calcareous, sandy muds with glauconite and granules of partly pyritised monosulphide in the prodelta acoustic facies; their heavy minerals indicate that these sediments have a mainly Continental provenance. Borehole 81/05 passed through only 2 m of fine-grained, shelly sand, of probable delta-top acoustic facies, from a British source (Figure 98).

The Winterton Shoal Formation records the first distinct uniting of the eastern and western delta fronts in the southern North Sea, and the concomitant northward swing of delta advance. These changes probably occurred during the Eburonian Stage. The formation is also the youngest in which sediment units of clearly British origin are discernible in shallow-seismic profiles.

The northward component in the direction of delta advance is well established in the Markham's Hole Formation (Figure 99). Only the delta-front acoustic facies, up to 105 m thick, has been sampled by BGS boreholes. The nonsigmoid subfacies in BGS borehole 82/20 comprises fine-grained, glauconitic, calcareous, muddy sand with partings of clay. In BGS borehole 89/05, the top 12 m of the formation are of the sigmoid subfacies; the sediments become slightly coarser grained, less calcareous and less glauconitic upwards, and are probably transitional to a nearby thin wedge of intertidal sediments belonging to the delta-top acoustic facies. Figure 99 shows that the early Pleistocene marine connection to East Anglia was severed during Markham's Hole times (Eburonian/Waalian?), by the end of which the coastline lay some 80 km to the north-east of the present-day East Anglian coast.

With deposition of the Outer Silver Pit Formation and the Aurora Formation, there was a markedly north-westward advance of the delta (Figures 100 and 101). The strongest downwarping of the basin was between 54° and 55°N, where these formations are of a delta-front acoustic facies that is more thickly bedded than the older formations farther south. This variant of the nonsigmoid subfacies merges imperceptibly with both prodelta and delta-top acoustic facies, making more difficult both internal facies delineation and differentiation of the two formations themselves. The Outer Silver Pit

izontal, rather discontinuous reflectors that are thought to indicate depositional surfaces controlled by local sea level. However, in the area of the Outer Silver Pit there are complicated arrangements of stacked, intersecting, concave-up seismic reflectors bounding local sets of inclined seismic reflectors; these are interpreted as channels and their fills. Element B comprises solely the Yarmouth Roads Formation, which attains a thickness of 160 m in the report area (Figure 102). It has been penetrated by numerous boreholes in several sectors of the North Sea; there is a widespread dominance of partly or wholly decalcified sands, commonly with scattered pebbles (including chalk), abundant plant debris, peat and wood clasts. Plant remains include those of the freshwater fern *Azolla filiculoides* Lambert. The sands are associated in places with intertidal sand/mud rhythmites studded with mud pebbles derived from intraformational hardgrounds. These features strongly support the seismic interpretation of the formation as a complex of delta-top deposits reflecting diverse depositional milieus.

In the area immediately south of the Dogger Bank, the Yarmouth Roads Formation can be divided into three acoustic members (BGS Silver Well Quaternary Geology sheet). The lowest member consists mainly of extensive lagoonal clays sampled in boreholes 89/05, 82/20 and 82/21 (Figure 102), whereas the middle and upper members are fine- or medium-grained sands with plant remains and well-worn shells and pebbles. These two members record an upward transition through beach deposits to more terrestrial conditions.

Towards the western edge of the basin in the Southern Bight, the base of the Yarmouth Roads Formation is sharply differentiated from the underlying fully marine formations of the Pleistocene delta by a strong reflector. This reflector probably corresponds to the lithological break recorded at 45.73 m depth in the Ormesby borehole of East Anglia (Harland et al., 1991), related to a sharp change from low-energy to moderate-energy conditions in an upward-shallowing sequence. This break can probably be attributed to contemporaneous tilting and erosion at the basin margin (Mathers and Zalasiewicz, 1988; Figure 93), because towards the centre of the basin, unbroken subsidence has produced a more gradational lithological transition, and the lower bounding seismic reflector is more discontinuous. The base of the formation therefore becomes less easily definable, although in places channelling has been noted (BGS Silver Well Quaternary Geology sheet).

The Yarmouth Roads Formation is strongly diachronous. In the Southern Bight its base is certainly coeval with the base of the Markham's Hole Formation. On geometric grounds (Figures 93 and 94), its stratigraphical base near the East Anglian coast could be expected to be coeval with the Praetiglian to Tiglian Westkapelle Ground Formation (BGS East Anglian Quaternary Geology sheet). Supporting evidence for this inference is provided some 8 km east of Lowestoft by two vibrocores of marine-lagoonal clays from near the base of the formation; these contain not only the typical early Pleistocene dinoflagellate cysts *Operculodinium israelianum* Rossignol and *Tectatodinium pellitum* Wall, but also indigenous *Amiculosphaera umbracula* Harland, *Protoperidinium* sp. nov. and *Tuberculodinium vancampoae* Rossignol, which are characteristic of the very earliest Pleistocene and the Pliocene (R Harland, written communication, 1987). Furthermore, some 40 km to the north-northwest of these vibrocores, rather coarse-grained littoral sands recovered in BGS borehole 79/07 (Figure 102) have yielded *Elphidium pseudolessonii* ten Dam and Reinhold (D

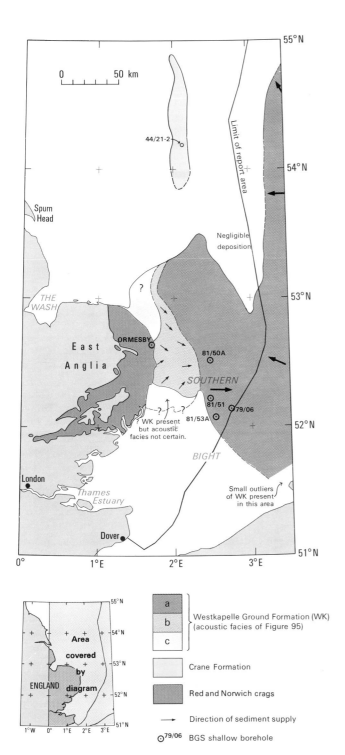

Figure 96 Distribution and acoustic facies of the Westkapelle Ground Formation and the crags of East Anglia. Adapted from Cameron et al. (1987) and Mathers and Zalasiewicz (1988).

Formation attains a thickness of about 100 m, and the Aurora Formation some 75 m. Only the Outer Silver Pit Formation has been sampled in the southern North Sea, where 11 m of delta-top acoustic facies in borehole 89/05 (Figure 100) consist of fine-grained, slightly pebbly and weakly calcareous sand.

Element B

Element B is characterised by a varied acoustic signature which appears generally rather chaotic, with sporadic, subhor-

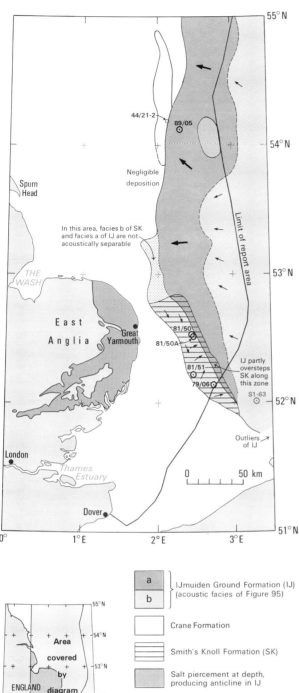

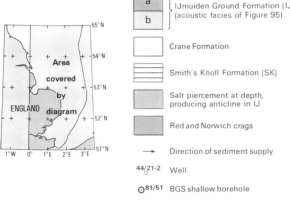

Figure 97 Distribution and acoustic facies of the IJmuiden Ground Formation and its correlatives. Adapted from Cameron et al. (1987).

Geology sheets), which occurred somewhat later than the beginning of the Cromerian Complex Stage (Figure 91). At about 54°N, the base of the formation is coeval with the reversely magnetised Aurora Formation (BGS Silver Well Quaternary Geology sheet). The polarity transition between 54°N and 56°N must therefore lie within the Yarmouth Roads Formation (Figure 94), a conclusion supported by somewhat fragmentary palaeomagnetic data from borehole 82/20. Hence, the age of much of the Yarmouth Roads Formation in the northern part of the report area can be established as Cromerian Complex (Figure 91). By early Cromerian Complex times, the whole southern North Sea area was thus a vast wetland complex of delta-top sediments extending in the UK sector to a shoreline in the vicinity of 55°N (Figure 103). Creation of this huge delta plain, which has been called Ur-Frisia, had taken some 1.6 million years (Jeffery and Long, 1989).

The uppermost parts of the Yarmouth Roads Formation have probably been eroded over much of the UK sector of the North Sea. Both the remaining parts and the missing upper parts are represented in East Anglia by fluvial and littoral deposits such as in the Kesgrave Formation (Zalasiewicz and Mathers, 1988). However, uncertainty about the length of time spanned by the boundary between the Kesgrave Formation and the underlying Norwich Crag Formation (time-equivalent to the Smith's Knoll/Westkapelle Ground formations) does not yet permit precise correlation of the base of the Kesgrave Formation with any particular level within the Yarmouth Roads Formation.

Shoreline changes in the early Pleistocene

To the north of the report area, there is seismic evidence of increasing marine influence on the deposition of the Yarmouth Roads Formation during the later part of the Cromerian Complex Stage (BGS Dogger and Swallow Hole Quaternary Geology sheets). Possibly this records the first decline in the growth of the Ur-Frisian delta. Certainly, a late Cromerian Complex basin-wide marine transgression reaching a similar geographical position to today's shoreline (Zagwijn, 1979) has been recognised in The Netherlands. In East Anglia, this transgression gave rise to the upper parts of the Cromer Forest Bed Formation (*sensu* Zalasiewicz and Mathers, 1988), that is, the lower members of the Cromer Forest Bed Formation *sensu* West (1980). Comparable deposits have not yet been identified offshore, even in the unglaciated parts of the Southern Bight, probably because of later erosion. The late Cromerian Complex transgression converted the delta plain of Ur-Frisia (Figure 103) into a sea which perhaps was never more than a few metres deep. An attractive explanation for the transgression is that the sediment supply to the still-subsiding basin was being gradually reduced. Zagwijn and Doppert (1978) have suggested that the reduction was a consequence of deteriorating climate in the source areas. Whatever the ultimate control, with subsequent regional sea-level fall caused by the trapping of water in extensive northern ice sheets, a ready-made, low-relief surface was available for the spread of Scandinavian and British ice sheets during the Elsterian (Anglian) Stage.

From the distribution of the early Pleistocene formations, and using the concepts of Figure 94, a picture emerges (Figure 104) of steady overall regression; any transgressions would have been local, resulting from differences in the balance between subsidence and sedimentation. The contemporaneous sea was generally warm-temperate (Long et al., 1988), and boreholes, especially in the Southern Bight (Cameron et al., 1984), record only upward-shallowing se-

M Gregory, written communication, 1980), a foraminifer last known from the Pre-Pastonian Stage (Tiglian — Gibbard et al., 1991) of East Anglia (Funnell, 1989).

Outside the report area, a little to the north of 56°N, the base of the Yarmouth Roads Formation lies just above the Brunhes–Matuyama transition from reversed to normal magnetic polarity (BGS Dogger and Swallow Hole Quaternary

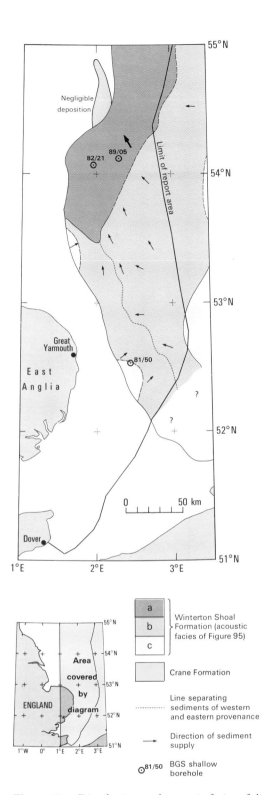

Figure 98 Distribution and acoustic facies of the Winterton Shoal Formation. Adapted from Cameron et al. (1987).

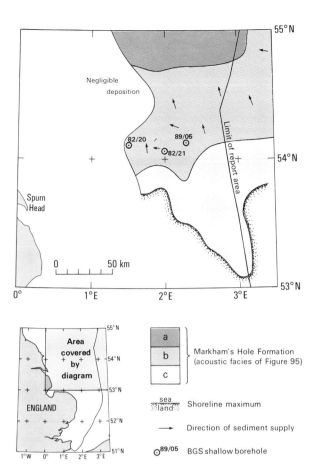

Figure 99 Distribution and acoustic facies of the Markham's Hole Formation. Adapted from Cameron et al. (1987).

quences. Glacial and glacioeustatic influences were evidently absent or insignificant. A similar series of depositional events is observable in the early Pleistocene of The Netherlands, where variations in pollen assemblages in deltaic sediments have been taken as proof of great climatic, and hence sea-level, changes (Zagwijn, 1989). From Figure 94, however, an initial conclusion based on the elevation of marine beds in East Anglia (West, 1980), is that although there was significant coastline migration in the early Pleistocene, regional sea levels remained similar to those of today.

NONDELTAIC DIVISION

Element C

Element C consists solely of the Swarte Bank Formation, the first unequivocal record of the invasion of ice into the Southern North Sea Basin. This formation fills a fan-like array of valleys (Figure 105), up to 12 km in width and 450 m in depth, that are cut into the Pleistocene deltaic and pre-Pleistocene strata (Cameron et al., 1987; 1989b). Boat shaped in plan, with an irregular thalweg, as if created by a series of giant scoops, they are generally considered to have been formed by subglacial meltwater under pressure (Ehlers et al., 1984; Boulton and Hindmarsh, 1987), although Wingfield (1990) has suggested a jökulhlaup origin. Because of their shape, the term 'scaphiform' is proposed as a descriptive, nongenetic term. In The Wash, these scaphiform valleys tend to be smaller and more amalgamated than farther offshore, probably as a result of more resistant substrata. Landwards, there is a gradual passage into the quasitabular geometry of the Anglian subglacial diamictons that are known collectively as the Chalky-Jurassic till.

The Swarte Bank Formation contains three members (Cameron et al., 1987, 1989b; Balson and Jeffery, 1991). The basal member has been penetrated by BGS boreholes in The Wash, and farther offshore in borehole 81/52A (Figure 105). It has been mapped as Chalky-Jurassic till (BGS East Anglia Quaternary Geology sheet). It comprises stiff, grey diamictons with, in some cases, lenses of coarse-grained glaciofluvial sand. It is overlain elsewhere by the spectacularly stratified and dominant middle member, which in boreholes in the Dutch sector consists of stiff, grey, glaciolacustrine muds

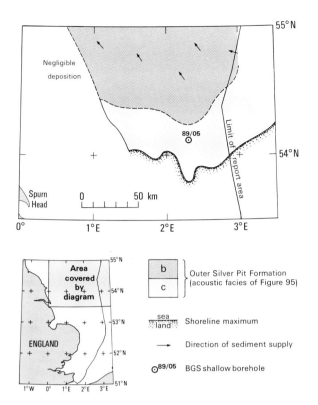

Figure 100 Distribution and acoustic facies of the Outer Silver Pit Formation.

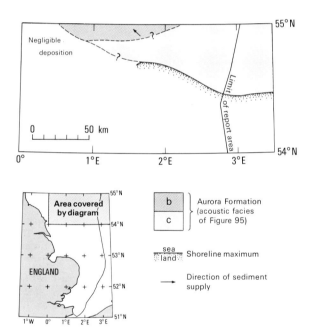

Figure 101 Distribution and acoustic facies of the Aurora Formation.

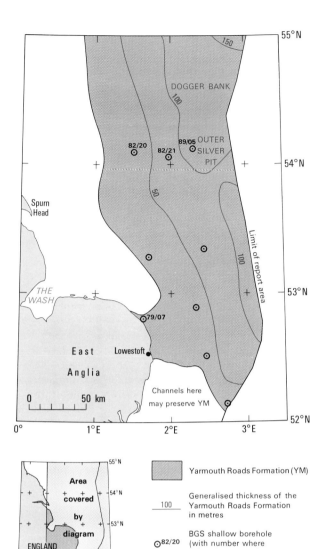

Figure 102 Distribution and generalised thickness, excluding effects of local erosion, of the Yarmouth Roads Formation.

(BGS Indefatigable Quaternary Geology sheet). BGS borehole 79/08 (Figure 105) records an upward transition from unfossiliferous, lacustrine clays into marine clays yielding a benthonic foraminiferal assemblage characteristic of waters so shallow that winter ice froze fast to the bottom (D M Gregory, written communication, 1980). The sporadic upper member comprises marine interglacial sediments (Jeffery, 1991).

Both from its stratigraphical position between the Cromerian Complex Yarmouth Roads Formation and sediments with interglacial Holsteinian fossil assemblages (Spurn Quaternary Geology sheet), and its apparent equivalence to the Elsterian Peelo Formation/Lauenburger Clay of The Netherlands and Germany (Ehlers et al., 1984), the Swarte Bank Formation is considered to be of late Elsterian to locally earliest Holsteinian age (Anglian — Gibbard et al., 1991).

The Swarte Bank Formation is the sole offshore relict of the Elsterian/ Anglian glaciation, and if the subglacial origin of the scaphiform valleys containing the formation is correct, then a line joining the present southern termination of each valley represents the minimum southern limit of a major still-stand of an ice sheet at that time. Such a line (Figure 105) agrees well with the inferred onshore limits of a late Anglian readvance which created the ice-pushed Cromer Ridge (Hart and Peglar, 1990). Grounded ice across the Southern Bight blocked northward-flowing drainage, and probably diverted it into the English Channel, creating a proto-Dover Strait (Gibbard, 1988). Meltwater drainage from the ice sheet must also have flowed into the English Channel, perhaps exploiting the same river channels (Hamblin et al., 1992).

Element D

With continued amelioration of the climate following the decay of the Elsterian ice sheet, rising sea level combined with continuing tectonic subsidence led, during the Holsteinian Stage, to the re-establishment of a shallow sea, the maximum limits of which lay partly landward of the present-day North Sea shorelines. The Sand Hole and Egmond Ground formations are the early and later deposits of that sea, which at its maximum development was probably deeper than the late Cromerian Complex sea.

The Sand Hole Formation is up to 20 m thick and is entirely confined to a bowl-like area around the Silver Pit (Figure 106). On seismic-reflection profiles it is characterised by closely spaced, parallel and even reflectors; borehole 81/52A (BGS Spurn Quaternary Geology sheet) returned 14 m of laminated clays which elsewhere in the vicinity have yielded abundant dinoflagellate cysts and a rich, diverse assemblage of interglacial, shallow-marine foraminifera (Fisher et al., 1969). It is inferred that the Sand Hole Formation was deposited during the early, warm part of the Holsteinian Stage in a quiet, restricted, marine environment.

This environment was subsequently displaced, through sea-level rise, by more open-marine conditions which allowed the widespread development of the Egmond Ground Formation (Figure 106), a lithologically variable deposit of locally gravelly sands interbedded with silt and clay (Cameron et al., 1989b), penetrated in the report area by a single borehole (81/52A). This formation is easily defined on seismic profiles by its persistent tabular geometry and, above all, by its conspicuous basal reflector which commonly truncates the upper parts of the Swarte Bank Formation, making it an excellent regional marker horizon. Some 8 m thick in the west and north, and some 20 m thick in the east, the formation is locally thicker in the tops of any incompletely filled Elsterian valleys. It contains typical shallow-water Holsteinian faunas,

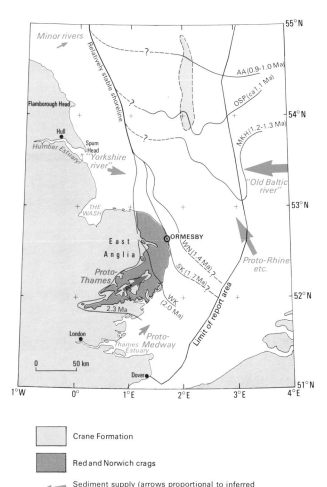

Figure 104 Inferred early Pleistocene shorelines. Speculative dates based on Cameron et al. (1984), palaeomagnetic data and unpublished reports. Shorelines on present-day land are interpreted from Mathers and Zalasiewicz (1985) and Harland et al. (1991).

and spores of the freshwater fern *Azolla filiculoides* Lambert, which became extinct in northern Europe after the Holsteinian. The faunas indicate a cool-temperate sea similar to northern parts of the present-day North Sea (Nilsson, 1983). Although the Holsteinian sea in the English Channel at its maximum invaded the Dover Strait (Hamblin et al., 1992), a connection of the North Sea to the Atlantic via the English Channel probably did not exist (Hinsch, 1985).

Element E

Two formations in this element, the Tea Kettle Hole and Cleaver Bank formations, record a reversion to glacially dominated sedimentation during the Saalian Stage.

The Cleaver Bank Formation (Figure 107) is a tabular body of stiff, laminated, dark grey clays with scattered angular granules of chert or chalk. The clays are intercalated with micaceous sands that are locally fine grained. It is only some 2 m thick in the west of the UK sector, and 8 m thick in the east. An arctic dinoflagellate-cyst assemblage is accompanied by abundant reworked Palaeogene cysts (R Harland, written communications, 1987). The formation is interpreted as a partly marine, proglacial diamicton of eastern provenance, a

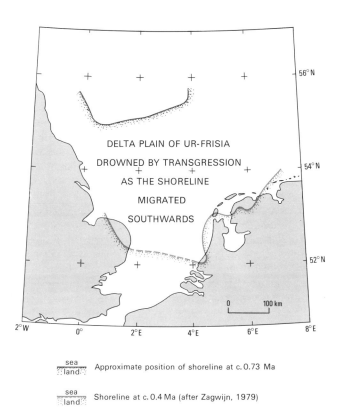

Figure 103 Inferred Cromerian Complex shorelines.

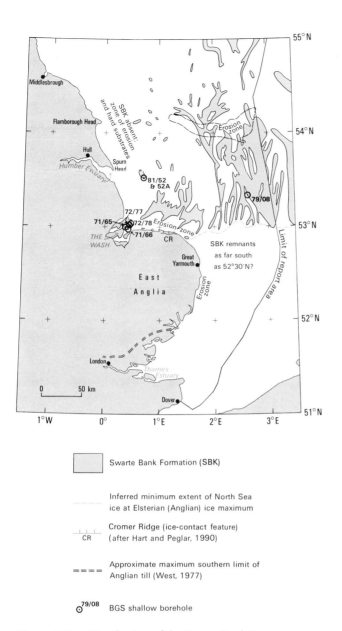

Figure 105 Distribution of the Swarte Bank Formation.

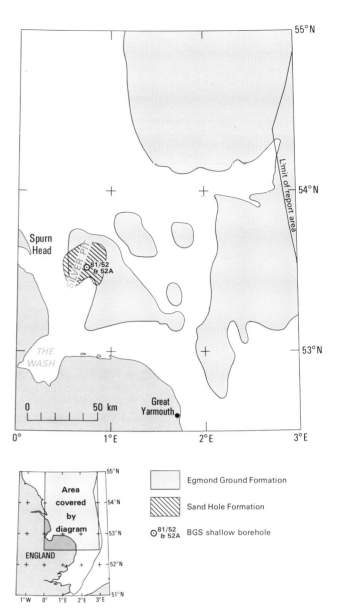

Figure 106 Distribution of the Egmond Ground and Sand Hole formations west of 3°E.

concept supported by its lateral transition east of 4°E into the subglacial, Saalian, Borkumriff Formation (Joon et al., 1990).

The sporadically distributed, periglacial, aeolian Tea Kettle Hole Formation (Cameron et al., 1989b) is so thin in the UK sector that its presence is generally undetectable in BGS seismic profiles. In boreholes in the Dutch sector it is up to 6 m thick and consists predominantly of fine-grained sands containing organic detritus. It is broadly contemporaneous with the Cleaver Bank Formation, and hence of Saalian age (Joon et al., 1990). It was probably originally much more extensive throughout the Southern North Sea Basin than it is now.

No unequivocal tills of Saalian age have been found in the report area, and the concomitant lack of scaphiform valleys suggests the area was never covered by a grounded ice sheet. Moreover, evidence in the Dutch sector (Joon et al., 1990) is for a Saalian ice front extending no more than 100 km northwest of the present-day Dutch coastline. In Britain, the very existence of Saalian subglacial deposits is disputed (Sumbler, 1983). The UK sector of the Southern North Sea Basin is therefore envisaged as having been a partly flooded, periglacial area during the Saalian glaciation, with the British ice sheet less extensive than either the earlier Elsterian or the succeeding Weichselian sheets (Balson and Jeffery, 1991).

Element F

During the Eemian Stage there was a return to higher sea levels, followed by a fall. This led to the development of element F, the base of which is locally marked by a gently undulating seismic reflector that is less well defined than the basal reflector of the Egmond Ground Formation. Element F consists of the Eem and Brown Bank formations (Figure 108), which in common with the Egmond Ground and Cleaver Bank formations do not occur to the north of the Dogger Bank owing to erosion during the subsequent Weichselian glaciation. In Figure 108, the Eem Formation is less widespread than shown on the BGS Indefatigable and Flemish Bight Quaternary Geology sheets, because the continuous seismic unit interpreted as Eem Formation on the western side of those sheets has been re-interpreted as largely Egmond Ground Formation (BGS Spurn Quaternary Geology sheet). The Eem Formation, if present, rests directly on the Egmond Ground Formation as discontinuous patches which are difficult to separate either lithologically or acoustically. Where it has been positively identified, the formation consists of up to 20 m of shelly sands passing westwards into muddy sands and muds of a more intertidal aspect.

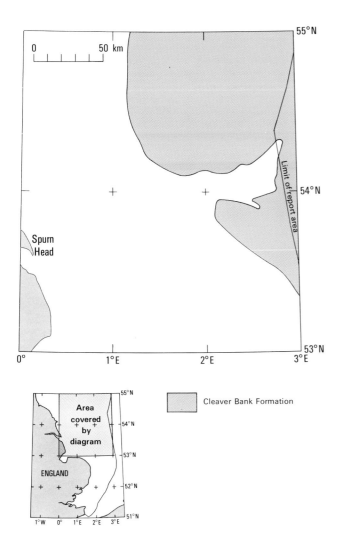

Figure 107 Distribution of the Cleaver Bank Formation west of 3°E.

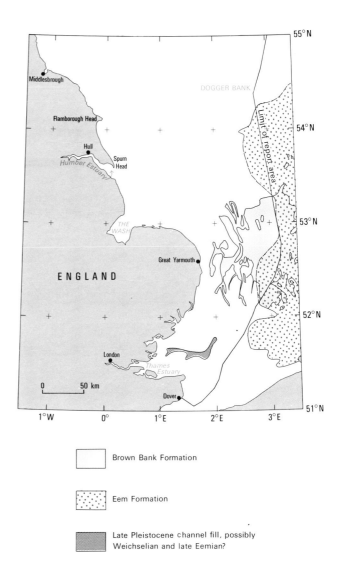

Figure 108 Distribution of the Eem and Brown Bank formations.

The Eem Formation is succeeded gradually in some places, sharply in others (Cameron et al., 1989b), by the late Eemian to early Weichselian brackish-water silty clays of the Brown Bank Formation (Figure 108). Generally 5 m thick, this latter formation is characterised by conspicuous sedimentary lamination, poverty of microfauna, and abundant bioturbation (Cameron et al., 1989b). A western marginal facies, developed in a series of north–south channels 2 to 15 km wide and up to 20 m deep (Figure 108), is sandier and lacks bioturbation. The formation records the former presence of a shallow, restricted, brackish lagoon that, in the UK sector, was supplied with sediments from the south-west. The lagoon resulted from a sea-level fall early in the Weichselian. Complete severing from marine influence is demonstrated by the presence in the Dutch sector of finely laminated lacustrine sediments at the top of the formation (Cameron et al., 1989b). By this stage, the Rhine and its tributaries had already ceased to flow into the North Sea Basin, having been diverted westwards down the English Channel (Oele and Schüttenhelm, 1979).

Element G

Reversion to glacially dominated sedimentary regimes in the closing stages of the Pleistocene is recorded by elements G and H.

Element G chronicles the growth, expansion and initial decay of ice sheets during regionally lowered sea levels of late Weichselian times. It consists largely of two laterally equivalent formations, the Bolders Bank and Dogger Bank formations (Figures 91 and 109). Minor components, commonly discernible only in seismic profiles of very high resolution, or in boreholes, are the Twente, Well Ground and Kreftenheye formations.

The Bolders Bank Formation typically has a chaotic to poorly ordered internal seismic-reflector configuration. A large suite of cores and BGS boreholes, notably 81/41, 81/43, 81/48, 81/52 and 81/52A, indicate that it characteristically consists of reddish to greyish brown, stiff diamictons that are generally massive but in places possess distinct, commonly arenaceous layering, and deformational structures. The majority of its pebbles, of which chalk is the most conspicuous component, are derived from the sedimentary rocks of eastern England. The pebble content tends to diminish eastwards. In general, the formation is less than 5 m thick, and commonly less than 1 m is preserved in deep-water areas west of the Dogger Bank, although to the east of Lincolnshire it may be 15 to 20 m thick (BGS Spurn Quaternary Geology sheet). Seismic-profile interpretation and a recent micromorphological study (Van der Meer and Laban, 1990) imply the formation is a composite of subglacial and supraglacial deposits. It resembles, and is almost in continuity with, the diamictons of Hunstanton and Holderness (BGS Spurn and East Anglia Quaternary Geology sheets), the latter perhaps thrust up in an ice-pushed ridge (Figure 110).

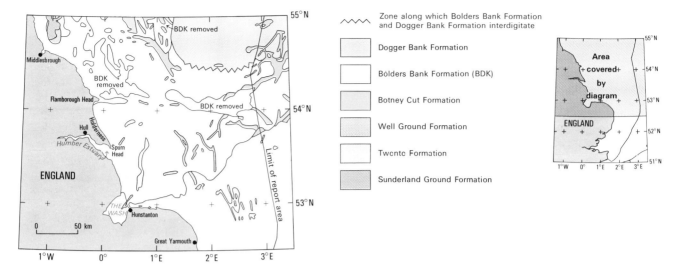

Figure 109 Distribution of deposits of the last glaciation (late Weichselian).

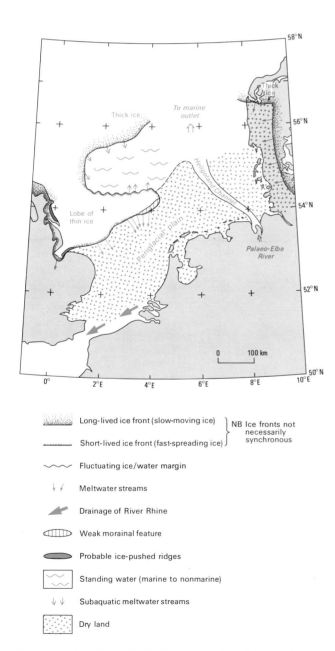

Figure 110 Generalised palaeogeography of the southern North Sea at the Weichselian ice maximum. Based on Eisma et al. (1979), BGS/RGD mapping, and Nilsson (1983).

The base of the formation is defined by a high-amplitude, remarkably flat or gently undulating reflector which is interrupted in places by the basal reflector of element H. Off Holderness, this base lies at about 32 m below sea level (BGS Spurn Quaternary Geology sheet), but slopes gently eastwards to a depth of about 55 m in the axis of the basin where it rests on Eemian or older sediments (BGS Silver Well Quaternary Geology sheet). The slope is probably largely due to tectonic subsidence.

Under the Dogger Bank, this basal seismic reflector continues uninterrupted; above it is an extensive, tabular depositional unit up to 42 m thick whose internal reflectors are better ordered than those of the Bolders Bank Formation. These features are suggestive of a proglacial, water-laid body. This unit is known as the Dogger Bank Formation; it interfingers with the Bolders Bank Formation and largely controls the morphology of the Dogger Bank, forming a high-standing core over and around which have been deposited Holocene sands reworked from glacial deposits. The formation is composed principally of clay-rich diamictons in which the pebbles are smaller, fewer, and less varied than in the Bolders Bank Formation, and well-developed stratification and lamination are more common. Reworked pre-Pleistocene dinoflagellate cysts in the top layers of the formation are mixed with indigenous cysts indicative of severe, cold, open-water, marine conditions (R Harland, written communication, 1987), although the lower parts may be glaciolacustrine in origin (BGS Silver Well Quaternary Geology sheet).

The shallow waters in which the Dogger Bank Formation was deposited probably occupied an ice-marginal fosse intermittently connected to a shrunken North Sea via a continuation of the Heligoland Channel (Figure 110). Farther south by contrast, the Weichselian ice front at its maximum extent lay on dry land. Meltwaters at various stages deposited the Well Ground Formation (Cameron et al., 1989b), which now forms discrete patches up to 5 m thick lying under, or around the limits of, the Bolders Bank Formation. Winds blowing around and over the ice sheet produced the Twente Formation, now preserved to the south of the Bolders Bank Formation in the UK sector as scattered outliers less than 1 m thick (Cameron et al., 1989b).

During the late Weichselian interval of ice invasion, the Rhine and its tributaries continued to flow down the English Channel, depositing the locally gravelly sands of the 8 m-thick, fluvial, Kreftenheye Formation in parts of the Southern Bight outside the UK sector (Cameron et al., 1989b).

Element H

Evidence of the final stages of decay of the late Weichselian ice sheets is preserved in element H. The widespread Botney Cut Formation occurs mainly in scaphiform valleys (Figure 109) within, and disposed roughly radially to, the original outer limits of the Bolders Bank Formation. In the report area, these scaphiform valleys are generally less than 100 m deep and less than 8 km wide; they are therefore shallower and smaller than those of the Elsterian glaciation, but are thought to have been formed in the same way. Their smaller dimensions are probably attributable to thinner ice. Several BGS boreholes indicate that a lower member of the Botney Cut Formation comprises stiff, reddish brown diamicton very like that of the Bolders Bank Formation, and that an upper member is composed of soft to stiff, laminated, sporadically pebbly glaciolacustrine to glaciomarine muds, commonly with the same distinctive reddish brown tints as the Bolders Bank Formation. This upper member also occupies wide, shallow, irregular hollows of uncertain genesis in the top of the glaciomarine Dogger Bank Formation (BGS Silver Well Quaternary Geology sheet).

Direct action of ice, perhaps augmented by dissolution, is more readily invoked to explain a trough some 75 km long, 15 km wide and up to 80 m deep, cut into weak Zechstein Group evaporites east of Sunderland. In this trough, up to 25 m of soft, reddish brown, proglacial, water-laid muds of the Sunderland Ground Formation (Figure 109) were deposited over the Botney Cut Formation as the ice front receded westwards in very late Weichselian times.

With rising sea levels offset by isostatic rise of land, deposition of glaciomarine muds in troughs and scaphiform valleys gave way in the early Holocene to the intertidal muds, silts and peats of the Elbow Formation (see Figure 116) at the base of element J (Figure 91), heralding the return to the fully marine conditions of the present, and the reversion of the mouth of the Rhine to its present position.

13 Holocene

Some of the earliest scientific investigations of Holocene sediments in the southern North Sea were made on samples collected from the nets of trawlers (e.g. Whitehead and Goodchild, 1909; Stather, 1912), but later, samples were taken specifically for fisheries research into spawning grounds (e.g. Borley, 1923; Davis, 1925). The development of the echosounder led to the first studies of bedforms (e.g. van Veen, 1935), since when commercial development has resulted in numerous studies of the deposits on, and just below, the sea floor. Early maps of the distribution of sediments on the floor of the southern North Sea were produced by Borley (1923), Lüders (1939) and Jarke (1956), although more rudimentary maps had been made earlier (e.g. Tesch, 1910). Since 1968, BGS has collected over 5000 sea-bed samples in the report area; these have been been combined with other data to produce maps at a scale of 1:250 000. A simplified version of the sea-bed sediment distribution is shown in Figure 111. Despite this wealth of data, relatively little is known either about the earliest Holocene sediments, which are locally buried under several metres of more recent deposits, or about certain nearshore areas that are too shallow for most survey vessels.

Holocene sediments in the report area generally form a thin veneer over Pleistocene or older formations. Exceptions are the thick coastal accumulations in areas such as the Humber, Wash and Thames estuaries, and where there are tidal sand ridges (sandbanks). The present-day bathymetry of the North Sea thus approximates to the morphology of the pre-Holocene land surface, but substantially modified in certain areas by sediment accretion, and elsewhere by erosion, particularly in coastal regions. The bathymetry shows a relatively flat surface with these Holocene modifications; water depths are generally between 20 and 40 m, deepening in the north-west to between 60 and 80 m (Figure 112).

Water depths shallower than 20 m are found around the coast and where there are major sandbanks, some of which have exposed crests at low tide. The Dogger Bank in the north-eastern part of the area is a large, broad, conspicuously shallow region with water depths of as little as 11 m near its south-western limit. The Dogger Bank is formed mostly of Pleistocene sediments, although early Holocene sediments are up to 10 m thick around part of its margin.

There are a number of elongate erosional deeps, the largest of which is the east–west-orientated Outer Silver Pit along the southern margin of the Dogger Bank. It is up to 15 km wide, over 100 km long, with water depths of up to 95 m. The Silver Pit, although smaller, is up to 98 m deep. These deeps form a radial pattern on the sea floor (Figure 112), and are believed to relate to the Weichselian ice limit (Balson and Jeffery, 1991). They may have formed by outbursts of glacial meltwaters normal to the ice margin (Wingfield, 1990), although modification by tidal currents during the early Holocene is likely (Donovan, 1973).

Tidal currents within the report area are locally strong. Average surface velocities at spring tide vary from less than 0.5 m/s in the north-east, to 1.5 or even 2 m/s in certain nearshore areas (Figure 113). Near-bottom velocities are lower, but only by a small amount, as the water is generally shallow. Measurements between the Sandettie and South Falls banks show near-bottom values up to approximately 1 m/s in a zone of active sand waves (Smith, 1988a).

Amplification of the tidal wave in the semienclosed southern North Sea means that large tidal ranges are experienced along the eastern coast of England. On the open coast, the maximum tidal ranges occur between the Humber and The Wash; at Skegness the mean spring range is 6.1 m and the mean neap range is 3.0 m. Minimum ranges occur on the East Anglian coast near Great Yarmouth and Lowestoft, where the mean spring tidal range is only 1.9 m. These variations in tidal ranges affect the width of the intertidal zone along the coastline.

Although the water is shallow enough for wave-induced currents with sufficient strength to move bottom sediments, particularly during northerly storms, these are generally of minor importance when compared to the stronger and more regular tidal streams (Caston, 1976; Morton, 1982), and are most effective only for the few hours when the tide and wind-induced currents act in the same direction (Jago, 1981). Currents may also be generated by seaward flow after major storm-surge events; core evidence from the German Bight suggests deposition from storm-induced currents (Morton, 1982).

THE HOLOCENE SEA-LEVEL RISE

As the Weichselian glaciers began their retreat, sea level began to rise from a eustatic low of around 120 m below present-day levels during the glacial maximum (Fairbanks, 1989). Regional sea levels were approximately 65 m below present at the beginning of the Holocene, 10 000 years ago (Figure 114), when the southern coast of the North Sea had probably only just encroached (Figure 115) into the report area (Jelgersma, 1979). Much of the area north of a line drawn approximately from The Wash to the southern edge of the Dogger Bank had been covered by lodgement tills or glacial outwash sands and gravels left behind during the retreat of the Weichselian ice sheets (Balson and Jeffrey, 1991). To the south, periglacial aeolian sands had been deposited. In late Weichselian to early Holocene times, fluviatile sands may have been deposited by rivers flowing across this land surface from eastern England or the European mainland.

Reconstructions of early Holocene sea levels in the southern North Sea are based mainly on radiocarbon dating of former coastal peat beds. Almost all of the data has come from peats now on land or in estuaries. In eastern England, peats have been dated from the Humber Estuary (Gaunt and Tooley, 1974), the fens surrounding The Wash (Shennan, 1986), the north Norfolk Coast (Funnell and Pearson, 1989), the Norfolk Broads (Shennan, 1987) and the Thames Estuary (Devoy, 1979; Greensmith and Tucker, 1971; 1973). Only a few peat samples from the offshore area have been dated (Godwin, 1960; Kirby and Oele, 1975), and because the North Sea was still some distance from its present coastline at the beginning of the Holocene, the earliest Holocene record is relatively poorly known. Jelgersma (1961; 1979) used data from Holocene deposits in The Netherlands and offshore to construct a Holocene sea-level curve for the North Sea (Figure 114). She used this curve, with known present-day

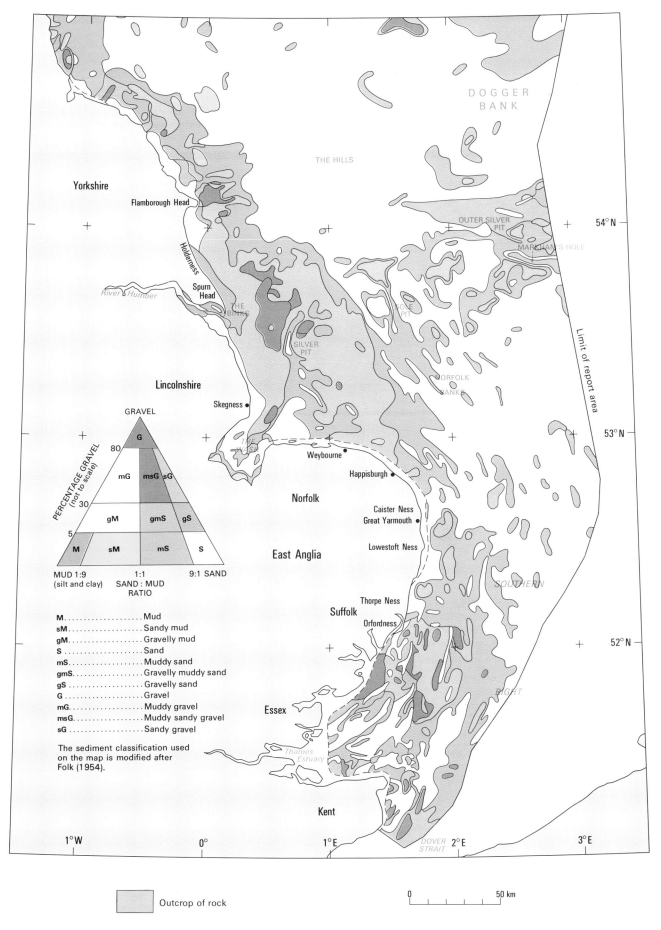

Figure 111 Sea-bed sediment distribution.

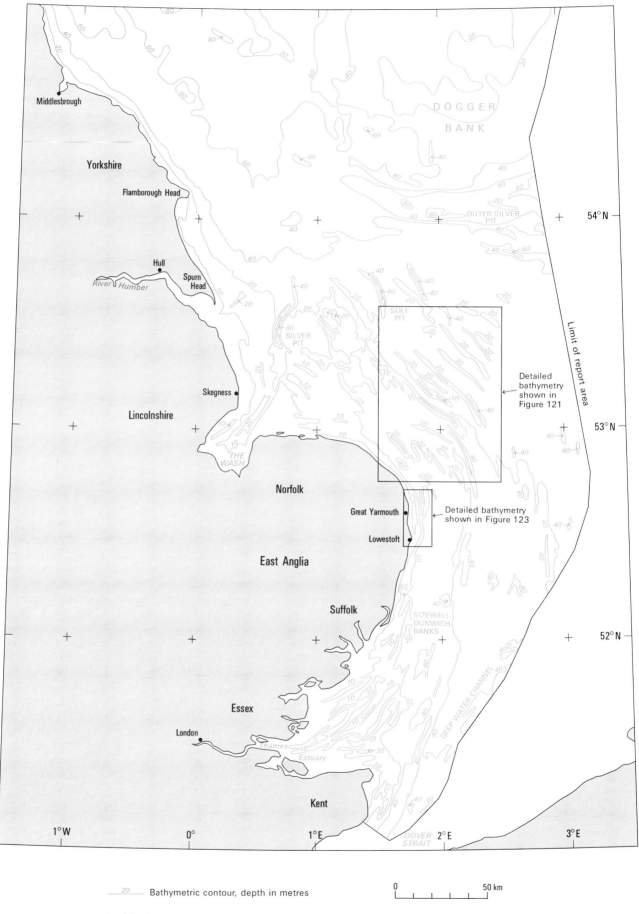

Figure 112 Generalised bathymetry.

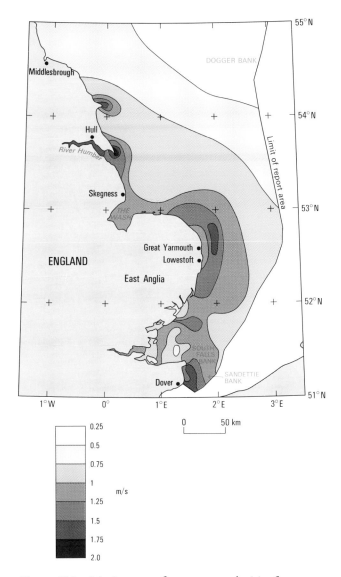

Figure 113 Maximum surface-current velocities for mean spring tides. After Sager and Sammler (1975).

bathymetry, to construct palaeocoastlines for various times throughout the Holocene (Figure 115).

Shells of molluscs typical of tidal flats are found at sites throughout the southern North Sea (Veenstra, 1965). Radiocarbon and stable-isotope analyses of such shells from sea-bed samples have been used to reconstruct the early Holocene palaeoenvironments (Eisma et al., 1981). The oldest radiocarbon ages found by Eisma et al. (1981) were between 9370 and 8260 BP in the Deep Water Channel area (Figure 115), which was initially flooded by the sea transgressing through the Dover Strait. Salinities at this time were brackish, around 13 parts per thousand. Younger samples indicate that the remainder of the southern North Sea became brackish between about 8000 and 7000 years BP, followed by fully marine conditions everywhere except in the Thames Estuary. Early Holocene mean water temperatures were around 10°C compared with 13.5° to 14°C at present. The flooding of the southern North Sea coincided with a sharp increase in average air temperatures in central England between 10 000 and 7000 years BP (Eisma et al., 1981).

In the early stages of the transgression, the Southern Bight contained a quiet, shallow sea of low tidal range. Once linkage between the English Channel and the North Sea was established (around 8300 years BP, Figure 115), there was probably a rapid transition to strong tidal currents and a large tidal range. Mathematical modelling suggests that this transition occurred when sea levels were approximately 10 to 15 m below present (Austin, 1991). This increase in tidal currents may have led to significant erosion of early Holocene tidal-flat sediments (Stride, 1989).

A simple picture of sea-level rise after the melting of Weichselian ice sheets is complicated by the effects of regional uplift and subsidence. In Scotland and northern England, isostatic uplift was responsible for the preservation of Holocene raised beaches, whereas in the southern North Sea region there has been regional subsidence of up to 2 mm/year during the last 4000 years (Shennan, 1989). Furthermore, detailed studies of sea-level change over the last 5000 years show a number of small oscillations (Devoy, 1979) rather than the smooth progressive rise shown in Figure 114; these may be linked to climatic fluctuations. For instance, palaeoecological and sedimentary changes at many coastal sites show the influence of a possibly worsening climate, and greater freshwater throughput in estuaries, after 3000 years BP (Devoy, 1982).

Sediment input from rivers may have varied through the Holocene, reducing as vegetation cover increased. More recently, man's activities have had a significant effect on Holocene sedimentation through tidal-flat and saltmarsh reclamation, coastal engineering projects and flood defences, farming practices, and offshore aggregate exploitation.

HOLOCENE SEDIMENTS

Dutch workers have given formation status to various early Holocene deposits in the Dutch sector of the southern North Sea. Where appropriate, this nomenclature has been used in the UK sector on BGS sea-bed sediment maps.

The Elbow Formation (Oele, 1969) was deposited during the early stages of the Holocene transgression; it consists of fine- or very fine-grained muddy sands with interbedded clay layers, and contains a mollusc assemblage characterised by *Spisula subtruncata*. A brackish-marine clay beneath the sands locally overlies a basal peat bed. The formation is mostly between 2 and 6 m thick, with a maximum thickness of 12 m; it is distributed widely within the Dutch sector but only occurs extensively in the eastern part of the report area (Figure 116). Similar nearshore and intertidal deposits with peat layers are found at sites around the eastern England coast close to the present coastline, such as off Lincolnshire (Swinnerton, 1931) and Suffolk. Peats recorded farther offshore near Leman Bank (Figure 116) yielded a radiocarbon age of 8425 ± 70 BP (Godwin, 1960), and contain a Mesolithic bone harpoon (Godwin and Godwin, 1933). Peat samples from the Deep Water Channel in the Southern Bight (Figure 115) gave even older ages of 9374 ± 90 and 9949 ± 125 BP (Kirby and Oele, 1975).

The Buitenbanken Formation is distributed mostly south of 53°N and traces laterally into, but may be locally younger than, the Elbow Formation. The formation consists of medium-grained, partly fine- or coarse-grained sand that is locally gravelly. It is mostly less than 2 m thick, with a maximum of 5 m, and commonly contains reworked Eemian (Ipswichian) or early Pleistocene mollusc shells.

The Well Hole Formation consists of locally laminated, marine, fine-grained sands and sandy muds that fill Pleistocene glacigene depressions. These deposits are usually between 5 and 20 m thick, but may be up to 25 m. In depressions, the formation unconformably overlies the Pleistocene Botney Cut Formation.

The Nieuw Zeeland Gronden Formation consists of two laterally equivalent members that occur north of 54°N. The

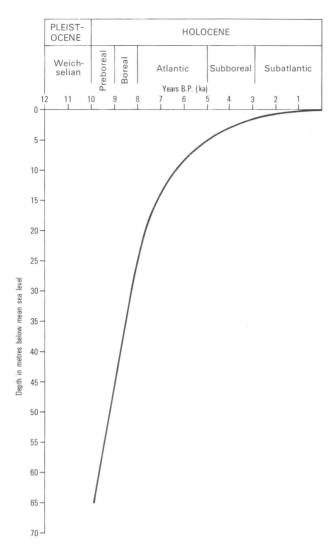

Figure 114 Generalised curve of relative sea level during the Holocene in the North Sea region. After Jelgersma (1979).

marine deposits are silty or sandy, and derived mainly from reworked periglacial or glacial deposits. The Terschellingerbank Member is a slightly muddy, fine-grained sand containing a sparse marine mollusc fauna. It is between 1 and 10 m thick, and is extensively distributed over and around the Dogger Bank. The Western Mud Hole Member consists of muddy, fine- or very fine-grained sands with a marine mollusc fauna including abundant *Turritella communis* Risso. This member is up to 7 m thick, but is restricted almost entirely to the region east of the median line.

The Indefatigable Grounds Formation consists of gravelly sand and sandy gravel that form a thin veneer over Pleistocene till deposits. It is generally less than 2 m thick, but is up to 5 m thick in places.

The Bligh Bank Formation comprises marine, fine- to medium-grained sands which form the 'mobile' marine sediment over much of the modern sea floor. The thickness varies from less than 1 m to over 35 m within the tidal sandbanks. This formation comprises most of the 'Young Sea-sands' of Dutch literature.

BEDFORMS

Three important bedforms are found on the floor of the southern North Sea. Sand ribbons and sandbanks are longitudinal bedforms parallel or subparallel to the dominant tidal flow, whereas sand waves are flow-transverse bedforms.

Sand ribbons

Sand ribbons vary greatly in size, but have lengths up to the order of several kilometres, and widths of tens of metres. Their thickness is always small, from a few grains up to about a metre (Belderson et al., 1982), and therefore they are best seen on sidescan sonar records, being generally too thin to be identified on seismic profiles. They may be composed of trains of small sand waves resting on a scoured bedrock or gravel pavement, and they commonly grade laterally into more sheet-like sands with large sand waves. They are therefore characteristic of areas of strong, rectilinear tidal currents with a limited supply of sand. Within this report area, sand ribbons have been recorded in the Dover Strait, off north-eastern East Anglia (McCave and Langhorne, 1982) and off Flamborough Head (Belderson et al., 1971; BGS Tyne-Tees and California Sea Bed Sediment sheets).

Other longitudinal bedforms within the area include obstacle marks formed by sand accumulation in the lee of wrecks (G F Caston, 1979).

Sand waves

Sand waves are conspicuous over large areas of the southern North Sea (Figure 117) both in the Southern Bight (Van Veen, 1938; McCave, 1971b; Terwindt, 1971) and to the south-west of the Dogger Bank (Dingle, 1965; G F Caston, 1979). They occur in water depths between 18 and 60 m. Their absence in water shallower than 18 m is attributed to the effects of storm-wave activity (Terwindt, 1971; McCave, 1971a), but in sheltered nearshore areas they occur within the intertidal zone. Many occur on the flanks of large tidal sand ridges but, where sand supply is sufficient, large areas of the sea floor may also be covered in sand waves. Their height and wavelength vary greatly; the largest at 16 m are found at the southern end of the Sandettie (Figure 118) and South Falls sandbanks, where the strongest and most prolonged tidal streams in the Dover Strait occur (Smith, 1988a). In the sand-wave field in the Southern Bight west of The Netherlands, the significant height (average of the largest third) of sand waves is over 7 m (McCave, 1971b). South-west of the Dogger Bank, the largest sand waves are approximately 4.5 m in height (Dingle, 1965).

Wavelengths vary between 50 and 500 m. In some areas where sand supply is limited, as for instance in the area north of Norfolk (BGS Spurn Sea Bed Sediment sheet), isolated sand waves may rest on a gravel-lag pavement.

Large sand waves may be covered by smaller sand waves (Terwindt, 1971; Langhorne, 1973) with wavelengths of 1 to 2 m, and crests orientated up to 60° relative to the main sand-wave crest (Langhorne, 1973). This may indicate that the crests of the large sand waves are oblique to the tidal flow, or they may indeed be relict. Migration of sand waves is believed to be in the direction in which their steeper, lee-slope faces, and to be in the direction of the dominant sand transport (Figure 119). Maximum lee slopes are of the order of 10° to 12°, with stoss slopes of 1° to 3° (Terwindt, 1971). It is not known how rapidly these bedforms migrate. In the southern North Sea, Terwindt (1971) was unable to detect movement within the inaccuracies of the navigation over a period of 3 years. Similarly, Burton (1977) found no systematic migration of the very large sand waves at the south-western end of Sandettie Bank. McCave (1971b) has estimated a migration rate of 15 m/year for southern North Sea sand waves.

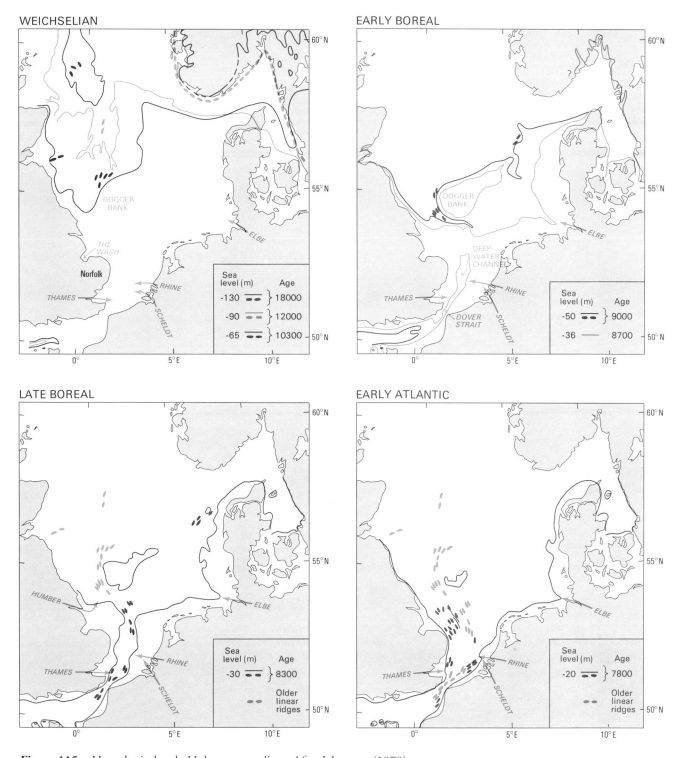

Figure 115 Hypothetical early Holocene coastlines. After Jelgersma (1979).

Sand waves in the southern North Sea generally consist of fine- to medium-grained sand. Very few studies of grain-size variation over the surface of a single sand wave have been performed in subtidal areas, but Houbolt (1968) showed a coarsening trend towards the crest of a sand wave off the Dutch coast. Using samples collected by divers, Terwindt (1971) found that whereas some sand waves show a similar coarsening towards the crest, others showed no variation, or had a fining trend towards the crest. The grain-size trend across a large sand wave may be complicated by individual trends within small sand waves migrating on its flanks.

In the gap between the southern end of the South Falls and Sandettie sandbanks, there are sand waves comprised of, or armoured with, coarse-grained sediments (Smith, 1988a; 1988b). These 'gravel waves' have characteristically very steep lee slopes of 35° to 41°, and are therefore believed to be active bedforms in an area where near-bottom current velocities reach between 0.7 and 0.82 m/s at spring tides (Smith, 1988b).

Sand waves on the flanks of sandbanks usually have their crests aligned at right angles to the ridge crest, particularly on the lower parts of the flanks, with a tendency to 'turn' towards the crest on higher parts of the bank. The direction of transport as indicated by the steep faces of the sand waves may be opposite on opposing sides of the banks, as on the North and South Falls (Houbolt, 1968; Caston, 1981). Around the ends of banks, sand waves of opposing asymmetry (Figure 118) may be separated by a zone of symmetrical for instance in The Wash (Evans, 1965; Wingfield et al., 1978; McCave and Geiser, 1978). These intertidal sand waves are much smaller in height than the large offshore bedforms. In a tidal inlet at Wells-next-the-Sea on the north Norfolk coast (Figure 117), intertidal sand waves with a height of less than 0.4 m have a threshold of movement of 0.6 m/s, and consequently move only during the highest spring tides (Allen and Friend, 1976).

Sandbanks (tidal sand ridges)

These are the largest bedforms on the continental shelf, and the sandbanks of the southern North Sea include some of the foremost examples in the world. The sandbanks in the report area are found to occur mostly parallel to one another in groups (Figure 120), the best known of which are the Norfolk Banks, off north-east Norfolk. Two other important groups are the Sand Hills between north-east England and the Dogger Bank, and the banks of the Thames Estuary region. Other isolated banks occur mostly in nearshore areas. The origins of the banks are, to a large extent, linked to the history of Holocene sea-level rise and the evolution of the Holocene coastline.

Sand Hills

The Sand Hills group probably began to form around 9000 years BP (Jelgersma, 1979), when the sea inundated the low ground between the Dogger Bank and the mainland of England (Figure 115). It is likely that at that time powerful tidal currents acted through a narrow strait, the 'Strait of Dogger' (Kenyon et al., 1981), which eventually isolated the Dogger Bank as an island (Figure 120). The Outer Silver Pit is an erosional relict of this strait. These tidal currents swept the available sand-sized sediment into sandbanks parallel or subparallel to the early Holocene coastline. As the transgression proceeded and water depths increased, current velocities diminished.

The present-day Sand Hills (Figure 120) are 12 to 21 m high, with symmetrical cross-sections (Dingle, 1965). The banks lie on a surface at around 50 m depth, and the crests of the banks lie at 25 to 35 m depth. To the west, the banks are mantled by large sand waves whose presence is thought to indicate that they are being maintained under present conditions, and are therefore active (Kenyon et al., 1981). However, in the east, the banks have smooth surface profiles devoid of sand waves; this is believed to indicate that the local currents are no longer able to maintain the banks, which are therefore considered to be moribund (Kenyon et al., 1981). South West Spit, the largest bank in the Sand Hills group, has sand waves on its western end but not on its eastern end, despite being relatively shallow in the east. This is perhaps due to the local current conditions related to the proximity of the Outer Silver Pit and Dogger Bank. To the north-west of the Dogger Bank, the East Bank Ridges (Figure 120) are sandbanks in water depths of 50 to 60 m, and are considered to be moribund.

Norfolk Banks

The Norfolk Banks rest on a relatively flat surface at 20 to 30 m depth (Figure 121). They consist of accumulations of sand on an erosional surface of Pleistocene deposits that in the interbank areas is either exposed or covered by a thin lag gravel. The largest of the Norfolk Banks, Well Bank, is over 50 km long, 1.7 km wide and rises to 38 m above the adjacent sea floor, although some banks are even higher at over 42 m (Caston, 1972).

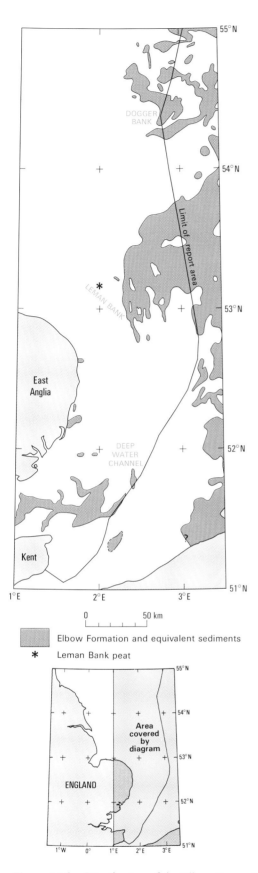

Figure 116 Distribution of the Elbow Formation and equivalent sediments.

sand waves (McCave and Langhorne, 1982; Caston, 1981). On nearshore sandbanks such as the Goodwin Sands, sand waves may be exposed at low water (Cornish, 1901).

Sand waves also occur within the intertidal zone, where sand-flat sediments are influenced by strong tidal currents, as

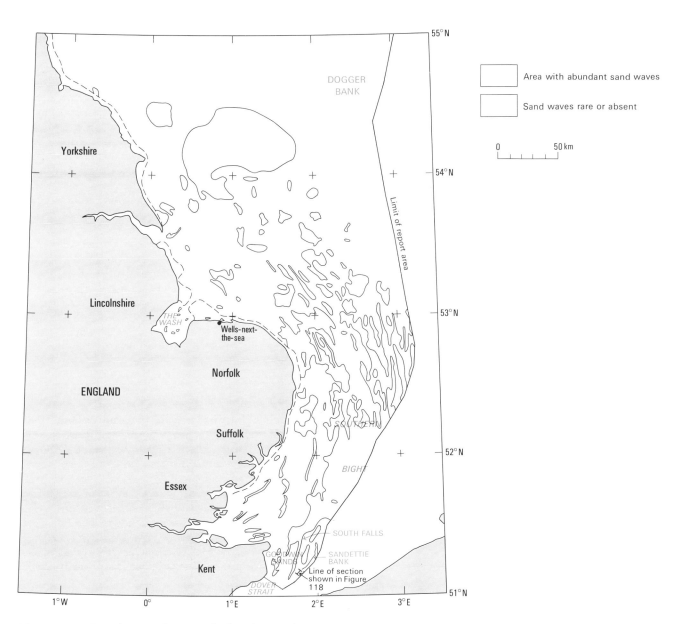

Figure 117 Distribution of areas with abundant sand waves.

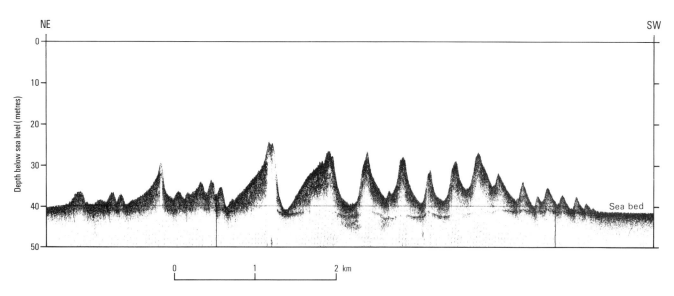

Figure 118 High-resolution seismic profile across the southern tip of Sandettie Bank, showing large sand waves with opposing asymmetry. For location see Figure 117.

The Norfolk Banks can be subdivided into a group of more nearshore, parabolic banks connected by low cols to form a zig-zag pattern; and an outer group of more linear banks. The innermost banks generally have sand waves on their flanks, the outermost banks tend to have a smooth profile. The parabolic form exhibited by the nearshore banks is believed by Caston (1972) to be a stage in bank development where a single bank may eventually split to form further banks. These banks are more active than the offshore linear banks, perhaps in part due to the greater tidal-current velocities closer to the coast. Many of the Norfolk Banks probably originated at around 7800 years BP (Jelgersma, 1979), although it is likely that the nearshore banks are more recent.

Net sand transport is in opposite directions on opposite sides of banks, as evidenced by the direction of asymmetry of sand waves on their flanks. Nevertheless, there is a dominant north-westerly transport direction (Caston and Stride, 1970). The linear banks are generally asymmetrical towards the north-east, with the steeper slopes up to 7°, and more gentle south-westerly slopes (Figure 122). Internal seismic reflectors within some of the banks dip to the north-east (Houbolt, 1968) (Figure 122), indicating a migration in that direction which may, or may not, be continuing at present. These internal reflectors probably represent boundaries between cross-bedded sets of sediments (e.g. McCave and Langhorne, 1982).

By studying hydrographic charts dated between 1851 and 1967, Caston (1972) found that many of the banks have elongated towards the north-west, the direction of net sand transport. Caston (1972) also found that during the same period, some of the more nearshore banks, such as Haisborough Sand, the Hewett Ridges and Smith's Knoll, had correspondingly migrated hundreds of metres to the north-east. The central part of Ower Bank had apparently migrated over 700 m to the south-west. However, after a similar study of hydrographic charts published between 1886 and 1950, McCave and Langhorne (1982) could detect no movement of Haisborough Sand to the north-east. The evidence for modern movement of these banks perpendicular to their long axes must therefore be regarded as equivocal; the evidence for migration in the direction of their long axes is better.

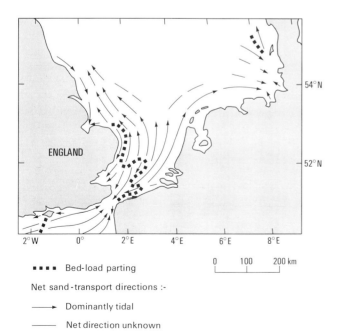

Figure 119 Net sand-transport directions in the southern North Sea region. After Johnson et al. (1982).

The banks to the east of Great Yarmouth (Figure 123) have shown measurable displacement almost entirely parallel to the coastline in the direction of their long axes. Since 1866, South Cross Sand may have moved several hundred metres, and the Scroby Sands over 1 km, to the north (Craig-Smith, 1972). South Cross Sand shows dipping internal seismic reflectors which reflect this migration (Figure 124). On the other hand, Corton Sand has moved over 1 km to the south during the same period. These nearshore banks are separated by channels which have been scoured to a level beneath the base of the bank sediments; Scroby Sands is effectively isolated on an elongate high. Similar highs are seen beneath some of the Thames Estuary banks, where D'Olier (1981) believed that such features predate bank formation. In the case of Scroby Sands, it is likely that at least some of the scouring is modern, and similar hollows interpreted as due to tidal scour occur adjacent to more linear Norfolk Banks (Donovan, 1973). These scoured deeps may serve to fix the banks on to the topographic highs, restricting lateral migration. Evidence from tidal-current data suggests that the Norfolk Banks formed at a greater angle to the prevailing tidal currents than they exhibit at present (Howarth and Huthnance, 1984), which may indicate that to some extent they are already moribund.

If the offshore migration of the banks is continuing at present, a continual supply of sand from the nearshore zone is required if the nearshore banks are not to decrease in size. This supply could come from rapid erosion of the East Anglian coast. It is unlikely that coastal erosion during the Holocene could have provided sufficient sediment to form the whole Norfolk Banks group. Clayton (1989) estimated coastal recession on the Norfolk coast to have been about 4.5 km during the last 5000 years, yielding about 2×10^9 m^3 of sand. The volume of Well Bank alone has been estimated to be 2.4×10^9 m^3 (Houbolt, 1968).

The banks consist of fine- to medium-grained sands which show a high degree of sorting. On Haisborough Sand, McCave and Langhorne (1982) found the sand to become finer grained across the bank from south-west to north-east, with the best sorting towards the bank crest. Shell content in the Norfolk Banks is low, less than 5 per cent on Well Bank (Houbolt, 1968). Very little is known of the internal composition of the Norfolk Banks, or indeed of any tidal sand ridge. Houbolt (1968) found no vertical gradation of grain size in a borehole through Ower Bank. A borehole through a moribund sandbank north-west of the Dogger Bank, just north of the report area, revealed only subtle vertical trends of grain size (Davis and Balson, 1992). Short cores taken on the steep flank of Well Bank indicate extensive bioturbation to depths of 55 to 60 cm from the surface, probably due to the burrowing echinoid *Echinocardium cordatum* (Pennant), although large populations of sand eels, which also burrow, are found on the same bank (Wilson, 1982).

THAMES ESTUARY BANKS

Farther south, the outermost Thames Estuary and the adjacent part of the southern North Sea contain a number of large sandbanks that can be conveniently divided into two subgroups (Figure 125). Firstly, there are those which form a radiate pattern within, and parallel to, the margins of the outer estuary coast. These banks, which include Long Sand, Sunk Sand and Gunfleet Sand, are generally broad, with their crests very close to sea level; some are exposed at low water.

The second subgroup consists of narrower, linear banks orientated approximately north–south in deeper water north of the Dover Strait. These form a pattern which radiates out

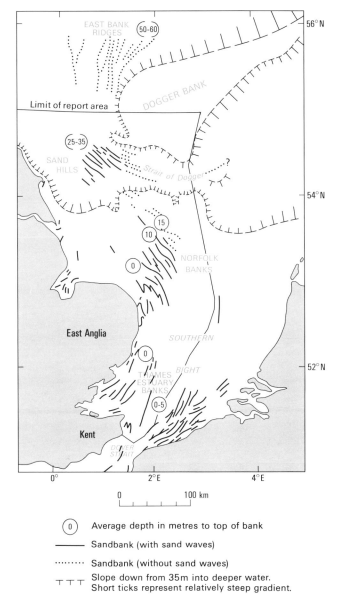

Figure 120 Distribution of major sandbanks in the southern North Sea. After Stride et al. (1982).

from the Dover Strait, and include the North Falls, South Falls, Galloper, and the Gabbard banks. The North and South Falls form a ridge 65 km long rising up to 35 m above the surrounding sea floor (Houbolt, 1968). The more offshore banks are asymmetrical in form, with steeper western slopes of around 3° to 4°, and shallower eastern slopes of approximately 1° (Caston, 1981). They are subparallel to the dominant tidal-flow direction, but slightly offset (Figure 125) in an anticlockwise sense (Kenyon et al., 1981). The Galloper sandbank appears to sit on a small elongate high of the underlying Tertiary bedrock (D'Olier, 1981). D'Olier (1981) believed other banks in this group may be sited on topographic highs which may initially have been the sites for beach deposits.

Most of the Thames Estuary sandbanks have sand waves on their surfaces, although on the estuarine banks they are restricted to the lower flanks in the deeper parts of the interbank channels. Sand-wave crest orientations on the outer banks are generally perpendicular to the bank crest and are dominantly asymmetric, with the steeper face to the south-south-west, the direction of net sediment transport. However, on the steeper western flanks of Inner Gabbard and South Falls, sand waves have their steep faces to the north-north-west (Caston, 1981). The upstream ends (heads) of the banks are generally blunt compared to the elongate downstream tails. Symmetrical sand waves may be present in a narrow zone at the ends of the banks, separating sand waves with opposing senses of asymmetry, implying transport towards, and accumulation at, these points. Caston (1981) believed that sand is added to the blunt upstream end, and therefore, if sand supply is plentiful, the bank should grow upstream.

The sediments of the Outer Gabbard sandbank are rich in calcium carbonate, which may comprise over 50 per cent of the sand. Lagaaij (in Houbolt, 1968) found Pliocene bryozoans in the sediments, and it is likely that much of the shelly material came from reworking of carbonate-rich Neogene formations on the sea floor. This implies that modern coastal erosion is not an important source of sediment for this sandbank.

The offshore sandbanks may have formed relatively early in the Holocene, as the sea transgressed from the south (Figure 115) through the Dover Strait (Jelgersma, 1979). The estuarine banks are more recent. The Goodwin Sands, off the Kent coast, are believed to be a banner bank formed in the lee of the headland of North Foreland (Stride, 1989).

DISTRIBUTION AND SOURCES OF SEA-BED SEDIMENT

The sediments of the floor of the southern North Sea (Figure 111) may be derived from a variety of sources. The contribution from each source varies geographically, temporally, and with the grain size of sediment being considered.

Some of the sediments presently exposed on the sea bed were deposited under conditions very different to those of the present; these include unlithified Pleistocene glacigenic or fluviatile sediments. These 'relict' sediments may provide a source for modern sea-bed sediments by erosion, or may be themselves altered in situ, as for instance by the infiltration of modern mud into the open framework of a Pleistocene gravel. Redistribution of the 'relict' deposits has been a very important process in the formation of the present-day sea-bed sediment distribution; much of this redistribution may have taken place during the early Holocene transgression over the land surface.

Coastline erosion has been an important source throughout the Holocene, and has yielded a large volume of sediment into the offshore area. The present coastline of eastern England south of Flamborough Head is dominated by low cliffs of unlithified Pleistocene or Tertiary formations. Coastal recession in these areas has been high, averaging 0.9 m/year for the last 5000 years in Norfolk (Clayton, 1989). Cliff recession of up to 2 m/year occurs along the Holderness coast (Figure 111) north of Spurn Head (Valentin, 1971). North of Flamborough Head, coastal erosion of relatively resistant Mesozoic formations is much slower (Jago, 1981).

The major rivers of eastern England presently input only silt- and clay-sized sediment into the southern North Sea, although their competence to transport sediment may have varied through the Holocene. Large amounts of fine-grained sediment may be transported into the area by water masses flowing from the North Atlantic, the Baltic, and particularly through the Dover Strait from the English Channel (Eisma and Kalf, 1979; Eisma, 1981).

Gravel

Gravelly sediments are widespread in the report area, but despite the strong tidal currents, they are not being transported

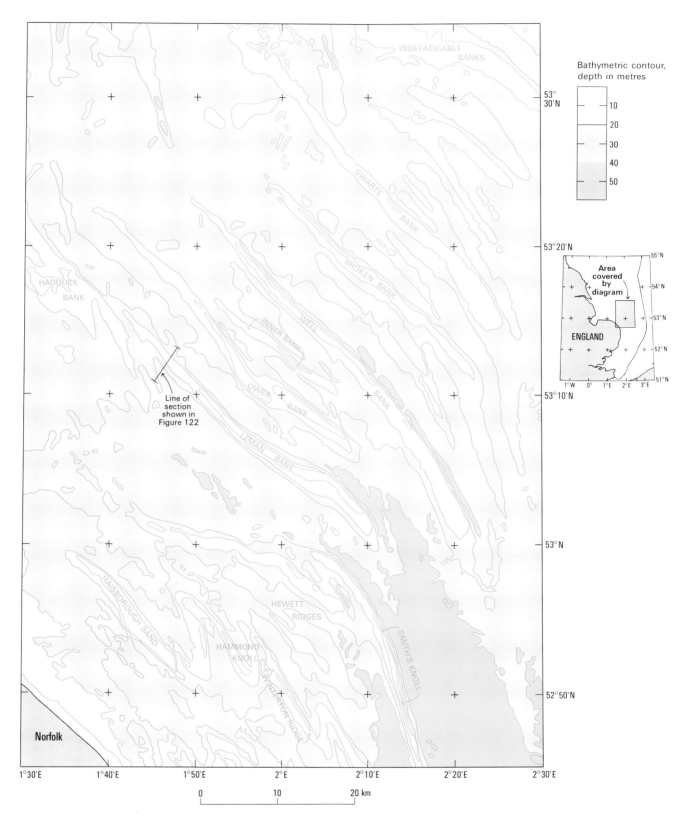

Figure 121 Morphology of the Norfolk Banks.

any significant distance at present. Exceptions are between the South Falls and Sandettie Banks, where 'gravel waves' have been described (Smith, 1988a; 1988b), and away from the coast, where wave-generated longshore drift can transport shingle. The distribution of gravelly sediments therefore reflects processes no longer active offshore, including glacial, fluvioglacial, fluvial and coastal processes. A small biogenic contribution to the gravel fraction may be made locally, mostly by mollusc shells.

The largest area of gravelly sediments occurs between the north Norfolk coast and Flamborough Head (Figure 111). These sediments are mostly less than a few tens of centimetres thick, and rest on an erosional surface of till or, near the Norfolk coast, on a bedrock of Chalk. This thin gravel sheet is overlain in places by deposits of sand, including the Norfolk Banks. In the area of Chalk outcrop, it is likely that a proportion of the gravel is derived directly from sea-floor erosion, as samples commonly include pebbles of chalk and flint.

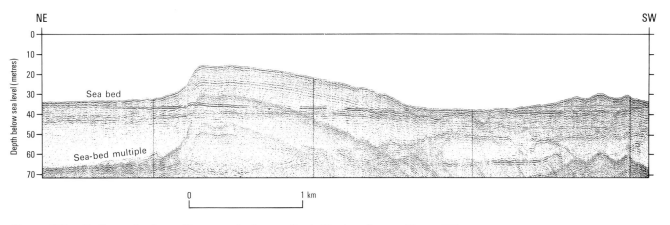

Figure 122 Shallow-seismic profile across the Leman Bank. For location see Figure 121.

Elsewhere the gravels contain a mixture of lithologies, including flint, but are dominated by Carboniferous sandstone and limestone, together with igneous rocks believed to have been derived from the Cheviots in north-east England (Veenstra, 1971). In general, the lithological assemblage indicates derivation from Palaeozoic formations of northern England and southern Scotland. These gravels may have arisen as lag deposits derived from moraines or glacial outwash fans (Robinson, 1968; Veenstra, 1971), parts of which may remain as topographic features or shoals to reflect the successive limits of the Weichselian ice sheet as it retreated northwards (Robinson, 1968).

Gravel-sized sediment may also have been derived by coastal erosion of Weichselian tills, which have provided considerable quantities of sand and mud to the sediment budget (McCave, 1973, 1987; O'Connor, 1987). However, the source sediments contain only a few per cent of gravel (Madgett and Catt, 1978). Gravel derived from modern coastal erosion along the coast between Spurn Head and Flamborough Head is transported southwards by longshore drift to form the sand and shingle spit of Spurn Head, and may be transported offshore in the area of The Binks (De Boer, 1964).

Patches of gravelly sediment on the Dogger Bank and to the south of the Outer Silver Pit have a similar composition to those to the west, and are believed to be similarly derived from Weichselian glacigenic deposits (Veenstra, 1969).

Gravelly deposits also occur on the sea floor off eastern East Anglia and in the Dover Strait. Off Great Yarmouth, the gravelly sediments rest unconformably on Pleistocene deposits, but farther south in the Outer Thames Estuary they rest on Tertiary formations. Considerable variation in pebble lithology is seen within this region, but flint is the dominant component. Quartz and quartzite are locally common, especially in the north. Most of the gravel originates from reworking by the marine transgression of deposits resting on the early Holocene land surface. These may have included fluvial-terrace deposits (D'Olier, 1975) or beach gravels formed from material of fluvial origin (Veenstra, 1971). Some of the flint pebbles show percussion marks on their surfaces, indicating beach processes (Veenstra, 1969).

Other gravel components show a relationship to the underlying geology. In the Dover Strait, rounded chalk pebbles are numerous, and in the vicinity of the Paleocene outcrop, very well-rounded, small, black, flint pebbles derived from Tertiary pebble beds are common. Gravels from areas underlain by the London Clay Formation contain large cobbles of calcareous mudstone derived from septarian concretions. In an area east of Orfordness, the gravels contain rounded phosphorite pebbles reworked from Pliocene lag deposits (Balson, 1989a).

Sand

Sandy sediments dominate over much of the offshore area. Much of the sand is mobile under present-day hydrodynamic conditions, and therefore its distribution relates to modern sand-transport processes. However, the heavy mineral content of the sands does show significant geographical variation (Baak, 1936), reflecting a variety of sources. Sands may also originate from fluvioglacial or fluvial deposits which were deposited before the area was inundated by the sea. It has been suggested that the large sand-wave field in the Southern Bight lies in an area previously occupied by an early Holocene Rhine–Meuse estuary complex (Nio, 1976), implying that this was the source of the sand. Cailleux (1942) identified quartz grains of aeolian origin that may be derived from patchily distributed Weichselian cover sands (Twente Formation).

At present, the rivers around the southern North Sea input very little sand; the main input is from coastal erosion. The rapidly eroding glacigenic sediments of the Holderness coast may supply $0.63 \times 10^6 \, m^3$ of sand per year (O'Connor, 1987), some of which is laid down on the spit of Spurn Head. Some enters the Humber Estuary to be deposited (Al-Bakri, 1986), and some is transported farther along the coast south of the Humber.

An important contribution from cliff erosion comes from the north Norfolk coast between Weybourne and Happisburgh (Figure 111), where cliffs of glacigenic sediments are receding at a rate of approximately 0.9 m/year (Clayton, 1989). Sediment derived from this erosion is transported to the west and south-east along the beaches by longshore drift, with the dominant transport to the east. A coarsening of sand grain-size on the beaches in the direction of transport has been noted by McCave (1978), and interpreted to be the result of winnowing of finer-grained sand from the beaches by wave action, followed by transport into the nearshore zone where the sand is removed by tidal currents. Vincent (1979) calculated potential sand-transport rates, and found that where the potential rate significantly exceeds the available supply of sand, the beaches tend to consist mostly of shingle.

The rate of transport along the beach varies both spatially and temporally, and can result in areas of sediment accretion which form the blunt headlands or nesses which characterise the East Anglian coast. These headlands were thought to be

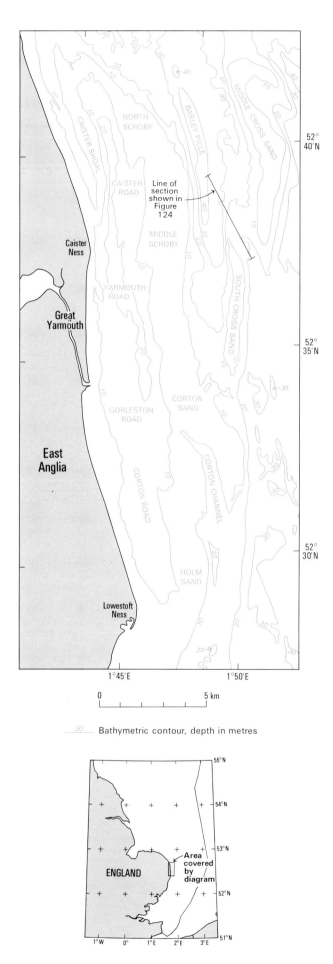

Figure 123 Morphology of nearshore sandbanks near the coast of East Anglia.

sites where sediment from offshore was being deposited (Robinson, 1966; Dugdale et al., 1978), but it is more likely that sand is lost from the coastline at the nesses (McCave, 1978; Carr, 1981), which are usually associated with a shore-attached sandbank. Lowestoft Ness and Caister Ness (Figure 111) both have shore-attached sandbanks which are connected through the complex of nearshore parabolic banks to the more offshore Norfolk Banks. Sand derived from coastal erosion may therefore be transported offshore along a complex transport path to maintain the offshore bank system tens of kilometres from the present coastline (Stride, 1988), although this has not been proven.

Clayton (1989) believed that a high rate of coastal retreat on the Norfolk coast has been maintained over the last 5000 years, and calculated that during this time approximately $2 \times 10^9 \, m^3$ of sand would have been eroded. This volume is of the same order as that presently represented by the nearshore banks Newcome, Holm, Corton and Scroby Sands (Clayton et al., 1983). Farther south, sediment budget calculations appear to confirm that coastal erosion is the main source of sand for the nearshore Sizewell–Dunwich Banks (Carr, 1981), which are connected to the coastline at Thorpe Ness.

The sand fraction of the sea-floor sediments is mostly medium grained. Fine-grained sand occurs in the mouths of estuaries such as The Wash, Humber and Outer Thames. Very fine-grained sands are found mostly on intertidal flats, or within pits like the Outer Silver Pit. Coarse sands are mostly limited to areas associated with winnowing by tidal currents, as for instance along the line of southern North Sea bed-load parting (Figure 119). Tidal currents decrease to the north, but increase to the south, of this line of sand-transport divergence. Consequently, sand grain sizes generally fine northward, but coarsen southward of the bed-load parting (Stride, 1989).

Mud

Muddy sediments in the southern North Sea have a restricted distribution (Figure 111), mostly confined to intertidal areas within estuaries and tidal inlets. At present, the waters of much of the report area are turbid, with suspended mud reaching >100 mg/l in nearshore water, and lower concentrations in the more saline central parts of North Sea where they may be <2 mg/l (Eisma and Kalf, 1979).

The fine-grained sediment may be derived from many potential sources, although Eisma and Kalf (1979) estimated that the most important is the inflow through the Dover Strait, which may supply 8 to 11.5 million tonnes/year. The major rivers (the Rhine, Scheldt, Thames and Humber) input 2.7 million tonnes/year, atmospheric dust 0.5 million tonnes/year, and coastal erosion 0.3 million tonnes/year. The input from organic production is considered to be negligible. The potential input from sea-floor erosion is not known, but was believed by McCave (1987) to be relatively unimportant. McCave (1987), however, gave an estimate for the contribution from coastal erosion which is much higher than that of Eisma and Kalf (1979); he presented a figure of 1.4 million tonnes/year from the Holderness cliffs alone, with a further 0.8 million tonnes/year from erosion of the East Anglian cliffs. Therefore, from coastal erosion in eastern England alone, McCave's (1987) figure is seven times Eisma and Kalf's (1979) figure for the total input to the southern North Sea by this process. This serves to indicate the difficulty of estimating the sediment budget from limited data.

The destinations of the mud are varied, although most of the fine-grained sediment released by coastal erosion remains

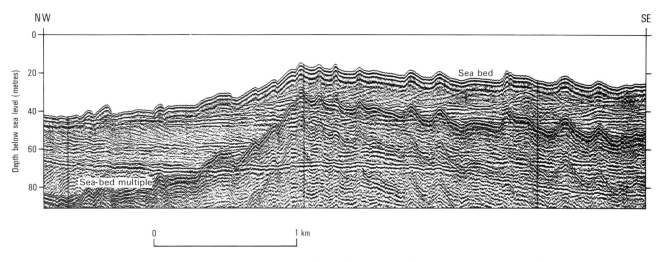

Figure 124 Seismic profile (sparker) at the northern end of South Cross Sand, showing dipping internal reflectors and surface sand waves. For location see Figure 123.

in the nearshore zone and is deposited mostly in estuaries (Kirby, 1987). The east coast estuaries have probably been major sinks for mud deposition throughout much of the Holocene to the present time. Holocene deposits, largely mud, exceed 15 m in thickness around the Humber Estuary, and more than 30 m in The Wash (Wingfield et al., 1978), for intertidal flats are zones of rapid sediment accretion. A plume of turbid water extending north-eastwards from the coast of East Anglia (Joseph, 1955) is related to the pattern of residual currents in the area (McCave, 1972). This plume lies across the south-eastern ends of the Norfolk Banks, and was believed by Stride et al. (1982) to be responsible for the deposition of thin mud laminae seen within sediment cores recovered from the surface of sandbanks in that area (Houbolt, 1968).

The large offshore depressions of the Outer Silver Pit and Markham's Hole are believed to have acted as sediment traps since early Holocene times (Zagwijn and Veenstra, 1966; Eisma, 1975), although it is not clear how much of the Holocene infill was deposited as intertidal flat sediments prior to the onset of modern marine conditions. On the basis of the presence of cinders at a depth of 0.95 m in a short core in the floor of the Outer Silver Pit, Zagwijn and Veenstra (1966) calculated a sedimentation rate of 7 mm/year, which they acknowledged as improbably high. Bioturbation by burrowing organisms may have mixed the sediments. The Outer Silver Pit contains approximately 20 m of Holocene sediment, which, assuming a constant rate of deposition, indicates a rate of 2 mm/year. Eisma (1975) believed a lower rate of 0.43 mm/year to be more likely. Eisma and Kalf (1979) pointed out that concentrations of suspended sediment are low in the vicinity of the Outer Silver Pit, perhaps indicating that mud deposition occurred mainly in the early Holocene; deposition rates at that time would therefore have been higher than 2 mm/year.

The Oyster Ground, east of the Outer Silver Pit in the Dutch sector (Figure 111), may be an area of modern mud accumulation which coincides with a zone of weak surface currents (Creutzberg and Postma, 1979). Muddy sediments are also found in nearshore areas such as off Great Yarmouth, where McCave (1981) believed deposition of mud to result from the influence on the nearshore zone of the turbidity maximum of the River Yare. Alternatively, the mud may originate from local erosion of early Holocene estuarine muds exposed on the sea floor, as found by Lees (1982) farther south.

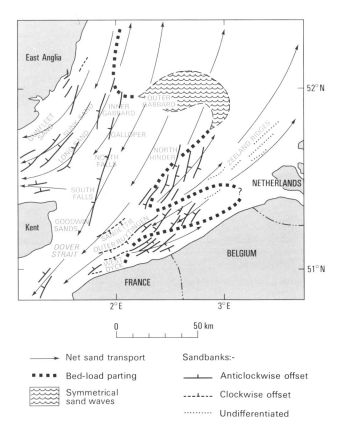

Figure 125 Net sand-transport directions in the southernmost part of the North Sea. Based on sand-wave and sandbank asymmetries, and tidal currents. After Kenyon et al. (1981).

Carbonate

The sea-floor sediments of the report area have a low carbonate content, mostly <10 per cent (Pantin, 1991), reflecting the dominance of glacigenic sources. The calcareous sediment grains may originate from modern, carbonate-secreting organisms, or from reworking of early Holocene shell-bearing sediments and older fossiliferous formations.

The southern North Sea is a very productive water mass, but the preservation potential for much of the biomass is low.

The contribution by planktonic organisms to fine-grained sediment is believed to be small (Eisma and Kalf, 1987); the contribution by benthic organisms may be greater, but is localised. Modern, marine, carbonate-secreting organisms in the area are dominated by benthic animals, including molluscs, barnacles, echinoderms, polychaetes and bryozoans. Calcareous foraminifera probably also make a contribution. The distribution of these benthic animals and their skeletal remains is to a large extent controlled by the nature of the substrate.

In regions of mobile sand within areas of large sand waves, echinoderms are abundant. *Donax vittatus* (da Costa) is the most significant mollusc species, with densities of up to 5 or 10 per square metre. This species grows rapidly and has a robust shell, and therefore makes an important contribution to the carbonate fractions in these regions (Wilson, 1982). In regions of gravelly or rocky substrates, encrusting organisms such as barnacles and tube-secreting polychaetes may dominate. In general, however, this region does not have carbonate-rich skeletal sediments such as are found in the Western Approaches (Evans, 1990) or to the north and west of Scotland (Pantin, 1991). This is partly due to dilution by lithic sediment, but is due also to the briefer time-span for Holocene marine deposition, and possibly to lower carbonate production rates.

Areas of high observed $CaCO_3$ values in sea-bed sediments of the southern North Sea commonly relate to derivation from older carbonate-rich sediments. In the Thames Estuary region, sediment rich in shells of *Cerastoderma* may be related to former early Holocene intertidal deposits (BGS Thames Estuary Sea Bed Sediments and Quaternary Geology sheet). Off East Anglia, erosion of shell-rich Neogene sediments has supplied fossils to the modern sediments; these include molluscs, brachiopods and bryozoans (Lagaaij in Houbolt, 1968; BGS Ostend Sea Bed Sediments sheet). Elsewhere, Palaeogene sediments have contributed abundant *Nummulites* (Kirby and Oclc, 1975). Where the Late Cretaceous Chalk is exposed on the sea floor, as for instance in the Dover Strait and off northern East Anglia, patchy sediments rich in rounded chalk fragments occur; these may be rapidly abraded by wave and current action to yield calcareous mud. Similarly, erosion of chalk-rich till may contribute calcareous mud.

Deposits very rich in $CaCO_3$ occur on tidal flats, as for instance on the coast of Essex where there are banks or ridges up to 3 m high formed of accumulations of mollusc shells (Greensmith and Tucker, 1969). These shell-rich bodies originated by wave accumulation, possibly after mass mortality events, and migrate landwards across the tidal flats. In The Wash, accumulations of living *Mytilus edulis* Linnaeus occur in large areas, known as mussel 'scalps', that are close to the low-water mark (Evans, 1965; Wingfield et al., 1978).

14 Economic geology

HYDROCARBONS

Over the last 25 years, commercial development of the gas reserves of the southern North Sea has helped to revitalise the fuel economy of the UK. Its many fields are currently providing about 29 billion cubic metres (bcm) of gas per annum, for sale mainly to British Gas and the petrochemical industry. This represents about 65 per cent of total gas production from UK waters (Department of Energy, 1990).

The first gas came ashore from the West Sole Gasfield (Figure 126) in 1967. By the end of 1989, 581 bcm of gas had been recovered from 28 separate fields (Department of Energy, 1990). Approval had been granted by the Department of Energy for the development of 7 additional fields, while plans were being considered for gas production from several more. The commercial potential of other discoveries has not been fully appraised, but the Department of Energy (1990) estimated that the Southern North Sea Gas Basin still has proven recoverable reserves of 299 bcm. They estimated that there may be between 155 and 755 bcm of gas in mapped, but untested, structural traps in the southern North Sea and Irish Sea, most of which are in the southern North

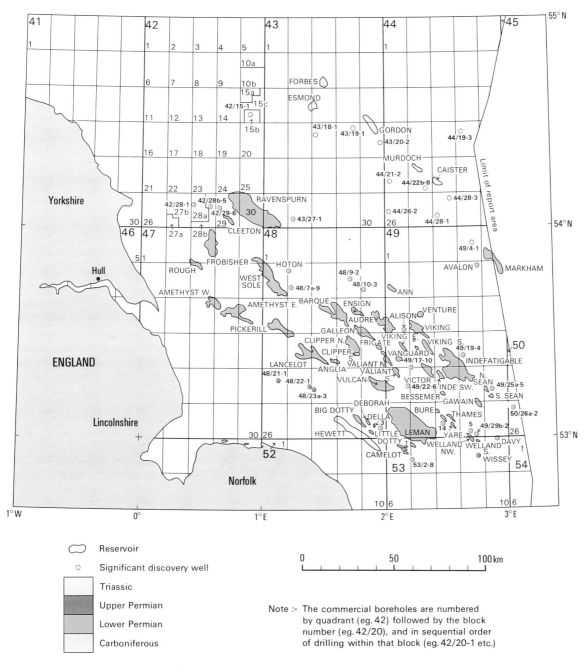

Figure 126 Locations of gasfields and other significant discovery wells in the southern North Sea. After Department of Energy (1991).

Sea. The operators' estimates of the original recoverable reserves in each of the fields in production or under development are listed in Figure 127.

Almost all major gas discoveries in the southern North Sea have been in Lower Permian and Triassic sandstone reservoirs. A number of wells have tested gas and oil shows from Upper Permian carbonate reservoirs, but appraisal drilling has indicated that their closures are relatively small, and they have only marginal economic potential. Since 1984, encouraging flow rates have been obtained from tests of Upper Carboniferous, Westphalian sandstones in the north of the area; some of these discoveries are likely to be commercially viable. There may also be potential for economic reserves of gas within Namurian, Dinantian, and perhaps even Devonian reservoirs in this northern region, although these plays are largely untested by drilling. There have been no significant shows of hydrocarbons within potential Middle Jurassic and Lower Cretaceous sandstone reservoirs adjacent to the Dowsing Fault Zone.

Carboniferous reservoirs

The early years of southern North Sea exploration established that there are prolific reserves of gas in Permian and Triassic reservoirs, but deeper hydrocarbon accumulations proved more difficult to locate. The Carboniferous came to be regarded as economic basement, important only for containing the source rocks from which the gas had migrated into the higher reservoirs. Government intervention in 1982 sharply increased the potential income from gas sales, stimulating the industry to explore pre-Permian reservoirs. Since 1984, there have been many discoveries of gas in Carboniferous sandstones in southern Quadrant 44 and in central Quadrant 43 (Figure 126). The most encouraging of these have been in the Murdoch and Caister gasfields, which press reports suggest may have combined reserves of about 28 bcm.

Exploration so far has indicated that the best economic potential for the Carboniferous occurs in fault-bounded and stratigraphical traps to the north of 54°N, especially where Westphalian sandstones are interbedded within mature, coal-measure source rocks. In this area, the overlying Lower Permian is composed of a lacustrine shale and evaporite sequence which, together with the Upper Permian evaporites, provides a regional seal to deep gas accumulations. Drilling has confirmed that the best reservoirs are the distributary-channel sandstones which comprise between 1 and 5 per cent of the Langsettian to early Bolsovian (Westphalian A to early Westphalian C) coal-measure sequence, and perhaps up to 25 per cent of the overlying Bolsovian redbeds. The sandstone which forms the principal reservoir in the Murdoch and Caister gasfields occurs near the base of the Duckmantian (Westphalian B); it is up to 40 m thick, and comprises a multistorey fill of a braided channel which may have been more than 10 km wide. Most of the early Westphalian sandstones which have tested gas nearby are between 8 and 25 m thick, whereas the late Westphalian sandstones are less than 15 m thick and may be less extensive laterally.

Cowan (1989), Besly (1990) and Leeder and Hardman (1990) have noted that the Westphalian sandstones of this northern region have undergone a complex diagenetic history that has caused a major reduction in their primary porosity, due mainly to the early precipitation of large volumes of quartz and dolomite cements. Virtually all porosity is secondary and patchily distributed, commonly with limited pore interconnection due to the presence of kaolinite. Potentially economic reservoirs seem to occur only in medium- and coarse-grained sandstones towards the centre of the largest

Gasfield	Operator's estimate of original recoverable reserves ($10^9 m^3$)	Operator's estimated peak production ($10^9 m^3$ per year)	Date of discovery
LEMAN	333.6	19.2	1966
INDEFATIGABLE	134.0	2.9	1966
HEWETT	115.0	8.6	1966
VIKING	80.1	6.4	1965
RAVENSPURN N & S	55.4	4.9	1983
WEST SOLE	53.8	2.25	1965
VALIANT N & S, VANGUARD, VULCAN	43.4	3.53	1970-1983
VICTOR	26.0	1.6	1972
BARQUE, CLIPPER	29.3	2.1	1983
AMETHYST E & W	23.9	1.9	1972
AUDREY	18.0	2.8	1976
SEAN N & S	15.9	0.7	1969
ROUGH	10.2	1.1	1968
THAMES, YARE, BURE	8.95	1.6	1969-1983
ESMOND	9.0	1.2	1982
WELLAND NW & S	8.6	1.0	1984
CLEETON	7.9	2.6	1983
CAMELOT (5 FIELDS)	5.7	0.7	1987
FORBES	1.4	0.5	1970
DELLA	1.7	0.2	1987
GORDON	4.1	0.4	1969

Figure 127 Gas reserves in the southern North Sea at fields in production or under development. After Department of Energy (1991). The Rough Gasfield has ceased production, and has been converted into a storage facility to meet peak seasonal demand for gas.

distributary channels. The log porosities of Westphalian reservoir sandstones in recently reported gas discovery wells range between 10 and 18 per cent (Besly, 1990). Tested flow rates of up to 1 million cubic metres of gas per day from some thicker Westphalian sandstones compare favourably with tests from the Lower Permian gasfields, but there are indications that many of the thinner Westphalian sandstones are more tightly cemented, have relatively poor porosity, and are less likely to yield commercial volumes of gas.

Westphalian source and reservoir rocks have been removed by post-Variscan erosion from over half of the area north of 54°N (Figure 18). Where these are absent, Namurian and Dinantian fluviodeltaic and Yoredale-cycle sequences subcrop beneath the Permian shales and evaporites. The comparatively few wells drilled to test their hydrocarbon potential encountered composite sandstone beds up to 90 m thick, appreciably thicker than in the Westphalian section. Besly (1990) observed that some of the Dinantian sandstones have log porosity maxima of over 20 per cent in well 44/2-1, and their measured core porosity is up to 14 per cent in a nearby Dutch well. Leeder and Hardman (1990) note that Namurian sandstones in wells adjacent to the Sole Pit Trough have much poorer reservoir properties than might be anticipated; they suggested that this may reflect severe, early, and irredeemable porosity loss due to mechanical compaction. Namurian sections in other areas of the basin, in which the grain framework was stabilised by silica and early carbonate cementation, may have more favourable reservoir qualities.

Between 53° and 54°N, there has been only limited exploration of the Carboniferous section beneath the many producing Lower Permian gasfields. Results from those few wells that have penetrated a significant thickness of Carboniferous sediments suggest that the Langsettian and Duckmantian coal measures become more shale-prone and less prospective southwards.

South of 53°N, Tubb et al. (1986) have suggested that there may be gas-filled Carboniferous reservoirs on the northern flank of the London–Brabant Massif, where distributary channel sandstones up to 30 m thick are intercalated within Bolsovian and Westphalian D redbeds; these rest unconformably on shaly, early Westphalian coal measures. The log porosities of the sandstones average between 14 and 20 per cent, and are locally up to 28 per cent. This play relies on intraformational sealing by Carboniferous shales, as the overlying Lower Permian sandstones are porous. The maturity of the early Westphalian source rocks provides another element of uncertainty, though Tubb et al. (1986) have deduced that these may have been locally mature for gas generation since the Jurassic.

Lower Permian reservoirs

It was the discovery in 1959 of huge gas reserves within the Lower Permian sandstones of the Groningen Gasfield in The Netherlands that first stimulated commercial interest in exploring for hydrocarbons in the southern North Sea. At Groningen, the reservoir sandstones are aeolian, and their west-dipping foresets suggested that they might continue through Dutch waters into the UK sector of the North Sea. Early drilling established that this is so, and that a thick Upper Permian evaporite seal extends across most of the offshore area. The underlying coal measures were also found to be almost everywhere mature for gas generation.

Reconnaissance seismic-reflection surveys carried out between 1962 and 1964 (Brennand and Van Hoorn, 1986) identified a number of promising anticlinal and fault-bounded structural traps. During 1965 and 1966, drilling of four of these structures led to the discovery of the Leman, Indefatigable, Viking and West Sole gasfields. Many smaller fields were discovered shortly afterwards, but unfavourable gas prices caused a lengthy hiatus in drilling between the discovery of the Audrey Gasfield in 1975 and of the Vanguard, Cleeton, Ravenspurn, Vulcan, Clipper, Barque and Bure fields (Figures 126 and 127) in 1982 and 1983. Most drilling activity since then has been aimed at appraising the commercial potential of such discoveries.

The drilling has proved that Lower Permian dune and fluvial sandstones are continuous for up to 150 km north of the London–Brabant Massif, before passing northward into desert-lake sediments with no hydrocarbon potential. Almost all gas in the offshore fields is being produced from the aeolian facies, for the fluvial sandstones are more argillaceous and mostly too well cemented to form good reservoirs. The Rough Gasfield is the principal exception, for it has produced from both fluvial and aeolian sediments (Goodchild and Bryant, 1986). In this field, the fluvial sandstones have an unusually high average porosity of 12.9 per cent, and their permeability varies between 0.1 and 180 mD (millidarcies), with an average of 15.7 mD. By contrast, the interbedded aeolian sandstones have porosities ranging between 6.3 and 22 per cent, with an average of 16 per cent, and their permeability varies between 0.05 and 1200 mD with an average of 156 mD (Goodchild and Bryant, 1986). These reservoir properties are very similar to those reported from aeolian sandstones in the Indefatigable and Sean gasfields (France, 1975; ten Have and Hillier, 1986). The latter authors and Conway (1986) have noted that the cross-bedded dune sediments tend to have better reservoir properties than the more argillaceous dune-base sands; the porosity and permeability of intercalated sabkha sediments are so poor that they may form permeability barriers in some fields.

Studies in the Leman and West Sole gasfields (van Veen, 1975; Butler, 1975) have indicated that the Leman Sandstone Formation is much thicker along the Sole Pit Trough than on its flanks, but has significantly inferior reservoir properties in the trough. Bulat and Stoker (1987) have deduced from interval velocities that the Lower Permian reservoirs have been buried about 1000 m deeper along the axis of the trough than on its flanks. The burial-related growth of authigenic minerals, especially illite, was the most important factor in damaging the reservoir properties of these sandstones (Marie, 1975). About 75 per cent of the gas in place will be recovered from Leman at the southern end of the Sole Pit Trough, compared with 80 per cent at Indefatigable on the eastern flank of the trough (Glennie, 1986a).

The uppermost 10 to 15 m of the Leman Sandstone Formation is tightly cemented by dolomite, and is nonproductive in many offshore fields. This dolomite cement is believed to have been derived from the overlying Zechstein carbonates. Porosity loss that continues for up to 65 m beneath the top of the dune sequence in some wells has been attributed by Glennie (1986c) to tighter grain packing caused by reworking of the sands during the Zechstein transgression.

The maturity of source rocks and the timing of gas generation relative to trap formation have been key elements in determining the location of gasfields in the southern North Sea. The Westphalian source rocks beneath the Sole Pit Trough may have generated their gas as early as the Jurassic, but most of this migrated into the contemporary highs which flanked the basin at that time (Glennie, 1986c). Subsequently, when the trough was inverted (Glennie and Boegner, 1981; Brennand and Van Hoorn, 1986), some of this gas re-migrated into the less-porous sandstones along the axis of the trough. The preservation of gas within the Lower Permian sandstones throughout this long migration history testifies to the efficiency of the Zechstein seal.

There have been no discoveries of gas in the porous Lower Permian sandstones to the south of the Camelot fields, and only one major discovery to the north-east of the Indefatigable and Viking gasfields (Figure 126). The former area is adjacent to the London–Brabant Massif, and has remained structurally high throughout its history. Here, the Zechstein seal is thin and breached by many faults which may have allowed most of the Carboniferous-sourced gas to migrate to the sea bed. Immaturity of the Carboniferous source rocks may be the main reason for the scarcity of gas in Permian reservoirs north-east of Indefatigable. Furthermore, Brennand and Van Hoorn (1986) have invoked an unfavourable relationship between the timing of trap formation in this area and the onset of gas migration from more distant, mature source rocks.

Upper Permian reservoirs

There are many oil and gas fields in Upper Permian carbonate reservoirs in The Netherlands, Germany and Poland. Gas production has also been obtained from Z2 and Z3 carbonates in the Lockton, Malton and Eskdale fields of north-east England. All of these fields are relatively small, and most would not be economically viable offshore. In the offshore Upper Permian, only the shelf and submarine slope carbonates offer potential hydrocarbon reservoirs, and these are confined to the southern and western margins of the Zechstein basin.

The early phase of exploration established that the Upper Permian carbonates are tightly cemented and nonproductive in most of the southern North Sea. Shows of oil or gas were recorded in a number of wells, but only the Wissey (formerly Scram) Gasfield (Figure 126) appears to contain potentially commercial volumes of gas, reported to be between 2.8 and 5.6 bcm (Clark, 1986). The gas is contained within both the Z1 and Z2 carbonates in Wissey, and nearby discovery well 53/5-1 tested gas from the Z3 reservoir. Subcommercial flow rates of gas have been tested from the Z2 and Z3 carbonates at Hewett (Cumming and Wyndham, 1975), and well 48/22-1 tested gas and 1584 barrels of oil per day from Z2 carbonates. Nearby appraisal well 48/22-2 encountered only shows of oil in the reservoir, which is tightly cemented by anhydrite (Clark, 1986), suggesting that this discovery is not commercially viable.

The porosity and permeability of the carbonates are very poor in almost all of the wells; this reflects the extensive plugging of their pore spaces by anhydrite and halite cements (Clark, 1980b). All gas in the carbonate reservoirs has migrated from mature Carboniferous source rocks, and geochemical studies suggest that the oil in the Z2 and Z3 reservoirs was derived from organic matter in the basin-floor and slope facies of these cycles (Taylor, 1986).

The results from the wells have been disappointing, but Clark (1986) noted that they have allowed the lithofacies and diagenetic history of each of the carbonates to be mapped across the basin in detail. These studies, integrated with new seismic techniques, may enable a better prediction as to where the optimum reservoir properties will be encountered by future drilling. The field size of Upper Permian discoveries will remain small by North Sea standards (Clark, 1986), and they are only likely to be developed offshore as satellites to more substantial reserves in other reservoirs.

Triassic reservoirs

The discovery of substantial reserves of gas in the Triassic reservoirs of the Hewett Gasfield provided the southern North Sea with one of its earliest exploration successes in 1966. The gas was being piped ashore from the deeper Hewett Sandstone reservoir by mid-1969, and from the Bunter Sandstone Formation by late 1973. Apart from the Esmond, Forbes and Gordon gasfields, which were developed jointly and came on-stream during 1985, the only discoveries of gas in Triassic reservoirs have been in block 44/23 and at the Little Dotty Gasfield adjacent to Hewett (Figure 126).

The Hewett Sandstone Bed is confined to a narrow belt along the northern flank of the London–Brabant Massif (Figure 49), where it is interbedded towards the base of the Bunter Shale Formation, which provides a seal to the reservoir. The sandstone is up to 61.5 m thick in the centre of the Hewett Gasfield, but is mostly less than 25 m thick. At Hewett, a relatively clean sand is sandwiched between two intervals of finer-grained, argillaceous sandstones and shales; the total reservoir section has an average porosity of 21.4 per cent and an average permeability of 1310 mD (Cumming and Wyndham, 1975).

The Bunter Sandstone Formation reservoir comprises a sheet-sand complex which is more than 50 m thick across most of the southern North Sea (Figure 50). Its porosity is between 20 and 25 per cent, and its permeability between 100 and 700 mD over most of the offshore area (Fisher, 1986), although there are relatively low porosity intervals containing appreciable quantities of halite cement in the reservoirs of the Forbes, Esmond and Gordon gasfields (Bifani, 1986). The claystones and evaporites of the Haisborough Group provide a regional seal for gas in the Bunter Sandstone Formation.

Despite excellent reservoir properties and the identification of many large closures, almost all tests of Triassic sandstones in the southern North Sea have been plugged and abandoned as dry. Only at a very few places have the underlying Upper Permian evaporites been breached by Carboniferous-sourced gas. In the centre of the evaporite basin, this seal has been breached where there has been very considerable diapiric withdrawal of the Zechstein halites, or along major faults. The Esmond, Forbes and Gordon gasfields are located above the crests of large Zechstein salt swells, and contain substantial reserves of gas only because they are adjacent to zones of extreme thinning of the Permian evaporites. At Hewett, Little Dotty, and along the northern flank of the London–Brabant Massif, the Permian halites are relatively thin, and many faults extend through the Permian sequence to offer migration routes for gas into the Triassic reservoirs. Large volumes of gas may have escaped to the sea bed along some of these faults.

The Triassic reservoirs contain a higher proportion of nitrogen and ethane in Hewett than do the nearby Lower Permian gasfields (Cumming and Wyndham, 1975), and the Bunter Sandstone Formation reservoirs contain between 8 and 16 per cent nitrogen in the northern group of fields (Bifani, 1986). Cumming and Wyndham (1975) have suggested that this is because the hydrocarbon component of the gas is absorbed preferentially, especially on the surfaces of clay minerals, during its migration from the Carboniferous source rocks. The higher concentrations of nitrogen simply reflect longer migration routes for gas into the Triassic, compared with the Lower Permian reservoirs.

NONHYDROCARBON RESOURCES

Sand and gravel aggregates

Extraction of marine-dredged sand and gravel from the floor of the southern North Sea has been making an increasingly major contribution to supplying the demand for raw materials by the UK construction industry. The aggregate is used mainly in the manufacture of concrete, but other uses include the production of asphalt, concrete blocks and constructional fill. From its early beginnings in the 1920s, marine aggregate production in the UK has risen steadily to more than 20 million tonnes per annum by 1989, representing about 16 per cent of the total requirement for sand and gravel in England and Wales (Ardus and Harrison, 1990). Marine supplies to London and south-east England are particularly important, amounting to around 11 million tonnes per annum, equivalent to 35 per cent of this region's sand and gravel consumption. About 30 per cent of the southern North Sea aggregate is exported to Belgium, The Netherlands and Germany (Nunny and Chillingworth, 1986).

Most of the dredging has been carried out in water depths of 18 to 30 m, although several ships have been commissioned recently with a capability to dredge from water depths of up to 50 m; this will extend the potential production zone across most of the southern North Sea (Ardus and Harrison, 1990). The areas from which the aggregate is currently being extracted are off East Anglia, the Outer Thames, and off the Humber Estuary (Figure 128). Production from the Outer Thames has begun to decline as its reserves diminish, and the dredging industry is looking increasingly to the East Anglian area to maintain the supply of marine aggregates to south-east England. Comments below on the quality and quantity of sand and gravel resources are based on studies carried out by

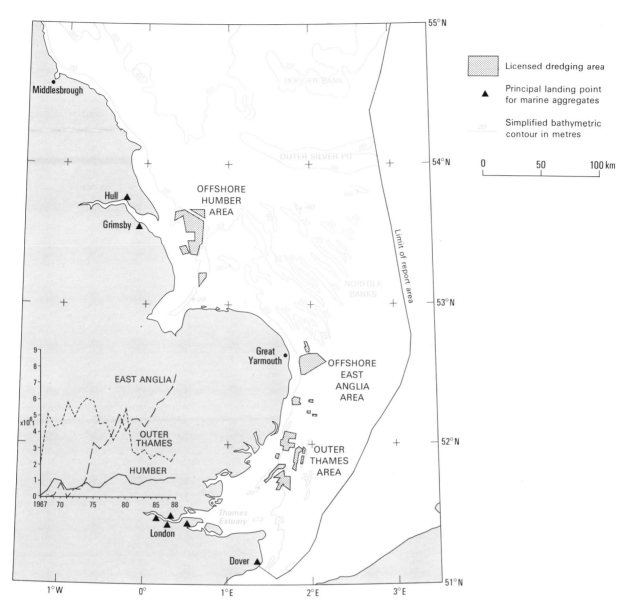

Figure 128 Areas of the southern North Sea currently licensed for marine aggregate production. Inset shows their annual production.

BGS on behalf of the Department of the Environment and the Crown Estate (Balson and Harrison, 1986; Harrison, 1988; Harrison, 1990).

The floor of the southern North Sea is dominated by sands and gravelly sands; muddy deposits occur only locally in the nearshore area and in a few sea-bed depressions. In large areas off the Humber Estuary, off East Anglia, and in parts of the Outer Thames, the surface layers contain up to 60 per cent gravel and only small amounts of silt and clay. These sandy gravels are generally less than 1 m thick, and in some areas form a veneer less than 0.1 m thick. Thicker gravel accumulations occur locally as the degraded remnants of former river channels and terraces. The present distribution of the gravels records the marine reworking of Pleistocene fluvial and glacial deposits during the early stages of the Holocene transgression across the area. Most of the gravels can be considered as relict deposits; overgrowths and encrustations by marine organisms indicate that they have been reworked to only a minor degree.

More than 90 per cent of the gravel fraction off East Anglia and in the Outer Thames is composed of flint. Quartzite, phosphorite and limestone pebbles are locally abundant, and shell debris and whole shells may constitute 30 per cent or more of the gravel fraction. Most of the flints are of pebble size, and only a small proportion of these is larger than 32 mm; shapes vary from well rounded to angular, although nodular and elongate forms also occur.

The gravels off the Humber Estuary are more varied in composition; Carboniferous sandstone and limestone are particularly common, but chalk, Jurassic mudstone, flint, and igneous and metamorphic rock types are also widespread. Clasts of Devonian sandstone and quartzite have also been noted. The gravels are believed to be lag deposits derived by marine winnowing of glacial moraines and outwash fans deposited during the Weichselian glaciation.

Most of the sandy sediment in the southern North Sea is medium grained (0.2–1.00 mm) and quartzose, with less than 10 per cent comminuted shell debris. Patches of coarse- and very coarse-grained sand occur in the more gravelly areas (Figure 111), and fine- to very fine-grained sands mantle sea-bed depressions. The superficial, Holocene sands have their maximum thickness of 40 m in the tidal sand ridges off the Norfolk coast; elsewhere, they are generally between 0.1 and 5 m thick.

Other minerals

Coal has been mined from beneath the North Sea for many years by extension of onshore coal workings for up to 7.5 km beyond the coastlines of Northumberland and Durham in north-east England. As conventional coal extraction is unlikely to prove rewarding in other areas, in-situ gasification may offer the only prospect for future development of the numerous Carboniferous coal seams known beneath much of the offshore area. On average, about 66 coal seams have been recorded per 1000 m of early Westphalian coal measures drilled in southern North Sea wells; they are mostly between 0.3 and 2 m thick, but are exceptionally up to 5 m thick. The proportion of coal varies from 2 to 12.5 per cent of the coal-measure sequence. The cumulative thickness of coal appears to be greatest under the Sole Pit Trough and the Anglo-Dutch Basin, where the coal measures are buried beneath 2500 to 4000 m of post-Carboniferous rocks. There may be better prospects for in-situ gasification off the East Anglian, Lincolnshire and Northumberland coasts where the coal measures are buried less deeply but are thinner.

There are enormous reserves of anhydrite, halite and potash salts in the Upper Permian and Triassic evaporites of the Southern North Sea Basin; equivalent sediments in north-east England have been mined for more than 100 years (Smith, 1974). There remain substantial reserves of these minerals onshore, and subsea extraction of evaporites is unlikely to be economic in the foreseeable future.

References

References to BGS offshore maps are not given here, but an index map showing their distribution in the report area is presented inside the back cover.

AGUIRRE, E, and PASINI, G. 1985. The Pliocene-Pleistocene Boundary. *Episodes,* Vol. 8, 116–120.

AL-BAKRI, D. 1986. Provenance of the sediments in the Humber Estuary and the adjacent coasts, eastern England. *Marine Geology,* Vol. 72, 171–186.

ALLEN, J R L, and FRIEND, P F. 1976. Changes in intertidal dunes during two spring-neap cycles, Lifeboat Station Bank, Wells-next-the-sea, Norfolk (England). *Sedimentology,* Vol. 23, 329–346.

ALLEN, P. 1976. Wealden of the Weald: a new model. *Proceedings of the Geologists' Association,* Vol. 86, 389–437.

— 1981. Pursuit of Wealden models. *Journal of the Geological Society of London,* Vol. 138, 375–405.

ALLSOP, J M. 1985. Geophysical investigations into the extent of the Devonian rocks beneath East Anglia. *Proceedings of the Geologists' Association,* Vol. 96, 371–379.

— 1987. Patterns of late Caledonian intrusive activity in eastern and northern England from geophysics, radiometric dating and basement geology. *Proceedings of the Yorkshire Geological Society,* Vol. 46, 335–353.

ANDERTON, R, BRIDGES, P H, LEEDER, M R, and SELLWOOD, B W. 1979. *A dynamic stratigraphy of the British Isles.* (London: George Allen & Unwin.)

ANDRÉ L, and DEUTSCH, S. 1984. Les porphyres de Quenast et de Lessines: géochronologie, géochimie isotopique et contribution au problème de l'âge du socle precambrien du Massif du Brabant (Belgique). *Bulletin de la Société Belge de Géologie,* Vol. 93, 375–384. [In French.]

ARDUS, D A, and HARRISON, D J. 1990. The assessment of aggregate resources from the UK continental shelf. 113–128 in *Ocean resources,* Vol. 1. Assessment. ARDUS, D A, and CHAMP, M A (editors). (Dordrecht: Kluver.)

ARTHUR, T J, PILLING, D, BUSH, D, and MACCHI, L. 1986. The Leman Sandstone Formation in UK Block 49/28: sedimentation, diagenesis and burial history. 251–266 in Habitat of Palaeozoic Gas in NW Europe. BROOKS, J, GOFF, G C, and VAN HOORN, B (editors). *Special Publication of the Geological Society of London,* No. 23.

AUBRY, M-P, HAILWOOD, E A, and TOWNSEND, H A, 1986. Magnetic and calcareous-nannofossil stratigraphy of the lower Palaeogene formations of the Hampshire and London basins. *Journal of the Geological Society of London,* Vol. 143, 729–735.

AUSTIN, R M. 1991. Modelling Holocene tides on the NW European continental shelf. *Terra Nova,* Vol. 3, 276–288.

BAAK, J A. 1936. *Regional petrology of the southern North Sea.* (Wageningen: H Veenman & Zonen.)

BADLEY, M E, PRICE, J D, and BACKSHALL, L C. 1989. Inversion, reactivated faults and related structures: seismic examples from the southern North Sea. 201–219 in Inversion tectonics. COOPER, M A, and WILLIAMS, G D (editors). *Special Publication of the Geological Society of London,* No. 44.

— — RAMBECK DAHL, C, and AGDESTEIN, T. 1988. The structural evolution of the northern Viking Graben and its bearing upon extensional modes of basin formation. *Journal of the Geological Society of London,* Vol. 145, 455–472.

BALSON, P S. 1980. The origin and evolution of Tertiary phosphorites from eastern England. *Journal of the Geological Society of London,* Vol. 137, 723–729.

— 1983. Temperate, meteoric diagenesis of Pliocene skeletal carbonates from eastern England. *Journal of the Geological Society of London,* Vol. 140, 377–385.

— 1987. Authigenic phosphorite concretions in the Tertiary of the Southern North Sea Basin: An event stratigraphy. *Mededelingen van de Werkgroep voor Tertiaire en Kwartaire Geologie,* Vol. 24, 79–94.

— 1989a. Tertiary phosphorites in the southern North Sea Basin: origin, evolution and stratigraphic correlation. 51–70 in *The Quaternary and Tertiary geology of the Southern Bight, North Sea.* HENRIET, J P, and DE MOOR, G (editors). (Brussels: Belgian Geological Survey.)

— 1989b. Neogene deposits of the UK sector of the southern North Sea (51°–53°N). 89–95 in *The Quaternary and Tertiary geology of the Southern Bight, North Sea.* HENRIET, J P, and DE MOOR, G (editors). (Brussels: Belgian Geological Survey.)

— 1990a. Episodes of phosphogenesis and phosphorite concretion formation in the North Sea Tertiary. 125–137 in Phosphorite research and development. NOTHOLT, A J G, and JARVIS, I (editors). *Special Publication of the Geological Society of London,* No. 52.

— 1990b. The 'Trimley Sands': a former marine Neogene deposit from eastern England. *Tertiary Research,* Vol. 11, 145–158.

— 1990c. The Neogene of East Anglia — a field excursion report. *Tertiary Research,* Vol. 11, 179–189.

— and CAMERON, T D J. 1985. Quaternary mapping offshore East Anglia. *Modern Geology,* Vol. 9, 221–239.

— and HARRISON, D J. 1986. Marine aggregate survey. Phase 1 – Southern North Sea. *British Geological Survey Marine Report,* No. 86/38.

— and HUMPHREYS, B. 1986. The nature and origin of fissures in the East Anglian Coralline and Red Crags. *Journal of Quaternary Science,* Vol. 1, 1–39.

— and JEFFERY, D H. 1991. The glacial sequence of the southern North Sea. 245–253 in *Glacial deposits in Great Britain and Ireland.* EHLERS, J, GIBBARD, P L, and ROSE, J (editors). (Rotterdam: Balkema.)

BARKER, R D, LLOYD, J W, and PEACH, D W. 1984. The use of resistivity and gamma logging in lithostratigraphical studies of the Chalk in Lincolnshire and South Humberside. *Quarterly Journal of Engineering Geology,* Vol. 17, 71–80.

BARTON, P, MATTHEWS, D H, HALL, J, and WARNER, M. 1984. Moho beneath the North Sea compared on normal incidence and wide angle seismic records. *Nature, London,* Vol. 308, 55–56.

— and WOOD, R. 1984. Tectonic evolution of the North Sea basin: crustal stretching and subsidence. *Geophysical Journal of the Royal Astronomical Society,* Vol. 79, 987–1022.

BELDERSON, R H, JOHNSON, M A, and KENYON, N H. 1982. Bedforms. 27–57 in *Offshore tidal sands, processes and deposits.* STRIDE, A H (editor). (London: Chapman and Hall.)

— KENYON, N H, and STRIDE, A H. 1971. Holocene sediments on the continental shelf west of the British Isles. *Report of the Institute of Geological Sciences,* No. 70/14, 161–170.

BESLY, B M. 1988. Palaeogeographic implications of late Westphalian to early Permian red-beds, central England. 200–221 in *Sedimentation in a synorogenic basin complex: the Upper Carboniferous of Northwest Europe.* BESLY, B M, and KELLING, G (editors). (London: Blackie & Son.)

— 1990. Carboniferous. 90–119 in *Introduction to the petroleum geology of the North Sea* (3rd edition). GLENNIE, K W (editor). (Oxford: Blackwell Scientific Publications.)

— and TURNER, P. 1983. Origin of red beds in a moist tropical climate (Etruria Formation, Upper Carboniferous, UK). 131–147 in Residual deposits. WILSON, R C L (editor). *Special Publication of the Geological Society of London,* No. 11.

BIFANI, R. 1986. Esmond Gas Complex. 209–221 in Habitat of Palaeozoic gas in NW Europe. BROOKS, J, GOFF, G C, and VAN HOORN, B (editors). *Special Publication of the Geological Society of London,* No. 23.

BISSON, G, LAMB, R K, and CALVER, M A. 1967. Boreholes in the concealed Kent Coalfield between 1948 and 1959. *Bulletin of the Geological Survey of Great Britain,* Vol. 26, 99–166.

BJØRSLEV NIELSEN, O, SØRENSEN, S, THIEDE, J, and SKARBØ, O. 1986. Cenozoic differential subsidence of the North Sea. *Bulletin of the American Association of Petroleum Geologists,* Vol. 70, 276–298.

BLESS, M J M, BOUCKAERT, J, CALVER, M A, GRAULICH, J M, and PAPROTH, E. 1977. Palaeogeography of Upper Westphalian deposits in NW Europe with reference to the Westphalian C north of the mobile Variscan belt. *Mededelingen van Rijks Geologische Dienst,* Vol. 28, 101–127.

BLONDEAU, A. 1972. *Les nummulites.* (Paris: Librairie Vuibert.) [In French.]

BLUNDELL, D J, and six others. 1991. Crustal structure of the central and southern North Sea from BIRPS deep seismic reflection profiling. *Journal of the Geological Society of London,* Vol. 148, 445–458.

BOIS, C, and five others. 1987. Operations and main results of the ECORS project in France. *Geophysical Journal of the Royal Astronomical Society,* Vol. 89, 279–286.

BORLEY, J O. 1923. The marine deposits of the southern North Sea. *Ministry of Agriculture and Fisheries, Fishery Investigations.* Series II, Vol. 4.

BOTT, M H P. 1967. Geophysical investigations of the northern Pennine basement rocks. *Proceedings of the Yorkshire Geological Society,* Vol. 36, 139–168.

— 1987. Subsidence mechanisms of Carboniferous basins in northern England. 21–32 in *European Dinantian environments.* MILLER, J, ADAMS, A E, and WRIGHT, V P (editors). (Chichester: John Wiley & Sons.)

BOULTON, G S, and HINDMARSH, R C A. 1987. Sediment deformation beneath glaciers: rheology and geological consequences. *Journal of Geophysical Research,* Vol. 92, 9059–9082.

BRADSHAW, M J, and PENNEY, S R. 1982. A cored Jurassic sequence from north Lincolnshire, England. *Geological Magazine,* Vol. 119, 113–134.

BRANDON, A, SUMBLER, M G, and IVIMEY-COOK, H I C. 1990. A revised lithostratigraphy for the Lower and Middle Lias (Lower Jurassic) east of Nottingham. *Proceedings of the Yorkshire Geological Society,* Vol. 48, 121–144.

BRENNAND, T P. 1975. The Triassic of the North Sea. 295–311 in *Petroleum and the continental shelf of North-west Europe.* WOODLAND, A W (editor). (London: Applied Science Publishers.)

— and VAN HOORN, B. 1986. Historical review of North Sea exploration. 1–24 in *Introduction to the petroleum geology of the North Sea* (2nd edition). GLENNIE, K W (editor). (London: Blackwell Scientific Publications.)

BROOKS, J R V. 1983. Geological information from hydrocarbon exploration on the United Kingdom Continental Shelf. 343–356 in Petroleum geochemistry and exploration of Europe. BROOKS, J (editor). *Special Publication of the Geological Society of London,* No. 12.

BROWN, L, and fourteen others. 1987. COCORP: new perspectives on the deep crust. *Geophysical Journal of the Royal Astronomical Society,* Vol. 89, 47–54.

BROWN, S. 1986. Jurassic. 133–159 in *Introduction to the petroleum geology of the North Sea* (2nd Edition). GLENNIE, K W (editor). (Oxford: Blackwell Scientific Publications.)

BRUNSTROM, R G W, and WALMSLEY, P J. 1969. Permian evaporites in the North Sea basin. *Bulletin of the American Association of Petroleum Geologists,* Vol. 53, 870–883.

BUCHARDT, B. 1978. Oxygen isotope palaeotemperatures from the Tertiary Period in the North Sea area. *Nature, London,* Vol. 275, 121–123.

BULAT, J, and STOKER, S J. 1987. Uplift determination from interval velocity studies, UK southern North Sea. 293–305 in *Petroleum geology of North West Europe.* BROOKS, J, and GLENNIE, K W (editors). (London: Graham and Trotman.)

BURTON, B W. 1977. An investigation of a sandwave field at the south western end of Sandettie Bank, Dover Strait. *International Hydrographic Review,* Vol. 54, 45–59.

BUTLER, J B. 1975. The West Sole gas-field. 213–221 in *Petroleum and the continental shelf of North-West Europe.* WOODLAND, A W (editor). (London: Applied Science Publishers.)

CAILLEUX, A. 1942. Les actions éoliennes périglacialires en Europe. *Mémoires de la Société Géologique de France,* No. 46.

CALVER, M A. 1969. Westphalian of Britain. *Compte Rendu du Congrès International de Stratigraphie et de Géologie du Carbonifère, Sheffield, 1967,* Vol. 1, 233–254.

CAMERON, T D J, BONNY, A P, GREGORY, D M, and HARLAND, R. 1984. Lower Pleistocene dinoflagellate cyst, foraminiferal and pollen assemblages in four boreholes in the southern North Sea. *Geological Magazine,* Vol. 121, 85–97.

— LABAN, C, and SCHÜTTENHELM, R T E. 1989a. Upper Pliocene and Lower Pleistocene stratigraphy in the Southern Bight of the North Sea. 97–110 in *The Quaternary and Tertiary geology of the Southern Bight, North Sea.* HENRIET, J P, and DE MOOR, G (editors). (Brussels: Belgian Geological Survey.)

— SCHÜTTENHELM, R T E, and LABAN, C. 1989b. Middle and Upper Pleistocene and Holocene stratigraphy in the southern North Sea between 52° and 54°N, 2° to 4°E. 119–135 in *The Quaternary and Tertiary geology of the Southern Bight, North Sea.* HENRIET, J P, and DE MOOR, G (editors). (Brussels: Belgian Geological Survey.)

— STOKER, M S, and LONG, D. 1987. The history of Quaternary sedimentation in the UK sector of the North Sea Basin. *Journal of the Geological Society of London,* Vol. 144, 43–58.

CARR, A P. 1981. Evidence for the sediment circulation along the coast of East Anglia. *Marine Geology,* Vol. 40, M9–M22.

CASEY, R. 1962. The ammonites of the Spilsby Sandstone, and the Jurassic–Cretaceous boundary. *Proceedings of the Geological Society of London,* No. 1598, 95–100.

— and GALLOIS, R W. 1973. The Sandringham Sands of Norfolk. *Proceedings of the Yorkshire Geological Society,* Vol. 40, 1–22.

CASTON, G F. 1979. Wreck marks: indicators of net sand transport. *Marine Geology,* Vol. 33, 193–204.

— 1981. Potential gain and loss of sand by some sand banks in the Southern Bight of the North Sea. *Marine Geology,* Vol. 41, 239–250.

CASTON, V N D. 1972. Linear sand banks in the southern North Sea. *Sedimentology,* Vol. 18, 63–78.

— 1976. A wind-driven near bottom current in the southern North Sea. *Estuarine, Coastal and Marine Science,* Vol. 4, 23–32.

— 1977. Quaternary sediments of the central North Sea. A new isopachyte map of the Quaternary of the North Sea. *Report of the Institute of Geological Sciences,* No. 77/11, 1–8.

— 1979. The Quaternary sediments of the North Sea. 195–270 in *The North-West European shelf seas: the sea bed and the sea in motion. 1. Geology and sedimentology.* BANNER, F T, COLLINS, M B, and MASSIE, K S (editors). (Amsterdam: Elsevier.)

— and STRIDE, A H. 1970. Tidal sand movement between some linear sand banks in the North Sea off northeast Norfolk. *Marine Geology*, Vol. 9, M38–M42.

CHADWICK, R A, PHARAOH, T C, and SMITH, N J P. 1989. Lower crustal heterogeneity beneath Britain from deep seismic reflection data. *Journal of the Geological Society of London*, Vol. 146, 617–630.

CHRISTIE-BLICK, N. 1991. Onlap, and the origin of unconformity-bounded depositional sequences. *Marine Geology*, Vol. 97, 35–56.

CLARK, D N. 1980a. The sedimentology of the Zechstein 2 Carbonate Formation of eastern Drenthe, The Netherlands. 131–165 *in* The Zechstein Basin with emphasis on carbonate sequences. FÜCHTBAUER, H, and PERYT, T M (editors). *Contributions to Sedimentology*, Vol. 9.

— 1980b. The diagenesis of Zechstein carbonate sediments. 167–203 *in* The Zechstein Basin with emphasis on carbonate sequences. FÜCHTBAUER, H, and PERYT, T M (editors). *Contributions to Sedimentology*, Vol. 9.

— 1986. The distribution of porosity in Zechstein carbonates. 121–149 *in* Habitat of Palaeozoic gas in NW Europe. BROOKS, J, GOFF, J C, and VAN HOORN, B (editors). *Special Publication of the Geological Society of London*, No. 23.

CLARKE, R H. 1973. Cainozoic subsidence in the North Sea. *Earth and Planetary Science Letters*, Vol. 18, 329–332.

CLAYTON, K M. 1989. Sediment input from the Norfolk Cliffs, eastern England — a century of coast protection and its effect. *Journal of Coastal Research*, Vol. 5, 433–442.

— MCCAVE, I N, and VINCENT, C E. 1983. The establishment of a sand budget for the East Anglian coast and its implication for coastal stability. 91–96 *in Shoreline protection.* (London: Thomas Telford.)

CLEMMENSEN, L. 1979. Triassic lacustrine red-beds and palaeoclimate: The Buntsandstein of Heligoland and the Malmros Klint Member of East Greenland. *Geologische Rundschau*, Vol. 68, 748–774.

COCKS, L R M, and FORTEY, R A. 1982. Faunal evidence for oceanic separations in the Palaeozoic of Britain. *Journal of the Geological Society of London*, Vol. 139, 465–478.

COLLETTE, B J. 1960. The gravity field of the North Sea. 47–96 *in* Gravity expeditions 1948–1958. *Publications of The Netherlands Geodetic Commission, Delft*, Vol. 5.

COLLINSON, J D. 1988. Controls on Namurian sedimentation in the Central Province basins of northern England. 86–101 in *Sedimentation in a synorogenic basin complex: the Upper Carboniferous of Northwest Europe.* BESLY, B M and KELLING, G (editors). (London: Blackie and Son.)

COLLINSON, M E. 1983. Fossil plants of the London Clay. *Palaeontological Association Field Guides to Fossils*, No. 1.

COLMAN, S M, and MIXON, R B. 1988. The record of major Quaternary sea-level changes in a large coastal plain estuary, Chesapeake Bay, eastern United States. *Palaeogeography, Palaeoclimatology, Palaeoecology*, Vol. 68, 99–116.

COLTER, V S, and REED, G E. 1980. Zechstein 2 Fordon Evaporites of the Atwick No. 1 borehole, surrounding areas of NE England and the adjacent southern North Sea. 115–129 *in* The Zechstein Basin with emphasis on carbonate sequences. FÜCHTBAUER, H, and PERYT, T M (editors). *Contributions to Sedimentology*, Vol. 9.

CONWAY, A M. 1986. Geology and petrophysics of the Victor Field. 237–249 *in* Habitat of Palaeozoic gas in NW Europe. BROOKS, J, GOFF, J C, and VAN HOORN, B (editors). *Special Publication of the Geological Society of London*, No. 23.

COOK, E E. 1965. Geophysical operations in the North Sea. *Geophysics*, Vol. 30, 495–510.

COPE, J C W, DUFF, K L, TORRENS, H S, WIMBLEDON, W A, and WRIGHT, J K. 1980a. A correlation of Jurassic rocks in the British Isles. Part 2: Middle and Upper Jurassic. *Special Report of the Geological Society of London*, No. 15.

— GETTY, T A, HOWARTH, M K, MORTON, N, and TORRENS, H S. 1980b. A correlation of Jurassic rocks in the British Isles. Part 1: Introduction and Lower Jurassic. *Special Report of the Geological Society of London*, No. 14.

CORNISH, V. 1901. On sand-waves in tidal currents. *Geographical Journal*, Vol. 18, 170–202.

COSTA, L I, DENNISON, C, and DOWNIE, C. 1978. The Paleocene/Eocene boundary in the Anglo-Paris Basin. *Journal of the Geological Society of London*, Vol. 135, 261–264.

— and DOWNIE, C. 1976. The distribution of the dinoflagellate *Wetzeliella* in the Palaeogene of north-western Europe. *Palaeontology*, Vol. 19, 591–614.

— and MANUM, S B. 1988. The description of the interregional zonation of the Paleogene (D1–D15) and the Miocene (D16–D20). 321–330 *in* The Northwest European Tertiary Basin. VINKEN, R (compiler). *Geologisches Jahrbuch*, Vol. A100.

COWAN, G. 1989. Diagenesis of Upper Carboniferous sandstones: Southern North Sea Basin. 57–73 *in* Deltas: sites and traps for fossil fuels. WHATELEY, M K, and PICKERING, K T (editors). *Special Publication of the Geological Society of London*, No. 44.

COX, B M, LOTT, G K, THOMAS, J E, and WILKINSON, I P. 1987. Upper Jurassic stratigraphy of four shallow cored boreholes in the UK sector of the southern North Sea. *Proceedings of the Yorkshire Geological Society*, Vol. 46, 99–109.

— and RICHARDSON, G. 1982. The ammonite zonation of Upper Oxfordian mudstones in the Vale of Pickering, Yorkshire. *Proceedings of the Yorkshire Geological Society*, Vol. 44, 53–58.

COX, F C, and five others. 1985. Palaeocene sedimentation and stratigraphy in Norfolk, England. *Newsletters on Stratigraphy*, Vol. 14, 169–185.

CRAIG-SMITH, S J. 1972. *The changing system.* East Anglian coastal study report, No. 3. (Norwich: University of East Anglia.)

CREUTZBERG, F, and POSTMA, H. 1979. An experimental approach to the distribution of mud in the southern North Sea. *Netherlands Journal of Sea Research*, Vol. 13, 99–116.

CRITTENDEN, S. 1981. The distribution of Palaeogene planktonic foraminiferida and the biostratigraphy of a borehole in the southern North Sea. *Newsletters on Stratigraphy*, Vol. 10, 34–44.

— 1982a. Lower Cretaceous lithostratigraphy NE of the Sole Pit area in the UK southern North Sea. *Journal of Petroleum Geology*, Vol.5, 191–202.

— 1982b. A note on Danian planktonic foraminiferida from the British southern North Sea. *Revista Española de Micropaleontología*. Vol. 14, 53–62.

— 1986. Planktonic foraminifera and the biostratigraphy of the Early Tertiary strata of the North Sea Basin; a brief discussion. *Newsletters on Stratigraphy*, Vol. 15, 163–171.

CUMMING, A D, and WYNDHAM, C L. 1975. The geology and development of the Hewett gas-field. 313–325 in *Petroleum and the continental shelf of North-West Europe*. WOODLAND, A W (editor). (London: Applied Science Publishers.)

CURRY, D, and SMITH, A J. 1975. New discoveries concerning the geology of the central and eastern parts of the English Channel. *Philosophical Transactions of the Royal Society of London*, Vol. 279A, 155–167.

DALEY, B. 1972. Some problems concerning the early Tertiary climate of southern Britain. *Palaeogeography, Palaeoclimatology, Palaeoecology*, Vol. 11, 177–190.

DAVIS, A G, and ELLIOTT, G F. 1957. The palaeogeography of the London Clay Sea. *Proceedings of the Geologists' Association*, Vol. 68, 255–277.

DAVIS, F M. 1925. Quantitative studies on the fauna of the sea bottom. No. 2: results of the investigations in the southern North Sea. *Ministry of Agriculture and Fisheries, Fishery Investigations*, Series II, Vol. 8.

DAVIS, R A, and BALSON, P S. 1992. Stratigraphy of a North Sea tidal sand ridge. *Journal of Sedimentary Petrology*. Vol 62, No. 1, 116–122

DE BATIST, M, DE BRUYNE, M, HENRIET, J P, and MOSTAERT, F. 1989. Stratigraphic analysis of the Ypresian off the Belgian coast. 75–88 in *The Quaternary and Tertiary geology of the Southern Bight, North Sea.* HENRIET, J P, and DE MOOR, G (editors). (Brussels: Belgian Geological Survey.)

DE BOER, G. 1964. Spurn Head: its history and evolution. *Transactions of the Institute of British Geographers,* Vol. 34, 71–89.

DEEGAN, C E, and SCULL, B J. 1977. A standard lithostratigraphic nomenclature for the central and northern North Sea. *Report of the Institute of Geological Sciences,* No. 77/25; *Norwegian Petroleum Directorate Bulletin,* No. 1.

DEPARTMENT OF ENERGY. 1990. *Development of the oil and gas resources of the United Kingdom: a report to Parliament by the Secretary of State for Energy,* April 1990. (London: HMSO.)

— 1991. *Development of the oil and gas resources of the United Kingdom: a report to Parliament by the Secretary of State for Energy,* April 1991. (London: HMSO.)

DESTOMBES, J P, and SHEPHARD-THORN, E R. 1971. Geological results of the Channel Tunnel site investigation 1964–65. *Report of the Institute of Geological Sciences,* No. 71/11.

DEVOY, R J N. 1979. Flandrian sea level changes and vegetational history of the Lower Thames Estuary. *Philosophical Transactions of the Royal Society of London,* Vol. 285B, 355–410.

— 1982. Analysis of the geological evidence for Holocene sea-level movements in south east England. *Proceedings of the Geologists' Association,* Vol. 93, 65–90.

DINGLE, R V. 1965. Sandwaves in the North Sea mapped by continuous reflection profiling. *Marine Geology,* Vol. 3, 391–400.

— 1971. A marine geological survey off the north-east coast of England (western North Sea). *Journal of the Geological Society of London,* Vol. 127, 303–338.

DIXON, R G. 1979. Sedimentary facies in the Red Crag (Lower Pleistocene), East Anglia. *Proceedings of the Geologists' Association,* Vol. 90, 117–132.

D'OLIER, B. 1975. Some aspects of Late Pleistocene–Holocene drainage of the River Thames in the eastern part of the London Basin. *Philosophical Transactions of the Royal Society of London,* Vol. 279A, 269–277.

— 1981. Sedimentary events during sea-level rise in the south-west corner of the North Sea. *Special Publication of the International Association of Sedimentologists,* No. 5, 221–227.

DONATO, J A, MARTINDALE, W, and TULLY, M C. 1983. Buried granites within the Mid North Sea High. *Journal of the Geological Society of London,* Vol. 140, 825–837.

— and MEGSON, J B. 1990. A buried granite batholith beneath the East Midlands Shelf of the Southern North Sea Basin. *Journal of the Geological Society of London,* Vol. 147, 133–140.

— and TULLY, M C. 1981. A regional interpretation of North Sea gravity data. 65–75 in *Petroleum geology of the continental shelf of North-West Europe.* ILLING, L V, and HOBSON, G D (editors). (London: Heyden and Son.)

DONOVAN, D T. 1963. *The geology of British seas.* (Hull: University of Hull.)

— 1968. *Geology of the shelf seas.* (London: Oliver and Boyd.)

— 1973. The geology and origin of the Silver Pit and other closed basins in the North Sea. *Proceedings of the Yorkshire Geological Society,* Vol. 39, 267–293.

— and DINGLE, R V. 1965. Geology of part of the southern North Sea. *Nature, London,* Vol. 207, 1186–1187.

— HORTON, A, and IVIMEY-COOK, H I C. 1979. The transgression of the Lower Lias over the northern flank of the London Platform. *Journal of the Geological Society of London,* Vol. 136, 165–173.

DUGDALE, R E, FOX, H R, and RUSSELL, J R. 1978. Nearshore sandbanks and foreshore accretion on the south Lincolnshire coast. *East Midland Geographer,* Vol. 7, 48–63.

DUNHAM, K C, and WILSON, A A. 1985. Geology of the Northern Pennine Orefield: Volume 2, Stainmore to Craven. *Economic Memoir of the British Geological Survey.*

EAMES, T D. 1975. Coal rank and gas source relationships — Rotliegendes reservoirs. 191–203 in *Petroleum and the continental shelf of North-West Europe.* WOODLAND, A W (editor). (London: Applied Science Publishers.)

EHLERS, J, MEYER, K D, and STEPHAN, H J, 1984. The pre-Weichselian glaciations of north-west Europe. *Quaternary Science Reviews,* Vol. 3, 1–40.

EISMA, D. 1975. Holocene sedimentation in the Outer Silver Pit area (southern North Sea). *Marine Science Communications,* Vol. 1, 407–426.

— 1981. Supply and deposition of suspended matter in the North Sea. *Special Publication of the International Association of Sedimentologists,* No. 5, 415–428.

— JANSEN, J H F, and WEERING, T C E VAN. 1979. Sea-floor morphology and recent sediment movement in the North Sea. 217–231 in The Quaternary history of the North Sea. OELE, E, SCHÜTTENHELM, R T E, and WIGGERS, A J (editors). *Acta Universitatis Upsaliensis: Symposium Universitatis Upsaliensis Annum Quingentesimum Celebrantis: 2.*

— and KALF, J. 1979. Distribution and particle size of suspended matter in the Southern Bight of the North Sea and the eastern Channel. *Netherlands Journal of Sea Research,* Vol. 13, 298–324.

— — 1987. Dispersal, concentration and deposition of suspended matter in the North Sea. *Journal of the Geological Society of London,* Vol. 144, 161–178.

— MOOK, W G, and LABAN, C. 1981. An early Holocene tidal flat in the Southern Bight. *Special Publication of the International Association of Sedimentologists,* No. 5, 229–237.

ELLIOTT, T. 1975. The sedimentary history of a delta lobe from a Yoredale (Carboniferous) cyclothem. *Proceedings of the Yorkshire Geological Society,* Vol. 40, 505–536.

ELLISON, R A. 1983. Facies distribution in the Woolwich and Reading Beds of the London Basin, England. *Proceedings of the Geologists' Association,* Vol. 94, 311–319.

EVANS, C D R. 1990. *United Kingdom offshore regional report: the geology of the western English Channel and its western approaches.* (London: HMSO for British Geological Survey.)

EVANS, G. 1965. Intertidal flat sediments and their environments of deposition in The Wash. *Quarterly Journal of the Geological Society of London,* Vol. 121, 209–245.

FAIRBANKS, R G. 1989. A 17 000–year glacio-eustatic sea level record: influence of glacial melting rates on the younger Dryas event and deep-ocean circulation. *Nature, London,* Vol. 342, 637–642.

FANNIN, N G T. 1989. Offshore investigations 1966–1987. *British Geological Survey Technical Report* WB/89/2.

FIELDING, C R. 1986. Fluvial channel and overbank deposits from the Westphalian of the Durham coalfield, NE England. *Sedimentology,* Vol. 33, 119–140.

FISHER, M J. 1986. Triassic 113–132 in *Introduction to the petroleum geology of the North Sea* (2nd edition). GLENNIE, K W (editor). (Oxford: Blackwell Scientific Publications.)

— FUNNELL, B M, and WEST, R G. 1969. Foraminifera and pollen from a marine interglacial deposit in the western North Sea. *Proceedings of the Yorkshire Geological Society,* Vol. 37, 311–320.

FLEISCHER, U. 1963. Surface ship gravity measurements in the North Sea. *Geophysical Prospecting,* Vol. 11, 535–549.

FOLK, R L. 1954. The distinction between grain size and mineral composition in sedimentary rock nomenclature. *Journal of Geology,* Vol. 62, 344–359.

FRANCE, D S. 1975. The geology of the Indefatigable Gas-Field. 233–239 in *Petroleum and the continental shelf of North-west Europe.* WOODLAND, A W (editor). (London: Applied Science Publishers.)

FREEMAN, B, KLEMPERER, S L, and HOBBS, R W. 1988. The deep structure of northern England and the Iapetus Suture zone from BIRPS deep seismic reflection profiles. *Journal of the Geological Society of London,* Vol. 145, 727–740.

FUNNELL, B M. 1972. The history of the North Sea. *Bulletin of the Geological Society of Norfolk,* Vol. 21, 2–10.

— 1987. Late Pliocene and early Pleistocene stages of East Anglia and the adjacent North Sea. *Quaternary Newsletter,* Vol. 52, 1–11.

— 1988. Foraminifera in the Late Tertiary and Early Quaternary Crags of East Anglia. 50–52 in *The Pliocene–Middle Pleistocene of East Anglia, field guide.* GIBBARD, P L, and ZALASIEWICZ, J A (editors). (Cambridge: Quaternary Research Association.)

— 1989. Quaternary. 563–569 in *Stratigraphical atlas of fossil foraminifera* (2nd edition). JENKINS, D G, and MURRAY, J W (editors). (Chichester: Ellis Harwood Ltd.)

— and PEARSON, I. 1989. Holocene sedimentation on the north Norfolk coast in relation to relative sea-level change. *Journal of Quaternary Science,* Vol. 4, 25–36.

FYFE, J A, ABBOTS, I, and CROSBY, A. 1981. The subcrop of the mid-Mesozoic unconformity in the UK area. 236–244 in *Petroleum geology of the continental shelf of North-West Europe.* ILLING, L V, and HOBSON, G D (editors). (London: Heyden and Sons.)

GALLOIS, R W. 1965. *British regional geology: the Wealden district* (4th edition). (London: HMSO for Institute of Geological Sciences.)

— and COX, B M. 1974. The stratigraphy of the Upper Kimmeridge Clay of The Wash area. *Bulletin of the Geological Survey of Great Britain,* Vol. 47, 1–16.

— — 1977. The stratigraphy of the Middle and Upper Oxfordian sediments of Fenland. *Proceedings of the Geologists' Association,* Vol. 88, 207–228.

GAUNT, G D, FLETCHER, T P, and WOOD, C J. 1992. Geology of the country around Kingston upon Hull and Brigg. *Memoir of the British Geological Survey,* Sheets 80 and 90 (England and Wales).

— and TOOLEY, M J. 1974. Evidence for Flandrian sea-level changes in the Humber Estuary and adjacent areas. *Bulletin of the Geological Survey of Great Britain,* No. 48, 25–40.

GEIGER, M E, and HOPPING, C A. 1968. Triassic stratigraphy of the Southern North Sea Basin. *Philosophical Transactions of the Royal Society of London,* Vol. 254B, 1–36.

GEORGE, T N, and six others. 1976. A correlation of Dinantian rocks in the British Isles. *Special Report of the Geological Society of London,* No. 7.

GIBBARD, P L. 1988. The history of the great northwest European rivers during the past three million years. *Philosophical Transactions of the Royal Society of London,* Vol. 318B, 559–602.

— and sixteen others. 1991. Early and early Middle Pleistocene correlations in the southern North Sea Basin. *Quaternary Science Reviews,* Vol. 10, 23–52.

GIBBS, A D. 1986. Strike-slip basins and inversion: a possible model for the southern North Sea gas areas. 23–35 *in* Habitat of Palaeozoic gas in NW Europe. BROOKS, J, GOFF, J C, and VAN HOORN, B (editors). *Special Publication of the Geological Society of London,* No. 23.

GLENNIE, K W. 1972. Permian Rotliegendes of northwest Europe interpreted in light of modern desert sedimentation studies. *Bulletin of the American Association of Petroleum Geologists,* No. 56, 1048–1071.

— 1983. Early Permian (Rotliegendes) palaeowinds of the North Sea. *Sedimentary Geology,* Vol. 34, 245–265.

— 1986a. Development of northwest Europe's southern Permian gas basin. 3–22 *in* Habitat of Palaeozoic gas in NW Europe. BROOKS, J, GOFF, J C, and VAN HOORN, B (editors). *Special Publication of the Geological Society of London,* No. 23.

— 1986b. The structural framework and the pre-Permian history of the North Sea area. 25–62 in *Introduction to the petroleum geology of the North Sea* (2nd edition). GLENNIE, K W (editor). (Oxford: Blackwell Scientific Publications.)

— 1986c. Early Permian–Rotliegend. 63–85 in *Introduction to the petroleum geology of the North Sea* (2nd edition). GLENNIE, K W (editor). (Oxford: Blackwell Scientific Publications.)

— and BOEGNER, P L E. 1981. Sole Pit Inversion tectonics. 110–120 in *Petroleum geology of the continental shelf of North-West Europe.* ILLING, L V, and HOBSON, G D (editors) (London: Heyden and Sons.)

— and BULLER, A T. 1983. The Permian Weissliegend of NW Europe: the partial deformation of aeolian dune sands caused by the Zechstein transgression. *Sedimentary Geology,* Vol. 35, 43–81.

GODWIN, H. 1960. Radiocarbon dating and Quaternary history in Britain. *Proceedings of the Royal Society of London,* Vol. 153B, 287–320.

— and GODWIN, M E. 1933. British Maglemose harpoon sites. *Antiquity,* Vol. 7, 36–48.

GOLDRING, R, BOSENCE, D W J, and BLAKE, T. 1978. Estuarine sedimentation in the Eocene of southern England. *Sedimentology,* Vol. 25, 861–876.

GOODCHILD, M W, and BRYANT, P. 1986. The geology of the Rough Gas Field. 223–235 *in* Habitat of Palaeozoic gas in NW Europe. BROOKS, J, GOFF, G C, and VAN HOORN, B (editors). *Special Publication of the Geological Society of London,* No. 23.

GRAY, I. 1975. Viking gas-field. 241–247 in *Petroleum and the continental shelf of North-West Europe.* WOODLAND, A W (editor). (London: Applied Science Publishers.)

GRAYSON, R F, and OLDHAM, L. 1987. A new structural framework for the northern British Dinantian as a basis for oil, gas and mineral exploration. 33–59 in *European Dinantian environments.* MILLER, J, ADAMS, A E, and WRIGHT, V P (editors). (Chichester: John Wiley and Sons.)

GREENSMITH, J T, and TUCKER, E V. 1969. The origin of Holocene shell deposits in the chenier plain facies of Essex (Great Britain). *Marine Geology,* Vol. 7, 403–425.

— — 1971. The effects of Late Pleistocene and Holocene sea-level changes in the vicinity of the River Crouch, East Essex. *Proceedings of the Geologists' Association,* Vol. 82, 301–321.

— — 1973. Holocene transgressions and regressions on the Essex coast, Outer Thames Estuary. *Geologie en Mijnbouw,* Vol. 52, 193–202.

GUION, P D, and FIELDING, C R. 1988. Westphalian A and B sedimentation in the Pennine Basin, UK. 153–177 in *Sedimentation in a synorogenic basin complex: the Upper Carboniferous of Northwest Europe.* BESLY, B M, and KELLING, G (editors). (London: Blackie & Son.)

HALLAM, A, and EL SHAARAWY, Z. 1982. Salinity reduction of the end-Triassic sea from the Alpine region into northwestern Europe. *Lethaia,* Vol. 15, 169–178.

HAMBLIN, R J O, and six others. 1992. *United Kingdom offshore regional report: the geology of the English Channel.* (London: HMSO for British Geological Survey.)

HANCOCK, J M. 1975. The sequence of facies in the Upper Cretaceous of northern Europe compared with that in the western interior. 83–118 *in* Cretaceous system in the western Interior of North America. CALDWELL, W G F (editor). *Geological Association of Canada Special Paper,* No. 13.

— 1976. The petrology of the Chalk. *Proceedings of the Geologists' Association,* Vol. 86, 499–535.

— 1986. Cretaceous. 161–178 in *Introduction to the petroleum geology of the North Sea* (2nd edition). GLENNIE, K W (editor). (Oxford: Blackwell Scientific Publications.)

— and KAUFFMAN, E G. 1979. The great transgressions of the Late Cretaceous. *Journal of the Geological Society of London,* Vol. 136, 175–186.

— and SCHOLLE, P. 1975. Chalk of the North Sea. 413–425 in *Petroleum and the continental shelf of North-west Europe.* WOODLAND, A W (editor). (London: Applied Science Publishers.)

HAQ, B U, HARDENBOL, J, and VAIL, P R. 1987. Chronology of fluctuating sea levels since the Triassic. *Science, New York*, Vol 235, 1156–1167.

HARDING, T P, and LOWELL, J D. 1979. Structural styles, their plate tectonic habitats, and hydrocarbon traps in petroleum provinces. *Bulletin of the American Association of Petroleum Geologists*, Vol. 68, 1016–1059.

HARLAND, R, BONEY, A P, HUGHES, M J, and MORIGI, A N. 1991. The Lower Pleistocene stratigraphy of the Ormesby borehole, Norfolk, England. *Geological Magazine*, Vol. 128, 647–660.

HARLAND, W B, and five others. 1982. *A geologic time scale.* (Cambridge: Cambridge University Press.)

— and GAYER, R A. 1972. The Arctic Caledonides and earlier oceans. *Geological Magazine*, Vol. 109, 289–314.

HARPER, M L, and SHAW, B E. 1974. Cretaceous–Tertiary carbonate reservoirs in the North Sea. *Offshore North Sea Technology Conference, Stavanger, Norway*, G–IV/4.

HARRISON, D J. 1988. The marine sand and gravel resources off Great Yarmouth and Southwold, East Anglia. *British Geological Survey Technical Report*, WB/88/9.

— 1990. Marine aggregate survey. Phase 3. East coast. *British Geological Survey Technical Report* WB/90/17.

HART, J K, and PEGLAR, S M. 1990. Further evidence for the timing of the, Middle Pleistocene glaciation of Britain. *Proceedings of the Geologists' Association*, Vol. 101, 187–196.

HART, M B, and BIGG, P J. 1981. Anoxic events in the Late Cretaceous chalk seas of north-west Europe. 177–185 in *Microfossils from Recent and fossil shelf seas.* NEALE, J W, and BRASIER, M D (editors). (Chichester: Ellis Horwood Ltd.)

HAVE, A TEN, and HILLIER, A. 1986. Reservoir geology of the Sean North and South gas fields, UK southern North Sea. 267–273 *in* Habitat of Palaeozoic gas in NW Europe. BROOKS, J, GOFF, J C, and VAN HOORN, B (editors). *Special Publication of the Geological Society of London,* No. 23.

HEMINGWAY, J E. 1974. Jurassic. 161–223 in *The geology and mineral resources of Yorkshire.* RAYNER, D H and HEMINGWAY, J E (editors). (Leeds: Yorkshire Geological Society.)

— and KNOX, R W O'B. 1973. Lithostratigraphical nomenclature of the Middle Jurassic of the Yorkshire Basin in north-east England. *Proceedings of the Yorkshire Geological Society*, Vol. 39, 527–535.

HENRIET, J P, and eight others. 1989. Preliminary seismic-stratigraphic maps and type sections of the Palaeogene deposits in the Southern Bight of the North Sea. 29–44 in *The Quaternary and Tertiary geology of the Southern Bight, North Sea.* HENRIET, J P, and DE MOOR, G (editors). (Brussels: Belgian Geological Survey.)

— DE BATIST, M, VAERENBERGH, W VAN, and VERSCHUREN, M. 1991. Seismic facies and clay tectonic features of the Ypresian Clay in the southern North Sea. *Bulletin van de Belgische Vereniging voor Geologie*, Vol. 97, 457–472.

HINSCH, W. 1985. Die Molluskenfauna des Eems-Interglazials von Offenbüttel-Schnittlohe (Nord-Ostsee-Kanal, Westholstein). *Geologisches Jahrbuch*, Vol. A86, 49–62. [In German.]

— 1986. The northwest German Tertiary Basin Miocene and Pliocene. 679–699 *in* Nordwestdeutschland im Tertiär. *Beitrage zur Regionale Geologie der Erde*, Vol. 18.

HIRN, A, DAMOTTE, B, TORREILLES, G, and ECORS SCIENTIFIC PARTY. 1987. Crustal reflection seismics: the contributions of oblique, low frequency and shear wave illuminations. *Geophysical Journal of the Royal Astronomical Society*, Vol. 89, 287–296.

HODGSON, A V. 1978. Braided river bedforms and related sedimentary structures in the Fell Sandstone Group (Lower Carboniferous) of north Northumberland. *Proceedings of the Yorkshire Geological Society*, Vol. 41, 509–522.

HOLLIGER, K, and KLEMPERER, S L. 1989. A comparison of the Moho interpreted from gravity data and from deep seismic reflection data in the northern North Sea. *Geophysical Journal*, Vol. 97, 247–258.

— — 1990. Gravity and deep seismic reflection profiles across the North Sea rifts. In *Tectonic evolution of the North Sea rifts.* BLUNDELL, D J, and GIBBS, A (editors). (Oxford: Clarendon Press.)

HOLMES, R. 1977. The Quaternary geology of the UK sector of the North Sea between 56° and 58°N. *Report of the Institute of Geological Sciences,* No. 77/14.

HOLMES, S C A. 1981. Geology of the country around Faversham. *Memoir of the Geological Survey of Great Britain.*

HOUBOLT, J J H C. 1968. Recent sediments in the Southern Bight of the North Sea. *Geologie en Mijnbouw*, Vol. 47, 245–273.

HOUSE, M R, and five others. 1977. A correlation of Devonian rocks of the British Isles. *Special Report of the Geological Society of London,* No. 8.

HOWARTH, M J, and HUTHNANCE, J M. 1984. Tidal and residual currents around a Norfolk sandbank. *Estuarine, Coastal and Shelf Science*, Vol. 19, 105–117.

HUGHES, M J. 1981. Contribution on the Oligocene and Eocene microfaunas of the southern North Sea. 186–195 in *Microfossils from Recent and fossil shelf seas.* NEALE, J W, and BRASIER, M D (editors). (Chichester: Ellis Horwood Ltd.)

IVIMEY-COOK, H C, and POWELL, J H. 1991. Late Triassic and early Jurassic biostratigraphy of the Felixkirk borehole, North Yorkshire. *Proceedings of the Yorkshire Geological Society,* Vol. 48.

JAGO, C F. 1981. Sediment response to waves and currents, North Yorkshire Shelf, North Sea. *Special Publication of the International Association of Sedimentologists,* No. 5, 283–301.

JANSEN, J H F, WEERING, T C E VAN, and EISMA, D. 1979. Late Quaternary sedimentation in the North Sea. 175–187 *in* The Quaternary history of the North Sea. OELE, E, SCHÜTTENHELM, R T E, and WIGGERS, A J (editors). *Acta Universitatis Upsaliensis: Symposium Universitatis Upsaliensis Annum Quingentesimum Celebrantis*: 2.

JANSSEN, A W. 1984. *Mollusken uit het Mioceen Van Winterswijk-Miste. Een inventarisatie met beschrijvingen en afbeeldingen van alle aangetroffen soorten.* (Amsterdam: KNNV, NGV, RGM.) [In Dutch.]

JARKE, J. 1956. Der Boden der südlichen Nordsee. *Deutsche Hydrographische Zeitschrift*, Vol. 9, 1–9. [In German.]

JEANS, C V. 1973. The Market Weighton structure: tectonics, sedimentation and diagenesis during the Cretaceous. *Proceedings of the Yorkshire Geological Society*, Vol. 39, 409–444.

JEFFERIES, R P S. 1963. The stratigraphy of the *Actinocamax plenus* Subzone (Turonian) in the Anglo-Paris Basin. *Proceedings of the Geologists' Association*. Vol. 74, 1–33.

JEFFERY, D H. 1991. Comment on: The origin of major incisions within the Pleistocene deposits of the North Sea — Wingfield (1990). *Marine Geology*, Vol. 96, 125–126.

— and LONG, D. 1989. Early Pleistocene sedimentation and geographic change in the UK sector of the North Sea. *Terra Abstracts*, Vol.1, 425.

JELGERSMA, S. 1961. Holocene sea level changes in The Netherlands. *Mededelingen van de Geologische Stichting*, Serie C, Vol. VI, No. 7.

— 1979. Sea-level changes in the North Sea basin. 233–248 *in* The Quaternary history of the North Sea. OELE, E, SCHÜTTENHELM, R T E, and WIGGERS, A J (editors). *Acta Universitatis Upsaliensis: Symposium Universitatis Upsaliensis Annum Quingentesimum Celebrantis*: 2.

JENKINS, D G, and HOUGHTON, S D. 1987 Age, correlation and paleoecology of the St Erth Beds and the Coralline Crag of England. *Proceedings of the Third Meeting of the Regional Committee of Northern Neogene Stratigraphy.* 147–156. JANSSEN, A W (editor). (Leiden: Rijkmuseum Geologiem Mineralogie.)

JENKINS, D G, and MURRAY, J W. 1989. *Stratigraphical atlas of fossil foraminifera* (2nd edition). (Chichester: Ellis Horwood Ltd.)

JENYON, M K. 1984. Seismic response to collapse structures in the southern North Sea. *Marine and Petroleum Geology*, Vol. 1, 27–36.

— 1985. Basin-edge diapirism and up-dip salt flow in Zechstein of southern North Sea. *Bulletin of the American Association of Petroleum Geologists,* Vol. 69, 53–64.

— 1986a. Some consequences of faulting in the presence of a salt rock interval. *Journal of Petroleum Geology,* Vol. 9, 29–51.

— 1986b. *Salt tectonics.* (Amsterdam: Elsevier Applied Science Publishers.)

— 1988. Overburden deformation related to the pre-piercement development of salt structures in the North Sea. *Journal of the Geological Society of London,* Vol. 145, 445–454.

— and CRESSWELL, P M. 1987. The Southern Zechstein salt basin of the British North Sea, as observed in regional seismic traverses. 277–292 in *Petroleum geology of North West Europe.* BROOKS, J, and GLENNIE, K W (editors). (London: Graham and Trotman.)

JOHNSON, M A, KENYON, N H, BELDERSON, R H, and STRIDE, A H. 1982. Sand transport. 58–94 in *Offshore tidal sands: processes and deposits.* STRIDE, A H (editor). (London: Chapman and Hall.)

JOON, B, LABAN, C, and MEER, J J M VAN DER. 1990. The Saalian glaciation in the Dutch part of the North Sea. *Geologie en Mijnbouw,* Vol. 69, 151–158.

JOSEPH, J. 1955. Extinction measurements to indicate distribution and transport of watermasses. *Proceedings of UNESCO Symposium on Physical Oceanography,* (1955, Tokyo). 59–75.

KENNEDY, W J, and GARRISON, R E. 1975. Morphology and genesis of nodular chalks and hardgrounds in the Upper Cretaceous of southern England. *Sedimentology,* Vol. 22, 311–386.

KENT, P E. 1947. A deep boring at North Creake, Norfolk. *Geological Magazine,* Vol. 84, 2–18.

— 1967. Outline geology of the Southern North Sea Basin. *Proceedings of the Yorkshire Geological Society,* Vol. 36, 1–22.

— 1975. Review of North Sea Basin development. *Journal of the Geological Society of London,* Vol. 131, 435–468.

— 1980a. Subsidence and uplift in East Yorkshire and Lincolnshire: a double inversion. *Proceedings of the Yorkshire Geological Society,* Vol. 42, 505–524.

— 1980b. *British regional geology: eastern England from the Tees to The Wash.* (London: HMSO.)

KENYON, N H, BELDERSON, R H, STRIDE, A H, and JOHNSON, M A. 1981. Offshore tidal sand-banks as indicators of net sand transport and as potential deposits. *Special Publication of the International Association of Sedimentologists,* No. 5, 257–268.

KIMBELL, G S, CHADWICK, R A, HOLLIDAY, D W, and WERNGREN, O C. 1989. The structure and evolution of the Northumberland Trough from new seismic reflection data and its bearing on modes of continental extension. *Journal of the Geological Society of London,* Vol. 146, 775–788.

KING, C. 1981. The stratigraphy of the London Clay and associated deposits. *Tertiary Research Special Paper,* No. 6.

— 1983. Cainozoic micropalaeontological biostratigraphy of the North Sea. *Report of the Institute of Geological Sciences,* No. 82/7.

— 1984. The stratigraphy of the London Clay Formation and Virginia Water Formation in the coastal sections of the Isle of Sheppey (Kent, England). *Tertiary Research,* Vol. 5, 121–160.

— 1989. Cenozoic of the North Sea. 418–489 in *Stratigraphical atlas of fossil foraminifera* (2nd edition). JENKINS, D G, and MURRAY, J W (editors). (Chichester: Ellis Horwood Ltd.)

— 1990. Eocene stratigraphy of the Knokke borehole (Belgium). *Toelichtende verhandelingen voor de Geologische en Mijnkaarten van België,* Vol. 29, 67–102.

KIRBY, G A, and SWALLOW, P W. 1987. Tectonism and sedimentation in the Flamborough Head region of north-east England. *Proceedings of the Yorkshire Geological Society,* Vol. 46, 301–309.

KIRBY, R. 1987. Sediment exchanges across the coastal margins of NW Europe. *Journal of the Geological Society of London,* Vol. 144, 121–126.

— and OELE, E. 1975. The geological history of the Sandettie–Fairy Bank area, southern North Sea. *Philosophical Transactions of the Royal Society of London,* Vol. 279A, 257–267.

KIRTON, S R, and DONATO, J A. 1985. Some buried Tertiary dykes of Britain and surrounding waters deduced by magnetic modelling and seismic reflection methods. *Journal of the Geological Society of London,* Vol. 142, 1047–1057.

KLEMPERER, S L. 1989. Deep seismic reflection profiling and the growth of the continental crust. *Tectonophysics,* Vol. 161, 233–244.

KNOWLES, B. 1964. The radioactive content of the Coal Measures sediments in the Yorkshire – Derbyshire Coalfield. *Proceedings of the Yorkshire Geological Society,* Vol. 34, 413–450.

KNOX, R W O'B. 1979. Igneous grains associated with zeolites in the Thanet Beds of Pegwell Bay, north-east Kent. *Proceedings of the Geologists' Association,* Vol. 90, 55–59.

— 1983. Volcanic ash in the Oldhaven Beds of southeast England, and its stratigraphical significance. *Proceedings of the Geologists' Association,* Vol. 94, 245–250.

— 1984a. Lithostratigraphy and depositional history of the Late Toarcian sequence at Ravenscar, Yorkshire. *Proceedings of the Yorkshire Geological Society,* Vol. 45, 99–108.

— 1984b. Nannoplankton zonation and the Palaeocene/Eocene boundary beds of NW Europe: an indirect correlation by means of volcanic ash layers. *Journal of the Geological Society of London,* Vol. 141, 993–999.

— 1989. The Palaeogene of the UK southern North Sea, 50°N–53°N: a survey of stratigraphical data. 17–28 in *The Quaternary and Tertiary geology of the Southern Bight, North Sea.* HENRIET, J P, and DE MOOR, G (editors). (Brussels: Belgian Geological Survey.)

— 1990. Thanetian and early Ypresian chronostratigraphy in south-east England. *Tertiary Research,* Vol. 11, 57–64.

— and FLETCHER, B N. 1978. Bentonites in the Lower D Beds (Ryazanian) of the Speeton Clay of Yorkshire. *Proceedings of the Yorkshire Geological Society,* Vol. 42, 21–27.

— MORIGI, A N, ALI, J R, HAILWOOD, E A, and HALLAM, J R. 1990. Early Palaeogene stratigraphy of a cored borehole at Hales, Norfolk. *Proceedings of the Geologists' Association,* Vol. 101, 145–151.

— and MORTON, A C. 1983. Stratigraphical distribution of early Palaeogene pyroclastic deposits in the North Sea Basin. *Proceedings of the Yorkshire Geological Society,* Vol. 44, 355–363.

— — 1988. The record of early Tertiary North Atlantic volcanism in sediments of the North Sea Basin. 407–419 *in* Early Tertiary volcanism and the opening of the NE Atlantic. MORTON, A C, and PARSON, L M (editors). *Special Publication of the Geological Society of London,* No. 39.

KOCKEL, F. 1988a. Assumed basin-wide relative sea-level changes in the NW European Tertiary Basin. Figure 35a *in* The Northwest European Tertiary Basin. VINKEN, R (compiler). *Geologisches Jahrbuch,* Vol. 100A.

— 1988b. Danian and Marine Montian, Palaeogeography. Map 1 *in* The Northwest European Tertiary Basin. VINKEN, R (compiler). *Geologisches Jahrbuch,* Vol. 100A.

KOOI, H, CLOETINGH, S, and REMMELTS, G. 1989. Intraplate stresses and the stratigraphic evolution of the North Sea Central Graben. *Geologie en Mijnbouw,* Vol. 68, 49–72.

LABAN, C, CAMERON, T D J, and SCHÜTTENHELM, R T E. 1984. Geologie van het Kwartair in de zuidelijke bocht van de Noordzee. *Mededelingen van de Werkgroep voor Tertiaire en Kwartaire Geologie, Natururhistorisch Museum, Leiden,* Vol. 21, 139-154.

LAKE, R D, ELLISON, R A, HENSON, M R and CONWAY, B W. 1986. Geology of the country around Southend and Foulness. *Memoir of the British Geological Survey,* Sheets 258 and 259 (England and Wales).

LAKE, S T, and KARNER, G D. 1987. The structure and evolution of the Wessex Basin, southern England: an example of inversion tectonics. *Tectonophysics,* Vol. 137, 347–378.

LAND, D H. 1974. Geology of the Tynemouth district. *Memoir of the Geological Survey of Great Britain.*

LANGHORNE, D N. 1973. A sandwave field in the Outer Thames Estuary, Great Britain. *Marine Geology,* Vol. 14, 129–143.

LARSONNEUR, C, HORN, R, and AUFFRET, J.P. 1975. Géologie de la partie méridionale de la Manche centrale. *Philosophical Transactions of the Royal Society of London,* Vol. 279A, 145–153. [In French.]

LE BAS, M J. 1972. Caledonian igneous rocks beneath central and eastern England. *Proceedings of the Yorkshire Geological Society,* Vol. 39, 71–86.

LEEDER, M R. 1974. Lower Border Group (Tournaisian) fluvio-deltaic sedimentation and palaeogeography of the Northumberland Basin. *Proceedings of the Yorkshire Geological Society,* Vol. 40, 129–180.

— 1982. Upper Palaeozoic basins of the British Isles: Caledonide inheritance versus Hercynian plate margin processes. *Journal of the Geological Society of London,* Vol. 139, 481–494.

— 1983. Lithospheric stretching and North Sea Jurassic sourcelands. *Nature, London,* Vol. 305, 510–514.

— 1987. Tectonic and palaeogeographic models for Lower Carboniferous Europe. 1–20 in *European Dinantian environments.* MILLER, J, ADAMS, A E, and WRIGHT, V P (editors). (Chichester: John Wiley & Sons.)

— and HARDMAN, M. 1990. Carboniferous of the Southern North Sea Basin and controls on hydrocarbon prospectivity. 87–105 *in* Tectonic events responsible for Britain's oil and gas reserves. HARDMAN, R F P AND BROOKS, J (editors). *Special Publication of the Geological Society of London,* No. 55.

— RAISWELL, R, AL-BIATTY, H, MCMAHON, A, and HARDMAN, M. 1990. Carboniferous stratigraphy, sedimentation and correlation of well 48/3-3 in the Southern North Sea Basin: integrated use of palynology, natural gamma/sonic logs and carbon/sulphur geochemistry. *Journal of the Geological Society of London,* Vol. 147, 287–300.

LEES, B J. 1982. Quaternary sedimentation in the Sizewell–Dunwich Banks area, Suffolk. *Bulletin of the Geological Society of Norfolk,* Vol. 32, 1–35.

LETSCH, W J, and SISSINGH, W. 1983. Tertiary stratigraphy of The Netherlands. *Geologie en Mijnbouw,* Vol. 62, 305–318.

LONG, D, LABAN C, STREIF, H, CAMERON, T D J, and SCHÜTTENHELM, R T E. 1988. The sedimentary record of climatic variation in the southern North Sea. *Philosophical Transactions of the Royal Society of London,* Vol. 318B, 523–537.

LOTT, G K. 1986. Late Triassic, Jurassic and early Cretaceous geology of the Southern North Sea Basin. Unpublished PhD thesis, University of Leicester.

— FLETCHER, B N, and WILKINSON, I P. 1986. The stratigraphy of the Lower Cretaceous Speeton Clay Formation in a cored borehole off the coast of north-east England. *Proceedings of the Yorkshire Geological Society,* Vol. 46, 39–56.

— KNOX, R W O'B, HARLAND, R, and HUGHES, M J. 1983. The stratigraphy of Palaeogene sediments in a cored borehole off the coast of north-east Yorkshire. *Report of the Institute of Geological Sciences,* No. 83/9.

— THOMAS, J B, RIDING, J B, DAVEY, R J, and BUTLER, N. 1989. Late Ryazanian black shales in the Southern North Sea Basin and their lithostratigraphical significance. *Proceedings of the Yorkshire Geological Society,* Vol. 47, 321–324.

— and WARRINGTON, G. 1988. A review of the latest Triassic succession in the UK sector of the Southern North Sea Basin. *Proceedings of the Yorkshire Geological Society,* Vol. 47, 139–147.

LOVELL, J P B. 1986. Cenozoic. 170–196 in *Introduction to the petroleum geology of the North Sea* (2nd edition). GLENNIE, K W (editor). (Oxford: Blackwell Scientific Publications.)

LUND, C-E, and HEIKKINEN, P. 1987. Reflection measurements along the EGT POLAR-profile, northern Baltic Shield. *Geophysical Journal of the Royal Astronomical Society,* Vol. 89, 361–364.

LÜDERS, K. 1939. Sediments of the North Sea. 322–342 in *Recent Marine Sediments, a symposium.* TRASK, P D (editor). (Tulsa: American Association of Petroleum Geologists.)

MADGETT, P A, and CATT, J A. 1978. Petrography, stratigraphy and weathering of Late Pleistocene till in East Yorkshire, Lincolnshire and North Norfolk. *Proceedings of the Yorkshire Geological Society,* Vol. 42, 55–108.

MALM, O A, and five others. 1984. The Lower Tertiary Balder Formation: an organogenic and tuffaceous deposit in the North Sea region. 149–170 in *Petroleum geology of the North European margin.* SPENCER, A M (editor). (London: Graham and Trotman Ltd.)

MARIE, J P P. 1975. Rotliegendes stratigraphy and diagenesis. 205–211 in *Petroleum and the continental shelf of North-West Europe.* WOODLAND, A W (editor). (London: Applied Science Publishers.)

MATHERS, S J, and ZALASIEWICZ, J A. 1985. Producing a comprehensive geological map. A case study — the Aldeburgh–Orford area of East Anglia. *Modern Geology,* Vol. 9, 207–220.

— — 1988. The Red Crag and Norwich Crag formations of southern East Anglia. *Proceedings of the Geologists' Association,* Vol. 99, 261–278.

MATTHEWS, D H, And CHEADLE, M J. 1986. Deep reflections from the Caledonides and Variscides west of Britain. 5–19 in *Reflection seismology: a global perspective.* BARAZANGI, M, and BROWN, L (editors). Geodynamics Series, No. 13. (Washington DC: American Geophysical Union.)

MCCAVE, I N. 1971a. Wave effectiveness at the sea bed and its relationship to bed-forms and deposition of mud. *Journal of Sedimentary Petrology,* Vol. 41, 89–96.

— 1971b. Sandwaves in the North Sea off the coast of Holland. *Marine Geology,* Vol. 10, 199–225.

— 1972. Transport and escape of fine-grained sediment from shelf areas. 225–248 in *Shelf sediment transport: process and pattern.* SWIFT, D J P, DUANE, D B, and PILKEY, O H (editors). (Stroudsberg, Pennsylvania: Dowden, Hutchinson and Ross Inc.)

— 1973. Mud in the North Sea. 75–100 in *North Sea science.* GOLDBERG, E D (editor). (Cambridge, Massachussets: MIT Press.)

— 1978. Grain-size trends and transport along beaches: example from eastern England. *Marine Geology,* Vol. 28, M43–M51.

— 1981. Location of coastal accumulations of fine sediments around the southern North Sea. *Rapport et procès-verbaux des réunions. Conseil Permanent International pour l'Exploration de la Mer,* Vol. 181, 15–27.

— 1987. Fine sediment sources and sinks around the East Anglian coast (UK). *Journal of the Geological Society of London,* Vol. 144, 149–152.

— and GEISER, A C. 1978. Megaripples, ridges and runnels on intertidal flats of The Wash, England. *Sedimentology,* Vol. 26, 353–369.

— and LANGHORNE, D N. 1982. Sand waves and sediment transport around the end of a tidal sand bank. *Sedimentology,* Vol. 29, 95–110.

MCQUILLIN, R, ARNOLD, S E, TULLY, M, and HULL, J H. 1969. Cruise report: Humber investigation, 1968. *Report of the Institute of Geological Sciences,* No. 69/3.

MEISSNER, R, MATTHEWS, D H, and WEVER, TH. 1986. The "Moho" in and around Great Britain. *Annales Geophysicae,* Vol. 4, 659–664.

MICHELSEN, O. 1978. Stratigraphy and distribution of Jurassic deposits of the Norwegian–Danish Basin. *Danmarks Geologiske Undersøgelse, Serie B,* No. 8, 1–133.

MILSOM, J, and RAWSON, P F. 1989. The Peak Trough — a major control on the geology of the North Yorkshire coast. *Geological Magazine,* Vol. 126, 699–705.

MITCHELL, M. 1981. The age of the Dinantian (Lower Carboniferous) rocks proved beneath the Kent Coalfield. *Geological Magazine,* Vol. 118, 703–711.

MORTIMORE, R N. 1983. The stratigraphy and sedimentation of the Turonian–Campanian in the Southern Province of England. *Zitteliana,* Vol. 10, 27–41.

— and POMEROL, B. 1987. Correlation of the Upper Cretaceous White Chalk (Turonian to Campanian) in the Anglo-Paris Basin. *Proceedings of the Geologists' Association,* Vol. 98, 97–143.

— and WOOD, C J. 1983. The distribution of flint in the English Chalk, with particular reference to the 'Brandon Flint Series' and the high Turonian flint maximum. 6–20 in *The scientific study of flint and chert.* SIEVEKING, G DE G, and HART, M B (editors). (Cambridge: Cambridge University Press.)

MORTON, A C, 1982. The provenance and diagenesis of Palaeogene sandstones of southeast England as indicated by heavy mineral analysis. *Proceedings of the Geologists' Association,* Vol. 93, 263–274.

— and KNOX, R W O'B. 1990. Geochemistry of late Palaeocene and early Eocene tephras from the North Sea Basin. *Journal of the Geological Society of London,* Vol. 147, 425–437.

MURRAY, K H. 1986. Correlation of electrical resistivity marker bands in the Cenomanian and Turonian Chalk from the London Basin to east Yorkshire. *Report of the British Geological Survey,* Vol. 17, No. 8.

NAM and RGD (Nederlandse Aardolie Maatschappij and Rijks Geologische Dienst). 1980. Stratigraphic nomenclature of The Netherlands. *Transactions of the Royal Dutch Geological and Mining Society of Delft.*

NEALE, J W. 1974. Cretaceous. 225–243 in *The geology and mineral resources of Yorkshire.* RAYNER, D H, and HEMINGWAY, J E (editors). (Leeds: Yorkshire Geological Society.)

NILSSON, T. 1983. *The Pleistocene: geology and life in the Quaternary ice age.* (Dordrecht: Reidel.)

NIO, S-D. 1976. Marine transgressions as a factor in the formation of sandwave complexes. *Geologie en Mijnbouw,* Vol. 55, 18–40.

NUNNY, R S, and CHILLINGWORTH, P C H. 1986. *Marine dredging for sand and gravel.* (London: HMSO.)

O'CONNOR, B A. 1987. Short and long term changes in estuary capacity. *Journal of the Geological Society of London,* Vol. 144, 187–195.

OELE, E. 1969. The Quaternary geology of the Dutch part of the North Sea, north of the Frisian Islands. *Geologie en Mijnbouw,* Vol. 48, 467–480.

— and SCHÜTTENHELM, R T E. 1979. Development of the North Sea after the Saalian glaciation. 191–215 *in* The Quaternary history of the North Sea. OELE, E, SCHÜTTENHELM, R T E, and WIGGERS, A J (editors). *Acta Universitatis Upsaliensis: Symposia Universitatis Upsaliensis Annum Quingentesimum Celebrantis: 2.*

OWEN, D E. 1987. Commentary: usage of stratigraphic terminology in papers, illustrations and talks. *Journal of Sedimentary Petrology,* Vol. 57, 363–372.

PANTIN, H M. 1991. The sea-bed sediments around the United Kingdom; their bathymetric and physical environment, grain size, mineral composition and associated bedforms. *British Geological Survey Research Report,* SB/90/1.

PARKER J R. 1975. Lower Tertiary sand development in the central North Sea. 447–453 in *Petroleum and the continental shelf of North-West Europe.* WOODLAND, A W (editor). (London: Applied Science Publishers.)

PENN, I E, COX, B M, and GALLOIS, R W. 1986. Towards precision in stratigraphy: geophysical log correlation of Upper Jurassic (including Callovian) strata of the Eastern England Shelf. *Journal of the Geological Society,* Vol. 143, 381–410.

PENNINGTON, J J. 1975. The geology of the Argyll Field. 285–291 in *Petroleum and the continental shelf of North-West Europe.* WOODLAND, A W (editor). (London: Applied Science Publishers.)

PHARAOH, T C, MERRIMAN, R J, WEBB, P C, and BECKINSALE, R D. 1987. The concealed Caledonides of eastern England: preliminary results of a multidisciplinary study. *Proceedings of the Yorkshire Geological Society,* Vol. 46, 355–369.

— and five others. 1991. Early Palaeozoic arc-related volcanism in the concealed Caledonides of southern Britain. *in* Proceedings of the International Meeting on the Caledonides of the Midlands and the Brabant Massif, Brussels 20–23 September 1989. ANDRÉ, L, HERBOSCH, A, VANGUESTAIN, M, and VERNIERS, J (editors). *Annales de la Société Géologique de Belgique,* Vol. 114.

PLANT, J A, and eight others. 1988. Metallogenic models and exploration criteria for buried carbonate-hosted ore deposits: results of a multidisciplinary study in eastern England. 321–352 in *Mineral deposits within the European Community.* BOISSONAS, J, and OMENETTO, P (editors). (Berlin: Springer-Verlag.)

POWELL, J H. 1984. Lithostratigraphical nomenclature in the Yorkshire Basin. *Proceedings of the Yorkshire Geological Society,* Vol. 45, 51–57.

QUINE, M L. 1988. Sedimentology of the Chalk of Coastal Haute Normandie, France. Unpublished PhD thesis, University of London.

RAMSBOTTOM, W H C. 1977. Major cycles of transgression and regression (mesothems) in the Namurian. *Proceedings of the Yorkshire Geological Society,* Vol. 41, 261–291.

— and six others. 1978. A correlation of Silesian rocks in the British Isles. *Special Report of the Geological Society of London,* No. 10.

RAWSON, P F, and seven others. 1978. A correlation of Cretaceous rocks in the British Isles. *Special Report of the Geological Society of London,* No. 9.

— and RILEY, L A. 1982. Latest Jurassic–early Cretaceous events and the "Late Cimmerian Unconformity" in the North Sea area. *Bulletin of the American Association of Petroleum Geologists,* No. 66, 2628–2648.

RAYNER, D H. 1981. *The stratigraphy of the British Isles* (2nd edition). (London: Cambridge University Press.)

READ, W A. 1979. A quantitative analysis of an Upper Westphalian fluviodeltaic succession and a comparison with earlier Westphalian deposits in the Kent Coalfield. *Geological Magazine,* Vol. 116, 431–443.

RESTON, T J, and BLUNDELL, D J. 1987. Possible mid-crustal shears at the edge of the London Platform. *Geophysical Journal of the Royal Astronomical Society,* Vol. 89, 251–258.

RHYS, G H. 1974. A proposed standard lithostratigraphic nomenclature for the southern North Sea and an outline structural nomenclature for the whole of the (UK) North Sea. A report on the joint Oil Industry — Institute of Geological Sciences Committee on North Sea Nomenclature. *Report of the Institute of Geological Sciences,* No. 74/8.

— 1975. A proposed standard lithostratigraphic nomenclature for the southern North Sea. 151–163 in *Petroleum and the continental shelf of North-West Europe.* WOODLAND, A W (editor). (London: Applied Science Publishers.)

RICHTER-BERNBURG, G. 1986. Zechstein 1 and 2 anhydrites: facts and problems of sedimentation. 157–163 *in* The English Zechstein and related topics. HARWOOD, G M, and SMITH, D B (editors). *Special Publication of the Geological Society of London,* No. 22.

RIDDLER, G P. 1981. The distribution of the Röt Halite Member in North Yorkshire, Cleveland and North Humberside. *Proceedings of the Yorkshire Geological Society,* Vol. 43, 341–346.

RIDING, J, and WRIGHT, J K. 1990. Palynostratigraphy of the Scalby Formation (Middle Jurassic) of the Cleveland Basin, north-east Yorkshire. *Proceedings of the Yorkshire Geological Society,* Vol. 47, 349–354.

Riley, L A, and Tyson, R V. 1981. Discussion of Fyfe et al. 244 in *Petroleum geology of the continental shelf of North-West Europe.* Illing, L V, and Hobson, G D (editors). (London: Heyden and Son.)

Robinson, A H W. 1966. Residual currents in relation to shoreline evolution of the East Anglian coast. *Marine Geology,* Vol. 4, 57–84.

— 1968. The submerged glacial landscape off the Lincolnshire coast. *Transactions of the Institute of British Geographers,* Vol. 44, 119–132.

Sager, G, and Sammler, R. 1975. *Atlas der Gezeitenströme für die Nordsee, den Kanal und die Irische See.* (Rostock: Seehydrographischer Dienst der Deutschen Demokratischen Republik.)

Sannemann, D. 1968. Salt-stock families in northwestern Germany. 261–270 *in* Dapirism and diapirs. *Memoirs of the American Association of Petroleum Geologists,* No. 8.

Savin, S M. 1977. The history of the earth's surface temperature during the past 100 million years. *Annual Review of Earth and Planetary Sciences,* Vol. 5, 319–355.

Schlanger, S O, and Jenkyns, H C. 1976. Cretaceous oceanic anoxic events — causes and consequences. *Geologie en Mijnbouw,* Vol. 55, 179–184.

Scholle, P A. 1974. Diagenesis of Upper Cretaceous chalks from England, Northern Ireland, and the North Sea. 177–210 *in* Pelagic sediments: on land and under the sea. Hsü, K J, and Jenkyns, H C (editors). *Special Publication of the International Association of Sedimentologists,* No. 1.

Scotese, C R, Gahagan, L M and Larson, R L. 1988. Plate tectonic reconstructions of the Cretaceous and Cenozoic ocean basins. *Tectonophysics,* Vol. 155, 27–48.

Selley, R C. 1976. The habitat of North Sea oil. *Proceedings of the Geologists' Association,* Vol. 87, 359–387.

Service Géologique de Belgique. 1964. Carte Aéromagnétique Résiduelle Polaire. 1:300 000.

— 1968. Massif du Brabant: Carte Géologique du Socle Paléozoïque. 1:300 000.

Shennan, I. 1986. Flandrian sea-level changes in Fenland, II. Tendencies of sea-level movement, altitudinal changes and local and regional factors. *Journal of Quaternary Science,* Vol. 1, 155–179.

— 1987. Holocene sea-level changes in the North Sea. 109–151 in *Sea-level changes.* Tooley, M J, and Shennan, I (editors). (Oxford: Blackwell.)

— 1989. Holocene crustal movements and sea-level changes in Great Britain. *Journal of Quaternary Science,* Vol. 4, 77–89.

Simon, J B, and Bluck, B J. 1982. Palaeodrainage of the southern margin of the Caledonian mountain chain in the northern British Isles. *Transactions of the Royal Society of Edinburgh: Earth Sciences,* Vol. 73, 11–15.

Skinner, A C, and Gregory, D M. 1983. Quaternary stratigraphy in the northern North Sea. *Boreas,* Vol. 12, 145–152.

Smith, D B, 1988a. Morphological development of the Sandettie–South Falls gap: a degeneration ebb dominated tidal passage in the southern North Sea. 51–64 in *Tide influenced sedimentary environments and facies.* De Boer, P L, Gelder, A Van, and Nio, S D (editors). (Dordrecht: D. Reidel Publishing Company.)

— 1988b. Bypassing of sand over sand waves and through a sand wave field in the central region of the southern North Sea. 39–50 in *Tide influenced sedimentary environments and facies.* De Boer, P L, Gelder, A van, and Nio, S-D (editors). (Dordrecht: D. Reidel Publishing Company.)

Smith, D B. 1974. Permian. 115–144 in *The geology and mineral resources of Yorkshire.* Rayner, D H, and Hemingway, J E (editors). (Leeds: Yorkshire Geological Society.)

— 1979. Rapid marine transgressions and regressions of the Upper Permian Zechstein Sea. *Journal of the Geological Society of London.* Vol. 136, 155–156.

— 1980. The evolution of the English Zechstein basin. 7–34 *in* The Zechstein Basin with emphasis on carbonate sequences. Füchtbauer, H, and Peryt, T M (editors). *Contributions to Sedimentology,* Vol. 9.

— 1981. The Magnesian Limestone (Upper Permian) reef complex of northeastern England. *Society of the Economic Palaeontologists and Mineralogists Special Publication,* No. 30, 161–186.

— 1989. The late Permian palaeogeography of north-east England. *Proceedings of the Yorkshire Geological Society,* Vol. 47, 285–312.

— Brunstrom, R G W, Manning, P I, Simpson, S, and Shotton, F W. 1974. A correlation of Permian rocks in the British Isles. *Special Report of the Geological Society of London,* No. 5.

— and Crosby, A. 1979. The regional and stratigraphical context of Zechstein 3 and 4 potash deposits in the British sector of the southern North Sea and adjoining land areas. *Economic Geology,* Vol. 74, 397–408.

— Harwood, G M, Pattison, J, and Pettigrew, T. 1986. A revised nomenclature for Upper Permian strata in Eastern England. 9–17 *in* The English Zechstein and related topics. Harwood, G M, and Smith, D B (editors). *Special Publication of the Geological Society of London,* No. 22.

Smith, K, and Smith, N J P. 1989. Deep geology. 53–64 in *Metallogenic models and exploration criteria for buried carbonate-hosted ore deposits — a multidisciplinary study in eastern England.* Plant, J A, and Jones, D G (editors). (Keyworth: British Geological Survey and Institution of Mining and Metallurgy.)

Smith, N J P. 1985. Structure contour and subcrop maps of the pre-Permian surface of the United Kingdom (south). (Keyworth: British Geological Survey.)

— 1987. The deep geology of central England: the prospectivity of the Palaeozoic rocks. 217–224 in *Petroleum geology of North West Europe.* Brooks, J, and Glennie, K W (editors). (London: Graham and Trotman.)

Soper, N J, Webb, B C, and Woodcock, N H. 1987. Late Caledonian (Acadian) transpression in north-west England: timing, geometry and geotectonic significance. *Proceedings of the Yorkshire Geological Society,* Vol. 46, 175–192.

Southworth, C J. 1987. Lithostratigraphy and depositional history of the Middle Triassic Dowsing Dolomitic Formation of the southern North Sea and adjoining areas. Unpublished PhD thesis, University of Oxford.

Staalduinen, C J van, and eight others. 1979. The geology of The Netherlands. *Mededelingen van Rijks Geologische Dienst,* Vol. 31–2, 9–49.

Stather, J W. 1912. Shelly clay dredged from the Dogger Bank. *Quarterly Journal of the Geological Society of London,* Vol. 68, 324–327.

Steele, R P. 1988. The Namurian sedimentary history of the Gainsborough Trough. 102–113 in *Sedimentation in a synorogenic basin complex: the Upper Carboniferous of Northwest Europe.* Besly, B M, and Kelling, G (editors). (Glasgow: Blackie and Son.)

Stewart, I J. 1987. A revised stratigraphic interpretation of the Early Palaeogene of the central North Sea. 557–576 in *Petroleum geology of North West Europe.* Brooks, J, and Glennie, K W (editors). (London: Graham and Trotman.)

Stoker, M S, Long, D, and Fyfe, J A. 1985. A revised Quaternary stratigraphy for the central North Sea. *Report of the British Geological Survey,* No. 17/2.

Stride, A H. 1988. Indications of long term episodic suspension transport of sand across the Norfolk Banks, North Sea. *Marine Geology,* Vol. 79, 55–64.

— 1989. Modern deposits, quasi-deposits and some Holocene? in the Southern Bight, North Sea. 149–159 in *The Quaternary and Tertiary geology of the Southern Bight, North Sea.* Henriet, J P, and De Moor, G (editors). (Brussels: Belgian Geological Survey.)

— BELDERSON, R H, KENYON, N H, and JOHNSON, M A. 1982. Offshore tidal deposits: sand sheet and sand bank facies. 95–125 in *Offshore tidal sands: processes and deposits.* STRIDE, A H (editor). (London: Chapman and Hall.)

SUMBLER, M G. 1983. A new look at the type Wolstonian glacial deposits of central England. *Proceedings of the Geologists' Association,* Vol. 94, 23–31.

SWINNERTON, H H. 1931. The post-glacial deposits of the Lincolnshire coast. *Quarterly Journal of the Geological Society of London,* Vol. 87, 360–375.

TAYLOR, B J, and five others. 1971. *British regional geology: northern England* (4th edition). (London: HMSO for Institute of Geological Sciences.)

TAYLOR, J C M. 1980. Origin of the Werraanhydrit in the southern North Sea — a reappraisal. 91–113 *in* The Zechstein Basin with emphasis on carbonate sequences. FÜCHTBAUER, H, and PERYT, T M (editors). *Contributions to Sedimentology,* No. 9.

— 1986. Late Permian — Zechstein. 87–111 in *Introduction to the petroleum geology of the North Sea* (2nd edition). GLENNIE, K W (editor). (Oxford: Blackwell Scientific Publications.)

— and COLTER, V S. 1975. Zechstein of the English sector of the Southern North Sea Basin. 249–263 in *Petroleum and the continental shelf of North-West Europe.* WOODLAND, A W (editor). (London: Applied Science Publishers.)

TAYLOR, S R. 1983. A stable isotope study of the Mercia Mudstones (Keuper Marl) and associated sulphate horizons in the English Midlands. *Sedimentology,* Vol. 30, 11–31.

TERWINDT, J H J. 1971. Sand waves in the Southern Bight of the North Sea. *Marine Geology,* Vol. 10, 51–67.

TESCH, J J. 1910. De physische gesteldheid der Noordsee. *Tijdschrift van het Kominklijk Nederlandsch Aardrijkskundig Genootschap,* Vol. 27, 702–740. [In Dutch.]

THOMPSON, P G. 1911. Note on the occurrence of stony beds underlying Harwich Harbour. *Essex Naturalist,* Vol. 16, 305–309.

THORNE, J A, and WATTS, A B. 1989. Quantitative analysis of North Sea subsidence. *Bulletin of the American Association of Petroleum Geologists,* Vol. 73, 88–116.

TOWNSEND, H A, and HAILWOOD, E A. 1985. Magnetostratigraphic correlation of Palaeogene sediments in the Hampshire and London Basins, southern UK. *Journal of the Geological Society of London,* Vol. 142, 957–982.

TUBB, S R, SOULSBY, A, and LAWRENCE, S R. 1986. Palaeozoic prospects on the northern flanks of the London–Brabant Massif. 55–72 *in* Habitat of Palaeozoic gas in NW Europe. BROOKS, J, GOFF, J C, and VAN HOORN, B (editors). *Special Publication of the Geological Society of London,* No. 23.

TURNER, J S. 1949. The deeper structure of central and northern England. *Proceedings of the Yorkshire Geological Society,* Vol. 27, 280–297.

VALENTIN, H. 1971. Land loss at Holderness. 116–137 in *Applied coastal geomorphology.* STEERS, J A (editor). (London: Macmillan.)

VAN DER MEER, J J M, and LABAN, C. 1990. Micromorphology of some North Sea till samples, a pilot study. *Journal of Quaternary Science,* Vol. 5, 95–101.

VAN HOORN, B. 1987. Structural evolution, timing and tectonic style of the Sole Pit Inversion. *Tectonophysics,* Vol. 137, 239–284.

VEEN, F R VAN. 1975. Geology of the Leman Gas-Field. 223–231 in *Petroleum and the continental shelf of North-West Europe.* WOODLAND, A W (editor). (London: Applied Science Publishers.)

VEEN, J VAN. 1935. Sand waves in the North Sea. *Hydrographic Review,* Vol. 12, 21–29.

— 1938. Die unterseeische sandwüste in der Nordsee. *Geologie der Meere und Binnengewässer,* Vol. 2, 62–86. [In German.]

VEENSTRA, H J. 1965. Geology of the Dogger Bank area, North Sea. *Marine Geology,* Vol. 3, 234–262.

— 1969. Gravels of the southern North Sea. *Marine Geology,* Vol. 7, 449–464.

— 1971. Sediments of the southern North Sea. *Report of the Institute of Geological Sciences,* No. 70/15, 9–23.

VINCENT, C E. 1979. Longshore sand transport rates; a simple model for the East Anglian coastline. *Coastal Engineering,* Vol. 3, 113–136.

VINKEN, R (compiler). 1988. The Northwest European Tertiary Basin. Results of the International Geological Correlation Programme Project No. 124. *Geologisches Jahrbuch,* Vol. A100.

WALKER, I M, and COOPER, W G. 1987. The structural and stratigraphic evolution of the northeast margin of the Sole Pit Basin. 263–275 in *Petroleum geology of North West Europe.* BROOKS, J, and GLENNIE, K W (editors). (London: Graham and Trotman.)

WALMSLEY, P J. 1983. The role of the Department of Energy in petroleum exploration of the United Kingdom. 3–10 *in* Petroleum geochemistry and exploration of Europe. BROOKS, J (editor). *Special Publication of the Geological Society of London,* No. 12.

WALTER, R. 1980. Lower Palaeozoic palaeogeography of the Brabant Massif and its southern adjoining areas. *Mededelingen van Rijks Geologische Dienst,* Vol. 32–2, 14–25.

WARRINGTON, G, and eight others. 1980. A correlation of Triassic rocks in the British Isles. *Special Report of the Geological Society of London,* No. 13.

WEST, R G. 1977. *Pleistocene geology and biology.* (London: Longman.)

— 1980. *The pre-glacial Pleistocene of the Norfolk and Suffolk coasts.* (Cambridge: Cambridge University Press.)

WHITEHEAD, H, and GOODCHILD, M H. 1909. Some notes on "moorlog", a peaty deposit from the Dogger Bank in the North Sea, with a report on the plant remains by C Reid and E M Reid. *Essex Naturalist,* Vol. 16, 51–60.

WHITTAKER, A (editor). 1985. *Atlas of onshore sedimentary basins in England and Wales: Post-Carboniferous tectonics and stratigraphy.* (Glasgow: Blackie and Son.)

— HOLLIDAY, D W, and PENN, I E. 1985. Geophysical logs in British stratigraphy. *Special Report of the Geological Society of London,* No. 18.

WIJHE, D H VAN. 1987. Structural evolution of inverted basins in the Dutch offshore. *Tectonophysics,* Vol. 137, 171–219.

WILLEMS, W. 1989. Foraminiferal biostratigraphy of the Palaeogene in the Southern Bight of the North Sea. 45–49 in *The Quaternary and Tertiary geology of the Southern Bight, North Sea.* HENRIET, J P, and DE MOOR, G (editors). (Brussels: Belgian Geological Survey.)

WILSON, J B. 1982. Shelly faunas associated with temperate offshore tidal deposits. 126–171 in *Offshore tidal sands: processes and deposits.* STRIDE, A H (editor). (London: Chapman and Hall.)

WINGFIELD, R T R. 1990. The origin of major incisions within the Pleistocene deposits of the North Sea. *Marine Geology,* Vol. 91, 31–52.

— EVANS, C D R, DEEGAN, S E, and FLOYD R. 1978. Geological and geophysical survey of The Wash. *Report of the Institute of Geological Sciences,* No. 78/18.

WOOD, C J, and SMITH, E G. 1978. Lithostratigraphical classification of the Chalk in North Yorkshire, Humberside and Lincolnshire. *Proceedings of the Yorkshire Geological Society,* Vol. 42, 263–287.

WOOLLAM, R, and RIDING, J B. 1983. Dinoflagellate cyst zonation of the English Jurassic. *Report of the Institute of Geological Sciences,* No. 83/2.

ZAGWIJN, W H. 1974. The Pliocene–Pleistocene boundary in western and southern Europe. *Boreas,* Vol. 3, 75–97.

— 1979. Early and Middle Pleistocene coastlines in the Southern North Sea Basin. 31–42 *in* The Quaternary History of the North Sea. OELE, E, SCHÜTTENHELM, R T E, and WIGGERS, A J (editors). *Acta Universitatis Upsaliensis: Symposium Universitatis Upsaliensis Annum Quingentesimum Celebrantis: 2.*

— 1985. An outline of the Quaternary stratigraphy of The Netherlands. *Geologie en Mijnbouw,* Vol. 64, 17–24.

— 1989. The Netherlands during the Tertiary and the Quaternary: a case history of Coastal Lowland evolution. *Geologie en Mijnbouw,* Vol. 68, 107–120.

— and DOPPERT, J W C. 1978. Upper Cenozoic of the Southern North Sea basin: palaeoclimatic and palaeogeographic evolution. *Geologie en Mijnbouw,* Vol. 57, 577–588.

— and VEENSTRA, H J. 1966. A pollen-analytical study of cores from the Outer Silver Pit, North Sea. *Marine Geology,* Vol. 4, 539–551.

ZALASIEWICZ, J A, and MATHERS, S J. 1988. The Pliocene to early Middle Pleistocene of East Anglia: an overview. 1–34 in *The Pliocene to Middle Pleistocene of East Anglia. Field guide.* GIBBARD, P L, and ZALASIEWICZ, J A (editors). (Cambridge: Quaternary Research Association.)

ZIEGLER, P A. 1982. *Geological atlas of Western and Central Europe.* (Amsterdam: Shell Internationale Petroleum Maatschappij BV.)

— 1987. Late Cretaceous and Cenozoic intra-plate compressional deformations in the Alpine foreland — a geodynamic model. *Tectonophysics,* Vol. 137, 389–420.

INDEX

Aalenian 70, 72, 73
Actinocamax plenus 90
Aegiranum Marine Band 34
Albian 80, 84, 86, 88
Aller Halite Formation 53, 54
Alpine compression/orogeny 15, 91, 101
Alston Block 24
Amethyst Gasfield 58
    Member 56, 58
Ampthill Clay 77, 78
Anglian 108, 109, 110
Anglo-Dutch Basin 6, 13, 17, 18, 24, 26, 27, 29, 30, 32, 34–36, 38, 39, 56, 58–61, 66, 136
Antian 105, 106
Antwerp Sands 99
Aptian 80, 86, 88
Argyll Oilfield 11
Askrigg Block 24
Audrey Gasfield 133
Aurora Formation 106, 107, 108
Azolla filiculoides 107, 111

Bacton Group 55, 56
Bajocian 68, 72, 73, 76
Bagshot Sands 97
Baltic 125
    craton 6
    rivers 94
    Shield 10
barnacles 130
Barque Gasfield 133
Barren Measures 32
    Red Beds 32
    Red Group 32
Bartonian 97
Basalanhydrit Formation 47, 48
base-Cretaceous unconformity 54
base-Permian unconformity 14, 27, 32, 35, 36
batholiths 8
Bathonian 68, 72, 73, 76
Baventian 106
bedforms 116, 120
Belgian sector 94, 97
Belgium 1, 3, 6, 9, 10, 11, 91, 94, 97–99, 134,
Billingham Formation 52
Binks, The 127
BIRPS 1, 5
Black Band 90
Blea Wyke Sandstone Formation 68, 73
Bligh Bank Formation 120
Blisworth Clay 76
Blue Anchor Formation 65
Blyth dyke swarm 8
Bolders Bank Formation 113–115
Bolsovian 14, 32, 34–36, 132, 133
Boom Clay Formation 98
Boreal Ocean 14, 43, 55
Borkumriff Formation 112
Botney Cut Formation 115, 119
Bouguer gravity anomaly 3, 6, 8, 12, 14
Boulby Formation 53

Brabourne borehole 56
Brent Group 72
Bröckelschiefer Member 56, 57
Brotherton Formation 50
Brown Bank Formation 112, 113
Brownmoor borehole 68, 77
Brunhes–Matuyama transition 101, 108
bryozoans 99, 130
Buitenbanken Formation 119
Bunter Sandstone Formation 14, 57, 58–60, 134
Bunter Shale Formation 56, 58, 134
Bure Gasfield 133
Burnham Chalk Formation 83, 89

Cadeby Formation 45
Cadomian 6
Caister Gasfield 132
    Ness 128
Calais 1
Calcinema permiana 50
Caledonian compression/deformation 6, 11, 13
    fold belt 6
    granite 8, 14, 24
    mountain chains 11
    shear zones 6
Callovian 68, 73, 76
Cambrian 9, 11
Camelot Gasfield 133
Campanian 89, 90
Carboniferous 1, 3, 5, 6, 10–14, 17, 18, 23–37, 39, 127, 131, 132, 133, 134, 136
Carcharocles megalodon 98
Carnallitic Marl 53
Carnallitic Marl Formation 53
Carstone Formation 80
Cementstone Group 24
Cenomanian 89, 90
Central Fracture Zone 18, 20
Central Graben 11, 14, 15, 67, 68, 72, 84, 91
Cerastoderma 130
Chalk/Chalk Group 1, 15, 20, 80, 83, 86, 88–91, 94, 95, 99, 100, 126, 130
    Sea 4, 89, 90
Channel Tunnel 88
Charnian 6, 11
Chattian 98
Cheviot Hills 127
Cinder Bed 83, 86
Cleaver Bank Formation 111, 112
    Bank High 14, 15, 54, 56, 58, 59, 61
Cleethorpes No 1 borehole 86
Cleeton Gasfield 133
Cleveland Basin 3, 6, 14, 15, 18, 56, 67, 68, 70, 73, 77, 78, 80, 89
    Ironstone Formation 68, 70, 72
Clipper Gasfield 133
Coal Measures 32
Corallian Formation 76, 77, 78
Coralline Crag 99
    Oolite Formation 77, 78
Cornbrash 73, 76

Corton Bank 128
Corton Sand 124
Crane Formation 103, 106
Cretaceous 1, 3, 4, 13–17, 18, 20, 43, 67, 80–91, 94, 95, 99, 132, 130
Cromer Forest Bed Formation 106, 108
    Knoll Group 83, 86
    Ridge 110
Cromerian Complex 108, 110, 111

Danian 89, 91, 94
Deckanhydrit Formation 49
Deep Water Channel 119
Deltaic division 101
Denmark 94
Department of Energy 1, 131
Derbyshire Coalfield 34
Deurne Sands 99
Devonian 3, 6, 8, 10–13, 17, 23, 24, 131, 135
diapirs/diapirism 3, 15, 20, 43, 56, 60, 91
Diest Sands 99
Dinantian 13, 14, 23, 24, 26, 27, 29, 30, 32, 132
dinoflagellate cyst 84, 94–97, 103, 105, 107, 111, 114
Dogger Bank 1, 107, 112–114, 116, 120, 122, 124, 127
    Bank Formation 113, 114, 115
    Fault Zone 20, 101
Donax vittatus 130
Dorset 73
Dotty Gasfield 58
Dover Strait 1, 11, 12, 15, 27, 32, 34, 36–38, 43, 56, 80, 86, 88, 89, 91, 111, 119, 120, 124, 125, 127, 128, 130
Dowsing Dolomitic Formation 59–61
    Fault Zone 4, 9, 15, 17, 18, 20, 24, 61, 65, 80, 132
Drill Stone 95
Duckmantian 14, 32, 34, 132
Dudgeon Saliferous Formation 61, 65
Durham 34, 38, 45, 136
    Coalfield 32
Dutch Central Graben 68
    sector 6, 34, 38, 58, 61, 72, 89, 94, 101, 109, 112, 113, 119, 121, 129, 133

East Anglia 1, 3, 6, 8, 11, 12, 15, 24, 27, 32, 34, 39, 101, 103, 105–109, 116, 120, 124, 127, 128, 129, 130, 134, 135, 136
    Bank Ridges 122
    Midlands 67, 71
    Midlands Shelf 3, 14–17, 18, 24, 39, 59–61, 65, 68, 71–73, 76–80, 84, 89
    Ruston borehole 27
Eastern Avalonia 10
Eburonian 106
Echinocardium cordatum 124
echinoderms 130
echinoids 99
Edale Gulf 24
Edlington Formation 48

Eem Formation 112, 113
Eemian 112–114, 119
Egmond Ground Formation 111, 112
Elbow Formation 115, 119
Elphidium pseudolessonii 107
Elsterian 101, 108, 110, 111, 112, 115
English Channel 80, 90, 91, 99, 110, 111, 113, 114, 119, 125,
Eocene 91, 94, 96–99
Esk Evaporite Formation 60
Eskdale Gasfield 133
Esmond Gasfield 58, 59, 134
Essex 130

Famennian 12
Fell Sandstone Group 24
Fennoscandia 13
Ferriby Chalk Formation 83, 86, 89
Firth of Forth 101
Flamborough Chalk Formation 83, 89
   Head 1, 80, 86, 90, 120, 125–127
Flemish Bight 24
foraminifera 89, 91, 93, 94, 105, 108, 111, 130
Forbes Gasfield 58, 59, 134
Fordon Formation 49
France 1, 5, 6, 14, 36, 38, 89, 94
Frodingham Ironstone 70

Gabbard Bank 125
Gainsborough Trough 24
Galloper Bank 125
Gault 83, 86
   Clay 88
Gauss magnetic epoch 101
German Bight 116
Germany 3, 14, 38, 40, 41, 43, 55, 56, 58, 59, 61, 67, 80, 94, 110, 133, 134
Goodwin Sands 122, 125
Gordon Gasfield 58, 59, 134
Grantham Formation 76
Grauer Salzton Formation 49, 50, 52
gravel waves 121, 126
Great Yarmouth 116, 124, 127, 129
Grenzanhydrit Formation 54
Groningen Gasfield 133
Gunfleet Sand 124

Haisborough Group 17, 56, 59, 66, 134
   Sand 124
Hales Clay 95
halokinesis 1, 3, 14–17, 20, 43, 49, 56, 91
Hampshire–Dieppe Basin 91, 97
Happisburgh 127
Hardegsen Unconformity 56
Harwich Harbour 95
   Member 95–97
Hastings Beds 86
Hauptanhydrit Formation 49, 52
Hauptdolomit Formation 47, 48
Heligoland Channel 114
Hettangian 68, 71
Hewett Gasfield 47, 57, 58, 134
   Ridges 124
   Sandstone Bed 57, 134
Holderness 113, 114, 125, 127, 128
Holm Bank 128
Holocene 1, 101, 114–130, 135
Holsteinian 110, 111
Humber Estuary 1, 86, 116, 127–129, 134, 135

Group 68, 76, 77, 80
Hunstanton 113
   Chalk Member 86
Hydraulic Limestones 68, 71
Hyperacanthum Zone 96

Iapetus Ocean 10, 11
   suture 10
IJmuiden Ground Formation 103, 105, 106
Indefatigable Gasfield 40, 133
   Grounds Formation 120
Inner Gabbard 125
Inoceramus 86
Intermediate Röt Mudstone Member 60
invertebrates 89
Ipswichian 119
Irish Sea 14, 131

Jet Rock Member 72
Jurassic 1, 3, 6, 13–17, 18, 20, 43, 56, 66, 67–80, 84, 86, 89, 94, 132, 133
Jurassic–Cretaceous boundary 78, 79, 83, 86

Kellaways Rock and Sand 73, 76
Kesgrave Formation 108
Kent 15, 27, 34, 36–38, 43, 56, 80, 86, 88, 94–96, 99, 125
   Coalfield 24, 27, 32, 34, 36, 86
Keuper Anhydritic Member 65
   Halite Member 61, 65
Kimmeridge Clay 76, 77
   Clay Formation 78, 79, 80, 86
Kimmeridgian 67, 68, 76, 77, 78
Kirkham Abbey Formation 48
Kreftenheye Formation 113, 114
Kupferschiefer Formation 45

Langsettian 27, 32, 34, 132
Laramide phase 91
Late-Cimmerian Unconformity 15, 80
Lauenburger Clay 110
Laurasia 23, 39
Laurentian landmass 13
   shield 10
Leine Halite Formation 49, 53
Leman Bank 119
   Gasfield 39–41, 47, 133
   Sandstone Formation 39–42, 133
Lenham Beds 99
Lias Group 65, 66, 68-72, 73, 76
Lilstock Formation 65
Lincolnshire 3, 32, 38, 58, 70, 73, 84, 86, 90, 113, 119, 136
   Limestone Formation 73, 76
Lista Formation 94, 95
Little Dotty Gasfield 134
Lockton Gasfield 133
London 134
   Basin 91, 95, 96
   Clay Formation 95–97, 100, 127
London–Brabant Massif 3, 5, 6, 10–17, 23, 24, 26, 27, 32, 34, 36, 38, 39, 43, 45, 48, 55–57, 59–61, 65, 68, 72, 76, 77, 80, 83, 84, 86, 88, 89, 133, 134
Long Sand 124
Low Countries 103
Lower Calcareous Grit Formation 77
   Chalk 83, 88, 90
   Greensand 80, 83, 86, 88
   Magnesian Limstone 45

Röt Mudstone Member 59
Werraanhydrit Formation 45, 47
lower-crustal reflectors 5, 6
Lowestoft 107, 116
   Ness 128
Lutetian 91, 97
Maastrichtian 89, 90
Main Bundsandstein Formation 58
   Coal 34
   Röt Evaporite Member 60
   Röt Halite Member 60
   Röt Mudstone Member 60
Malton Gasfield 133
Market Weighton Block 68, 70, 86
   structure 77
Markham's Hole 1, 129
   Formation 106, 107
Marl Slate Formation 45
Marlstone Rock Bed 70, 72
Marsdenian 29, 30
Matuyama magnetic epoch 101, 103
Meetjesland Formation 97
Mercia Mudstone Group 56
Mesolithic bone harpoon 119
Meuse 127
Mid North Sea High 3, 5, 8, 10–12, 14, 18–20, 23, 24, 38, 39, 43, 52, 55, 58, 68, 73, 80, 86
Middle Chalk 83, 88, 90
   Marls Formation 48
Middlesbrough 43
Midlands Microcraton 11
Millstone Grit 30
Miocene 15, 91, 94, 98, 99, 101
Mohorovičić discontinuity 5, 6, 14
molluscs 99, 120, 126, 130
Murdoch Gasfield 8/9, 32, 34, 42, 132
Muschelkalk Dolomitic Member 60
   Evaporite Member 60
   Evaporites 61
   Halite Member 60
   Mudstone Member 60
Mytilus edulis 130

Namurian 14, 23, 24, 27, 29, 30, 32, 34, 132
nannoplankton 94, 96
Neogene 4, 94, 98–101, 125, 130
Neogloboquadrina atlantica 100
Netherlands, The 1, 4, 15, 27, 40, 42, 56, 58, 60, 67, 94, 97, 98, 101, 102, 103, 106, 108–110, 116, 120, 133, 134
Nettleton Bottom borehole 68, 76, 77
Newcome Bank 128
Nieuw Zeeland Gronden Formation 119
Nodule Bed 84
Nondeltaic division 101, 109
Norfolk 1, 45, 80, 84, 86, 89, 94, 95, 116, 120, 122, 124, 125–128, 135
   Banks 122, 124, 126, 128, 129
   Broads 116
North Creake borehole 11
   European Triassic Basin 56, 61, 65
   Falls Bank 121, 125
   Foreland 1, 125
Northampton Sand 76
Northern North Sea Basin 68
   Permian Basin 38
Northumberland 24, 136
   Trough 11, 12, 14, 23, 24
Norwegian Sea 91

sector 96
Norwegian–Danish Basin 67, 68
Norwegian–Greenland Sea 91
Norwich Crag Formation 106, 108
Nummulites 91, 97, 130

Old Red Sandstone continent 11, 23
Oldhaven Beds 96
Olduvai magnetic event 99, 106
Oligocene 15, 91, 94, 97, 101
Ordovician 6, 10, 11, 32, 80
Orfordness 127
Ormesby borehole 94, 105, 107
    Clay 94, 95
Osgodby Formation 73
Outer Gabbard sandbank 99, 125
Outer Silver Pit 1, 101, 107, 116, 122, 127, 128, 129
    Silver Pit Fault 18
    Silver Pit Formation 106, 107
Outer Thames 127, 128, 134, 135
Ower Bank 124
Oxford Clay Formation 73, 76, 77
Oxfordian 68, 73, 77, 78
Oyster Ground 129

Palaeogene 4, 15, 91, 94–99, 100, 111, 130
Paleocene 89–91, 94–97, 127
Pastonian 106, 108
Pecten Ironstone 72
Peelo Formation 110
Pegmatitanhydrit 53, 54
    Formation 53
Pegwell Bay 94
Penarth Group 56, 65, 66, 68, 71
Penhurst borehole 56
Pennine Basin 32, 34
    High 14, 39, 43, 55, 56, 59
Permian 1, 3, 5, 6, 14–17, 18, 20, 23, 24, 38–57, 89, 91, 131, 132, 133, 134, 136
Permo-Triassic 15, 41
Phrygiana faunal belt 34
Plattendolomit Formation 49, 50, 52
Pleistocene 1, 4, 16, 20, 39, 91, 99, 101–115, 119, 120, 122, 125, 135
Plenus Marl 90
Pliensbachian 68, 70, 72
Plio-Pleistocene boundary 99, 100
Pliocene 15, 94, 99–101, 105, 107, 127
pollen 102, 105
polychaetes 130
Portlandian 83
Portlandian/Volgian 76
Praetiglian 101, 105, 106, 107
Precambrian 6, 9, 10, 11
Priabonian 97
Proterozoic 10
proto-Atlantic fracture zone 43
proto-Dover Strait 110
Proto-Tethys Ocean 11
Psiloceras 66
Purbeck Beds 83, 86

Quaternary 1, 4, 15, 20, 56, 67, 80, 89, 94, 96, 101–130

radiocarbon and stable-isotope analyses 119
radiolaria 97
Ravenscar Group 73
Ravenspurn Gasfield 133

Reading Formation 95
Red Chalk Formation 80, 86
Red Crag 99, 100
    Crag Formation 99, 100, 105
Redcar Formation 68

Réunion magnetic event 105
Rhabdammina biofacies 91
Rhaetian 65, 67, 71
Rhaetic Sandstone Member 66
Rhine 94, 113–115, 127, 128
Riff Sandstone Member 59
Rogenstein Member 56, 58
Röt Evaporites 61
    Halite Member 60
Roter Salzton Formation 49, 53
Rotliegendes Group 38, 39, 42
Rough Gasfield 40, 133
Rupel Formation 98
Rupelian 97, 98
Ryazanian 83, 84, 86

Saalian 111, 112
salt pillows/piercement 3, 5, 15, 20, 68, 80, 94
Sand Hills 122
Sand Hole Formation 111
sand ribbons 120
    wave 1, 116, 120–122, 124, 125, 127
sandbank 116, 120, 121, 122, 124, 125, 128
Sandettie Bank 116, 120, 121, 126
Sandringham Sands 78, 80, 84
Saxthorpe borehole 11
Scalby Formation 73
Scandinavian ice sheets 108
Scheldt 128
Scotland 91, 119, 127, 130
Scottish Highlands 10
Scram Gasfield 134
Scremerston Coal Group 24
Scroby Sands 124, 128
Seal Sands borehole 26
Sean Gasfield 133
Sele Formation 95
Sherburn Anhydrite 53
Sherwood Sandstone Group 56, 58
Shipwash 95
Silurian 11, 13
Silver Pit 1, 80, 111, 116
Silverpit Formation 39–42
    Halite Member 42
Sinemurian 70, 72
Sizewell–Dunwich Banks 128
Skegness 116
Smith's Knoll 124
    Knoll Formation 101, 105, 106, 108
Sneaton Halite Formation 54
Sole Pit 1
    Pit Inversion axis 17, 32, 94, 133
    Pit Inversion 68, 76, 94
    Pit Trough 3, 6, 14–17, 18, 20, 38–40, 56, 58, 59, 61, 65, 68, 72, 73, 76–78, 80, 84, 89, 132, 133, 136
Solling Mudstone Member 59
Somerton borehole 11, 26
South Creake borehole 11
    Cross Sand 124
    Falls Bank 116, 120, 121, 125, 126
    Foreland 1
    Hewett Fault Zone 17

West Spit 122
Southern Bight 3, 101, 103, 105–107, 108, 110, 114, 119, 120, 127
    North Sea Gas Basin 131
    Permian Basin 38, 42
Speeton Clay 80
    Clay Formation 79, 80, 84, 86, 88
    Cliffs 80, 86
Spilsby Sandstone 78, 80
    Sandstone Formation 80, 83, 84
Spisula subtruncata 119
Spurn Head 125, 127
Staintondale Group 53
Staithes Formation 68, 70, 72
Stassfurt Halite Formation 47, 49, 50
Stephanian 23, 32, 34
Stinkkalk Member 48
Stinkschiefer Member 48
Stradbroke Trough 99
Strait of Dogger 122
Subcrenatum Marine Band 32, 34
Suffolk 119
Sunderland 115
    Ground Formation 115
Sunk Sand 124
Swarte Bank Formation 109, 110, 111
    Bank Hinge Zone 17, 18, 56
Symon Unconformity 36

Tea Kettle Hole Formation 111, 112
Terebratula 99
Terebratula maxima 99
Terschellinger-bank Member 120
Tertiary 1, 4, 5, 8, 9, 13, 15, 17, 20, 80, 89, 90, 91–100, 125, 127
Tethyan province 67
Tethys 14, 56, 61, 80, 91
Thames 128
Thames Estuary 1, 80, 96, 97, 116, 119, 122, 124, 125, 130
    Estuary banks 124, 125
Thanet Formation 94, 95
Thanetian 94
Thorpe Ness 128
Thurnian 105
tidal sand ridges 1, 120, 122
Tiglian 105, 106, 107, 108
Toarcian 68, 70, 72, 73
Tongeren Formation 97
Tornquist Sea 10
Tournaisian 24, 27
Trent Formation 65
Triassic 3, 6, 11, 14–17, 18, 20, 39, 43, 55–66, 68, 70, 80, 86, 89, 131, 132, 134, 136
Trimley Sands 99
Triton Anhydritic Formation 61, 65
Turonian 80, 89, 90
Turritella communis 120
Twente Formation 113, 114, 127

Upper Calcareous Grit Formation 78
    Chalk 83, 88, 89
    Estuarine Series 76
    Greensand 83, 86, 88
    Magnesian Limestone 50
    Röt Evaporite Member 60
    Röt Halite Member 60
    Röt Mudstone Member 60
    Rotliegend Group 38
    Werraanhydrit Formation 45

Werraanhydrit Member 45
Upware Limestone 77
Ur-Frisia 108

Valanginian 80, 84
Vale of Pickering–Flamborough Head Fault Zone 15, 17, 18, 20
Valhall Formation 86
Vanderbeckei Marine Band 34
Vanguard Gasfield 133
Variscan compression/deformation 14, 16, 23, 38
  faults 11, 56
  foreland 14
  foreland basin 3, 14, 43, 55
  highlands 39, 40
  orogeny 3
  structures 14, 34
Victor Gasfield 40
Viking Gasfield 40, 133
  Graben 14, 15, 67, 80, 84, 91
Virginia Water Formation 97
Viséan 24, 26, 27
Vulcan Gasfield 133

Waalian 106
Waltonian 105
Wash, The 1, 8, 11, 76, 78, 79, 84, 86, 101, 109, 116, 122, 128, 129, 130
Weald Basin 15, 86
  Clay 86

Weald–Artois Axis 91
Wealden Beds 80, 83, 86, 88
Weichselian 101, 112–116, 119, 127, 135
Weissliegend facies 41
Well Bank 122, 124
  Ground Formation 113, 114
  Hole Formation 119
Wells-next-the-Sea 122
Welsh Massif 80
Welton Chalk Formation 83, 89
Werra Halite Formation 45
Werraanhydrit Formation 45, 47
Wessex-Channel Basin 56
West Netherlands Basin 67, 68, 72, 77
  Sole Gasfield 40, 42, 70, 78, 131, 133
  Sole Group 68, 70, 72, 73, 76
  Walton Beds 77
Westbury Formation 65
Western Approaches 130
  Mud Hole Member 120
Westkapelle Ground Formation 100, 101, 105–108
Westphalian 6, 14, 23, 24, 27, 32, 34–37, 132, 133, 136
  A 27, 32, 132
  B 14, 32, 132
  C 14, 32, 132
  D 14, 32, 34, 35, 36, 133
Weybourne 127
Widmerpool Gulf 24
Whitby Mudstone Formation 68

Wiltshire 77
Winterton Formation 56, 65
  Shoal Formation 106
Wissey Gasfield 134
Woolwich Formation 95

X magnetic event 105

Yare river 129
Yarmouth Roads Formation 103, 107, 108, 110
Yeadonian 29, 30, 32
Yoredale cyles 24, 26, 132
Yorkshire 1, 3, 5, 8, 14, 34, 38, 48–50, 53, 54, 66, 67, 70, 73, 77, 79, 84, 89
  Coalfield 34
Young Sea-sands 120
Ypresian 91, 96, 97

Z1 Group 43, 45, 47, 48, 134
Z2 Group 43, 45, 47, 48, 49, 50, 134
Z3 Group 43, 49, 50, 52, 53, 134
Z4 Group 53, 54
Z5 Group 45, 54
Zechstein 38–40, 41, 42, 43–55, 68, 80, 115, 133, 134
Zechsteinkalk Formation 45
Zelzate Formation 97
Zomergem Member 97

BRITISH GEOLOGICAL SURVEY

Keyworth, Nottingham NG12 5GG
(0602) 363100

Murchison House, West Mains Road, Edinburgh EH9 3LA
031-667 1000

London Information Office, Natural History Museum
Earth Galleries, Exhibition Road, London SW7 2DE
071 589 4090

The full range of Survey publications is available through the Sales Desks at Keyworth and at Murchison House, Edinburgh, and in the BGS London Information Office in the Natural History Museum Earth Galleries. The adjacent bookshop stocks the more popular books for sale over the counter. Most BGS books and reports are listed in HMSO's Sectional List 45, and can be bought from HMSO and through HMSO agents and retailers. Maps are listed in the BGS Map Catalogue, and can be bought from Ordnance Survey agents as well as from BGS.

The British Geological Survey carries out the geological survey of Great Britain and Northern Ireland (the latter as an agency service for the government of Northern Ireland), and of the surrounding continental shelf, as well as its basic research projects. It also undertakes programmes of British technical aid in geology in developing countries as arranged by the Overseas Development Administration.

The British Geological Survey is a component body of the Natural Environment Research Council.

HMSO publications are available from:

HMSO Publications Centre
(Mail, fax and telephone orders only)
PO Box 276, London SW8 5DT
Telephone orders 071-873 9090
General enquiries 071-873 0011
Queueing system in operation for both numbers
Fax orders 071-873 8200

HMSO Bookshops
49 High Holborn, London WC1V 6HB
(counter service only)
071-873 0011   Fax 071-873 8200
258 Broad Street, Birmingham B1 2HE
021-643 3740   Fax 021-643 6510
Southey House, 33 Wine Street, Bristol BS1 2BQ
0272-264306   Fax 0272-294515
9 Princess Street, Manchester M60 8AS
061-834 7201   Fax 061-833 0634
16 Arthur Street, Belfast BT1 4GD
0232-238451   Fax 0232-235401
71 Lothian Road, Edinburgh EH3 9AZ
031-228 4181   Fax 031-229 2734

HMSO's Accredited Agents
(see Yellow Pages)

And through good booksellers

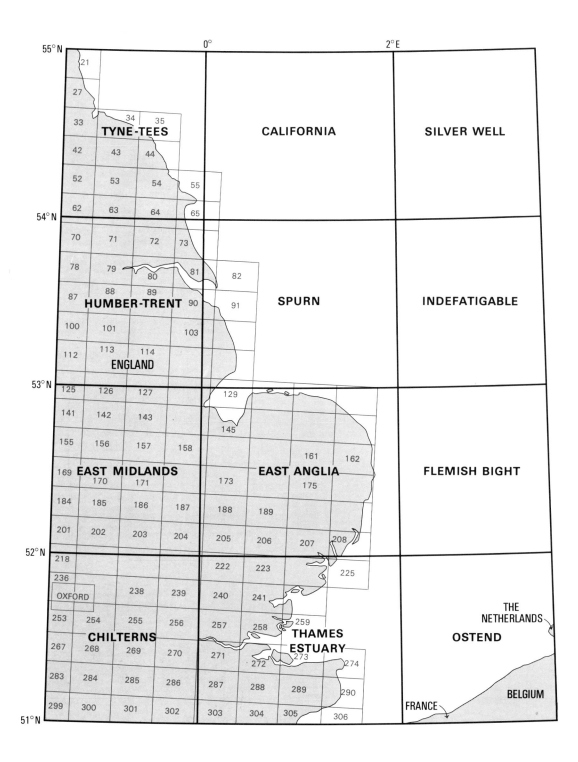